PRECISION AGRICULTURE TECHNOLOGY FOR CROP FARMING

PRECISION AGRICULTURE TECHNOLOGY FOR CROP FARMING

Edited by
Qin Zhang
Washington State University
Prosser, Washington, USA

CRC Press
Taylor & Francis Group
6000 Broken Sound Parkway NW, Suite 300
Boca Raton, FL 33487-2742

First issued in paperback 2021

Version Date: 20150731

ISBN 13: 978-1-03-209827-2 (pbk)
ISBN 13: 978-1-4822-5107-4 (hbk)

Publisher's Note
The publisher has gone to great lengths to ensure the quality of this reprint but points out that some imperfections in the original copies may be apparent.

Visit the Taylor & Francis Web site at
http://www.taylorandfrancis.com

and the CRC Press Web site at
http://www.crcpress.com

Contents

Foreword

Precision agriculture technologies have developed over the last three decades to aid plant agriculture. This book reviews what has happened in the past, what the current situation is, and predicts what the future may hold for these technologies. Top experts who have contributed to the development of precision agriculture provide the information.

Agriculture must provide an ever-increasing amount of quality food, fiber, feed, and fuel for humankind. And it must do this in a manner that is environmentally, economically, and sociopolitically sustainable. This will become even more challenging in the future as there is no single technology that can solve this problem. The development and proper implementation of precision agriculture therefore can be a great help toward achieving this very important task.

NEED FOR PRODUCTION

The Food and Agriculture Organization of the United Nations (UN) estimates that we will need 60% more food by 2050. This is partially due to increasing population. In 2012, the world's population passed thc 7 billion mark. Although the population growth rate has halved since its peak, the UN predicts that the world's population will increase to 9.6 billion by 2050. Those extra mouths need to be fed.

But a bigger cause of the need for the increased production is the changing diets of many consumers, especially those in high-population emerging economies. For example, the consumption of meat increased about 800% in China from 1978 to 2008. The move from diets heavy in staple crops to diets that include substantial amounts of animal products and fruits and vegetables demands much more production from plant agriculture. At the same time, the average total calories consumed per capita have increased from 2250 in 1961 to 2750 in 2007 and are predicted to reach 3070 in 2050. The confluence of more people and more per capita demands leads to the need for great increase in plant agriculture production.

The world depends heavily upon fossil fuels for its fuel, chemicals, and fibers. However, easily accessible supplies of oil and gas are finite and we will soon reach their limit. In addition, the extraction of fossil fuels from below the earth's surface unfortunately brings carbon to the surface and into the atmosphere, thereby increasing the greenhouse effect. Plant agriculture removes carbon dioxide from the atmosphere. More plant production is needed to replace fossil fuels and consequently to provide the raw materials for biofuels, chemical feedstock, and natural fibers. However, increasing such uses of plants could consume agricultural resources that could contribute to food production. This again shows the need for increased productivity.

As discussed above, it is obvious that more agricultural production is needed to support increased populations and changing diets while reducing fossil fuel dependence. By properly responding to spatial and temporal variability in soils, crops, and pests, precision agriculture technologies help increase the productivity and efficiency of plant agriculture. The best way to meet the production needs of the future is to use precision agriculture in combination with the best genetics, cultural practices, equipment, and agronomic management, to achieve maximum production.

NEED FOR SUSTAINABILITY

Humans are now dominating the earth. In order to ensure the health and happiness of future generations, we must live in a sustainable manner. There needs to be environmental, economic, and sociopolitical sustainability. This sustainability can be improved through the use of precision agriculture.

The growing and harvesting of crops remove nutrients from the soil, which must be replaced for long-term environmental sustainability. In low-income countries where the supplies of fertilizers are limited, they should be applied in the areas where they will do the most good. In high-income countries, fertilizers are often uniformly overapplied to avoid the economic consequences of nutrient deficiencies in any area. The mobility of the overapplied nitrogen and phosphorus can then cause those nutrients to be removed from agricultural fields and subsequently lead to drinking water contamination or excessive algae growth. Applying the right fertilizers in the right place at the right time is important to maintaining a proper crop-growing environment without pollution.

Water is a similar environmental issue. Agricultural irrigation represents about 70% of humans' water usage. There are competing demands for our limited water resources. Precision agriculture irrigation can help maximize water use efficiency.

Traditionally, pesticides are applied uniformly in an agricultural field. However, insect, disease, and weed pests tend to be spatially variable. Therefore, the uniform application of pesticides often results in pesticides being released into environments where they are not needed. It would promote environmental sustainability if precision agriculture was utilized and pesticides were applied just where and when they are needed.

Economic sustainability is also promoted by precision agriculture. Inputs such as water, fertilizers, and pesticides contribute very significantly to the costs of production. Reductions in those inputs and increases in quality production from precision agriculture can make farming more economically rewarding. There are also secondary economic benefits in the increased input use efficiency, thus reducing embedded energy costs and environmental costs.

Over half of the world's population now lives in urban, rather than rural, environments as the migration to cities continues. A disproportionate percentage of the migrants are young adults in search of better economic opportunities and more rewarding jobs. Their migration from rural areas has a detrimental effect on sociopolitical sustainability in both rural and urban areas. Although the effect may be small, the introduction of advanced precision agriculture technologies may prove attractive enough to some potential migrants to encourage them to remain in rural communities. The infrastructure, personnel, and experience of precision agriculture may help reduce the digital divide between rural and urban populations.

POTENTIAL OF PRECISION AGRICULTURE TO HELP MEET THESE NEEDS

Meeting the production and sustainability needs for plant agriculture in the future is a difficult task. The greatest chance of meeting those needs is if there is an integration of advances in many areas. There need to be better technologies in genetics, cultural practices, weather prediction, equipment, and farm management. Existing and to-be-developed precision agriculture technologies must be effectively and efficiently integrated into the crop production systems to contribute to increased production and sustainability.

The purpose of this book is to facilitate that integration by conveying information on precision agriculture technology to other researchers and practitioners. The chapters are written by experts who have contributed significantly to the development of precision agriculture technologies. They discuss the developments of the past, describe the current situation, and provide some predictions of the likely future.

ORGANIZATION OF THIS BOOK

The first chapter gives a brief review of the history of precision agriculture to establish a background to the discussion of particular technologies and applications. The next chapters provide details on technologies for sensing, data handling, modeling, and control. The technologies, when integrated, are the vital tools needed for precision agriculture to be successful. The following chapters show how precision agriculture can be used in large-scale agriculture, community agriculture, diversified farming, and as a good agricultural practice. Finally, the needs for the future are proposed.

Of course, there is much more information on precision agriculture than can be included in one book. The authors have utilized their vast experience and knowledge to select the most important and relevant information. I hope you find the book as interesting and informative as I have.

John K. Schueller
Mechanical and Aerospace Engineering Department
University of Florida
Gainesville, Florida

Editor

Dr. Qin Zhang is the director of the Center for Precision and Automated Agricultural Systems and a professor of agricultural automation in the Department of Biological Systems Engineering at Washington State University (WSU). His research interests include agricultural automation, intelligent agricultural machinery, agricultural robotics, and precision agriculture. Before joining the faculty at WSU, he was a professor at the University of Illinois at Urbana-Champaign, working on the development of agricultural mechanization and automation solutions. He has authored 2 textbooks and 6 separate book chapters, edited 2 technical books and 2 conference proceedings, published 125 peer-reviewed journal articles, presented more than 200 papers at national and international professional conferences, and been awarded 10 U.S. patents. He is currently the editor in chief of *Computers and Electronics in Agriculture* and Section III (Plant Production) Chair of CIGR (International Commission of Agricultural and Biosystems Engineering). Dr. Zhang has been invited to present numerous seminars and teach short courses at 18 universities, 9 research institutes, and 11 industry companies in North America, Europe, and Asia. He has also been invited to give keynote speeches at 14 international technical conferences.

Contributors

Hermann Auernhammer
Department of Life Science Engineering
Munich University of Technology (TUM)
Freising, Germany

Josse De Baerdemaeker
Department of Biosystems
KU Leuven—University of Leuven
Leuven, Belgium

Liping Chen
National Engineering Research Center for Intelligent Equipment in Agriculture (NERCIEA)
Beijing, China

Markus Demmel
Institute for Agricultural Engineering and Animal Husbandry
Bavarian State Research Center for Agriculture LfL
Freising, Germany

David Franzen
Department of Soil Science
North Dakota State University
Fargo, North Dakota

Shufeng Han
Deere & Company
Urbandale, Iowa

Won Suk Lee
Agricultural and Biological Engineering Department
University of Florida
Gainesville, Florida

Minzan Li
College of Information and Electrical Engineering
China Agricultural University
Beijing, China

David Mulla
Department of Soil, Water, and Climate
University of Minnesota
St. Paul, Minnesota

William R. Raun
Department of Plant and Soil Sciences
Oklahoma State University
Stillwater, Oklahoma

Wouter Saeys
Department of Biosystems
KU Leuven—University of Leuven
Leuven, Belgium

Sakae Shibusawa
Faculty of Agriculture
Tokyo University of Agriculture and Technology
Tokyo, Japan

Xiaoyu Song
National Engineering Research Center for Information Technology in Agriculture (NERCITA)
Beijing, China

Brian L. Steward
Agricultural and Biosystems Engineering Department
Iowa State University
Ames, Iowa

Marvin L. Stone
Biosystems and Agricultural Engineering Department
Oklahoma State University
Stillwater, Oklahoma

Ruixiu Sui
USDA-ARS
Stoneville, Mississippi

Hong Sun
College of Information and Electrical Engineering
China Agricultural University
Beijing, China

Lie Tang
Agricultural and Biosystems Engineering Department
Iowa State University
Ames, Iowa

Chenghai Yang
USDA-ARS
College Station, Texas

Guijun Yang
National Engineering Research Center for Information Technology in Agriculture (NERCITA)
Beijing, China

Qin Zhang
Center for Precision and Automated Agricultural Systems
Washington State University
Prosser, Washington

Chunjiang Zhao
National Engineering Research Center for Information Technology in Agriculture (NERCITA)
Beijing, China

1 A History of Precision Agriculture

David Franzen and David Mulla

CONTENTS

1.1 INTRODUCTION

There is little evidence that the ancients, although they recognized production differences between fields, considered within-field variation to be worthy of concern (Cato, 160 BC). Romans bought land based on their impression of the care of the farm, its location on the landscape, and soil characteristics. Farms were fertilized using a variety of manures, composts, and the liquid left over after olive pressing. Perhaps, owing to the intense workload of performing basic farming practices, little thought appears to be given to within-field variability. Often the landowners concerned themselves much more with slave or freeman management than within-soil differences. Much more attention was given to acquisition of land rather than dealing with deficient areas of individual fields (Slavin, 2012). In colonial America and the new United States, similar practices were adopted as that of ancient peoples with regard to certain crops on certain soils, crop rotation, and the use of manures on worn-out soils (Jefferson, 1824).

1.2 BRIEF REVIEW OF PRECISION AGRICULTURE HISTORY

When scientists from the new U.S. land-grant colleges first met to discuss agricultural school objectives, the very first experimental subject proposed was dealing with variability in plot crop yield due to soil heterogeneity (Hatch, 1967). Despite efforts to find the most uniform areas possible to conduct field experiments, the problem of field heterogeneity continued to confound researchers (Harris, 1920). Serious steps forward in improving decisions confounded with spatial variability on a small spatial scale began in the early 1920s. Robert A. Fisher started his breakthrough work on the foundation of experimental design at the Rothamsted Experiment Station in Harpenden, Hertfordshire, England in 1919 (Box, 1978). Over the following 7 years, he developed a series of statistical tools used as a foundation for most small-plot and even full-field experiments (Fisher, 1935). The use of principles established by Fisher and the expansion of statistical tools to include problems associated with slopes and systematic differences in soils, such as a Latin-square design or the use of replication blocks, have been greatly useful in decreasing the effect of spatial variability in small-plot experiments for generations of field researchers. However, none of these tools is particularly helpful for managing field variability of nutrients, weeds, insects, seeding rate, or other management inputs.

Most of the work in site-specific agriculture conducted since 1920 has concentrated in crop nutrient management. Soil testing has received a great deal of attention, since it was identified as a means of determining the nutrient supplying capacity of soil since the writings of Sprengel (1839). In soil analysis, a soil sample is taken from the field, processed to take out small stones, then usually subjected to mixing with a liquid extractant, filtered and the nutrient of interest is determined so that a relative amount of the nutrient can be compared to the amount correlated with some degree of possible crop response. The response can be beneficial to the crop or it might sometimes be toxic, depending on the extracted element or compound and the amount in the soil (Melsted, 1967). Although laboratory errors are possible, and minimized commercially through a system of laboratory checks and blanks (SSSA, 2004), the greater source of error in determining the usefulness of a soil analysis recommendation is from sampling error (Cline, 1944; Reed and Rigney, 1947; Hemingway, 1955; Graham, 1959).

The first known recommendation for soil sampling to address field heterogeneity was published by Linsley and Bauer in 1929. Figure 1.1 shows the sampling strategy that was recommended for a 12.5-ha field. The inspiration for advising farmers to soil sample to 15 cm depth and analyze in a 0.4-ha grid, with additional sample cores to a 30 cm depth was related to the effort it took to spread agricultural limestone onto an acidic farm field. At that time, the ground limestone was delivered by rail car. The transport, usually a horse-drawn wagon, was filled with shovels from the rail car by hand, driven back to the farm where the limestone was scooped out of the back of the wagon onto the field by hand. With recommended limestone rates well over a T ha^{-1}, the labor of sampling was far less than that of applying limestone to areas of the field where it was not required. However, the practicality of these recommendations quickly disappeared by the development of mechanical, self-propelled fertilizer application equipment. By 1938, there were many fertilizer application machines available to farmers and many were regularly used, including broadcast, hill-placed, and near-seed-banded equipment (Salter, 1938). Salter references several sources where studies of over 20 different types of machines were compared. These early machines were small, but they were great labor-saving tools that made laborious soil sampling more difficult than addressing field variability.

The basic soil sample from the 1950s to the present day in many areas is the composite sample that represents a field (Melsted and Peck, 1973). Although researchers familiar with crop nutrient spatial variability included cautions to only include relatively uniform, similar soils in a composite sample, sample cores were typically taken from multiple areas of fields described by farmer field boundaries and not by soils within them. Melsted and Peck (1973) describe two fields that were periodically sampled in a 24.3-m (80-ft) grid pattern from 1961. Figure 1.2 shows images of Melsted and Peck, the visionaries who helped build the foundation for

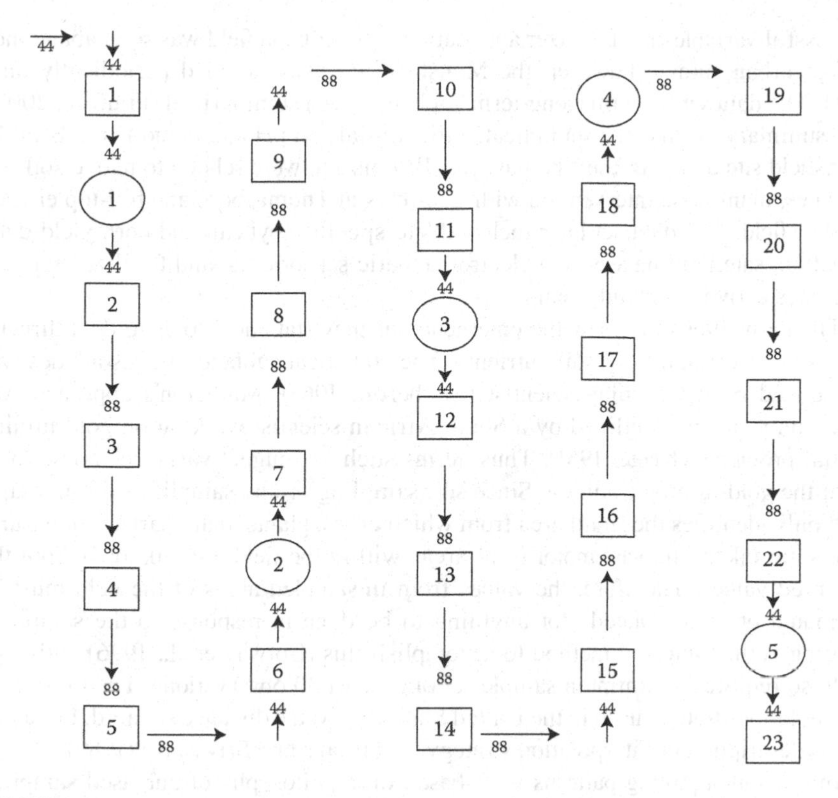

FIGURE 1.1 Recommended sampling strategy for a 12.5-ha square field. Rectangles are for 0–15 cm surface cores, and circles denote locations for a deeper, 0–30 cm core. The 44 or 65 designations are for steps between sampling points, because that was the only reasonably efficient location producer at the time. (Adapted from Linsley, C.M. and F.C. Bauer. 1929. *Test Your Soils for Acidity*. University of Illinois Circular 346, Urbana, IL.)

FIGURE 1.2 S.W. Melsted (a) and T.R. Peck (b), the visionaries who helped build the foundation for successful variable-rate fertilizer application.

successful variable-rate fertilizer application. The Urbana field was soon abandoned due to urban sprawl; however, the Mansfield site was sampled periodically until 1994. The data set from this long-term study has been summarized (Franzen, 2007). The summary of this data set indicated that initial soil pH and P and K levels in the Mansfield site and later the Thomasboro, Illinois site were related to native soil differences, man-made intervention with tree-rows at Thomasboro and hill-top erosion at Mansfield. The data set later included site-specific soybean and corn yield data, elevation, satellite imagery, soil electromagnetic sensor data, and fertilizer application rates early in the study years.

The early 1960s also saw the emergence of new statistical tools to deal directly with spatial variability of soil nutrients. The statistical subfield of geostatistics was introduced by a Canadian scientist (Matheron, 1963). Matheron's approach was based on principles outlined by a South African scientist working on gold mining spatial problems (Krige, 1951). Thus, terms such as "nugget variance" come to us from the gold-mining tradition. Since soil sampling or any sampling within a farm field only identifies the small area from which cores, plants, plant parts, or measurements are taken, the vast majority of areas within the field are unknown from the observed values. Therefore, the values from unsampled areas of the field must be estimated or "interpolated" for anything to be done in response to the sampling. Kriging is the preferred method to accomplish this (Gotway et al., 1996), although to do so requires a minimum sample set of at least 30 observations. The most used interpolation strategy used in the United States is inverse distance squared. For a layman's description of interpolation strategy, see Isaaks and Srivastava (1989).

Initial soil sampling patterns were based on a philosophy of unbiased sampling and the lack of locating instruments within fields that are taken for granted today, such as radar and especially global positioning satellite (GPS) receiving devices. The unbiased sampling approach discouraged taking sample cores from unusual areas or in a random manner over the field. Rather, regular grid sampling, with equal distance between sampling locations was most recommended, with the Melsted/Peck data set being most extensive product of that philosophical approach. The regular grid sampling pattern was the most widely researched and recommended approach until the 1990s (Cline, 1944; Yates, 1948; McIntyre, 1967; Peck and Melsted, 1973; Burgess et al., 1981; McBratney and Webster, 1983; Webster and Burgess, 1984; Petersen and Calvin, 1986; Sabbe and Marx, 1987).

Variability of soil nutrients can be the result of natural soil processes, parent material differences, organic matter differences, and erosion patterns, but they can also be caused by systematic fertilizer application errors (Jensen and Pesek, 1962; Franzen, 2007) and land leveling artifacts (James and Dow, 1972; Knighton and James, 1985) used to prepare for irrigation.

Owing to the variability of crop nutrients within fields and the varying degree of variation of those properties, the density of soil sampling should reflect soil nutrient patterns. Depicting soil nutrient patterns became particularly important as a practical issue rather than an academic question with the advent of the variable-rate fertilizer applicator in the late 1980s. AgChem Equipment Company of Jackson, Minnesota was the first successful variable-rate broadcast granular fertilizer spreader, with commercial equipment used in 1988 in central Illinois. The founder and president of the company, Al McQuinn, attended the first International Conference of Precision Agriculture in 1992 in Minneapolis, Minnesota and gave an impassioned scolding directed at the general academic community for not providing adequate direction and recommendations for soil sampling to direct the new equipment (Franzen, 2014, personal experience). The required research was already being conducted in at least Illinois (Franzen and Peck, 1992), Wisconsin (Wollenhaupt et al., 1994), and Nebraska (Gotway et al., 1996) at this time; however, the data generated was very large in these experiments and computational power was not nearly as great or as rapid as what is available today. In analyzing data in Illinois, for example, generating just one map of a 12.5-ha test field with 256 sampling points took 2 h to calculate and print. The same process today takes about 10 s from start to finished printed page if the sampling results are already uploaded. The publication of results from the Illinois and Wisconsin experiments and the Nebraska experiments that followed therefore took longer than most fertilizer rate experiments conducted at the time. The results from the Illinois (Franzen and Peck, 1995), Wisconsin (Wollenhaupt et al., 1994), and Nebraska studies (Gotway et al., 1996) all indicated that to reveal soil P, K, and soil pH patterns adequately to direct a meaningful variable-rate fertilizer or lime application, a 0.4-ha grid (one sample per acre) was necessary.

Although most researchers that have sampled whole fields at small spatial scale (less than 0.4 ha sampling density) have found that 0.4 ha is a minimum density to sample if nutrient variability is in a range where variable-rate fertilizer application would make a positive difference to the farmer, most grid soil sampling today is conducted in a 1 sample ha^{-1} grid (2.5 acre grid). The reason that this is acceptable in the central Corn Belt of the United States today is that most fields and areas within those fields have soil P and K levels in the high range. Although there is considerable variability of P and K levels within the field, most of the field, regardless of soil test level, will result in the same P and K fertilizer recommendation. Therefore, a less dense grid is acceptable, because failure to represent the P and K nutrient level still results in the proper recommendation as provided by university soil fertility experts (Bullock, 2002; Wittry and Mallarino, 2004).

With advances in grid sampling density, considerable energy was given to determine the best manner to map grid-sampled data and how to generate improved soil sampling strategies. Using large soil sampling data bases, several researchers concluded that maps should be developed using kriging estimation rather than interpolation procedures such as inverse distance (Russo, 1984; Laslet et al., 1987; Laslet and McBratney, 1990; Laslet, 1994; Gotway et al., 1996; Kravchenko and Bullock, 1999; Kravchenko, 2003). Several publications were offered to aid practitioners in understanding the principles and uses of geostatistical methods and sampling strategies that might help support improved variable-rate fertilization strategies (Mulla, 1991; Wollenhaupt, 1996; Mulla and McBratney, 2002). Although kriging is the preferred interpolation method with which to map soil nutrients, it is necessary to have a sample size of over 30 points, and it is also particularly helpful if a systematic unaligned grid was used. Both of these parameters result in sufficient variogram points to produce the relationship regression curve/line between variogram and distance between points. Use of a regular grid, such as that used in Franzen (2007) produces limited points to construct the regression relationship, in the Illinois case, perhaps 30 distances. If a systematic unaligned grid was used, the number of distance possibilities would exceed 100 for the same number of sampling points.

Grid sampling strategies other than a regularly spaced grid design, and subsequent within-field variable-rate fertilizer application would not be practical without reliable locating devices that could be linked with input rate. Before automatic locating systems, any within-field application relied on limiting it to smaller field boundary marked with posts or flags. Some elementary within-field lime or fertilizer application was possible using brightly marked flagging (Franzen, personal experience, 1980s). The first automatic locating devices worked on radar. The U.S. Navy first utilized radio positioning devices in the 1920s. With satellite technology, the Navy deployed a radio-directed locating system called NAVSAT in 1964 (Danchik, 1998). In agriculture, with the emergence of commercial variable-rate input application equipment, the availability of application equipment preceded the GPS network available from the U.S. Department of Defense. For a short time, radar positioning systems were used. These systems were cumbersome, with the radar posts needing deployment to define field boundaries before an applicator could begin work (Tillet, 1991).

The U.S. Department of Defense was granted congressional funding for a satellite positioning system that became known as GPS. The GPS idea was tested in phases, with 11 satellites launched up to 1985. Deployment of the remaining GPS satellites was delayed by the space shuttle Challenger explosion in 1986, which was the only launch platform for the satellites at the time. In 1989, satellite launches resumed with enhanced satellites compared with the first 11 previously launched, including greater longevity probability. The GPS satellite network of 24 satellites was completed in 1994. In 1993, a joint Department of Defense and Department of Transportation agreement was signed that allowed civilian use of the GPS system. The availability of GPS for agriculture was a huge development for precision agriculture, and before long several companies offered GPS for agricultural use. The positioning of the original systems would not allow location within a few meters of an intended location, but real-time kinetic correction towers were built by agribusinesses and farmers so that better location could be provided (Allen et al., 2004). GPS signals directly from satellite signals have inherent errors associated with atmospheric layer differences. Corrections are made using corrective satellite differential receivers and transmitters, such as those used in the John Deere GreenStar™ system (Brimeyer, 2005), and through subscription to ground-based real-time kinematics differential towers often managed by third parties, such as the North Dakota/Minnesota Rural Tower Network (http://www.ruraltowernetwork.com/).

The activity of grid sampling research stimulated thinking into alternative methods of determining spatial nutrient patterns in soil and plants. One of the earlier alternative methods was to use soil survey as a delineation tool (Carr et al., 1991; Mausbach et al., 1993; Wibawa et al., 1993) with some success. Others documented that landscape, or landscape position had an influence on crop yield and crop nutrient availability (Malo et al., 1974; Malo and Worcester, 1975). Canadian researchers in particular documented the influence of topography on soil nutrients and yield early in the process (Spratt and McIver, 1972; Pennock et al., 1987, 1992; Nolan et al., 1995; Penney et al., 1996). Others observed that terrain position was related to differences in crop yield and quality (Fiez et al., 1994a,b; Kravchenko and Bullock, 2000). A breakthrough in sampling design occurred when intensively sampled fields for nitrogen exhibited similar patterns of residual nitrate in successive years. A general understanding of soil nitrate was previously that it would change in value from year to year. Soil nitrate levels indeed change between years in values, but the relative amounts are present in relatively stable areas or zones. Figure 1.3 shows a field sampled in 1994 and 1995 in a 33-m grid. The field was in spring wheat in 1994, fertilized with a uniform rate of N in spring 1995, and a substantial sunflower crop was harvested in 1995. The images from each year depict residual nitrate following the harvest of each crop.

Since soil P and K levels vary little between years, it is difficult to attribute stability of P and K patterns to some logical, underlying reason. However, for residual soil nitrate patterns to remain stable over years, there must be an underlying reason for this result. One of the five soil forming factors is topography (Jenny, 1941). Ruhe (1969) explained that water (and presumably any solutes it contains) moves through and within a landscape, but it always moves to the same places (Figure 1.4). The Valley City residual nitrate is strongly related to topography patterns in the field (Figure 1.5). From the mid-1990s, delineation of fields into nutrient management zones became a strategy that makes variable-rate nutrient application in many regions practical, including nitrogen in the northern Great Plains of the United States (Franzen et al., 1998; Fleming et al., 2000; Fridgen et al., 2000; Inman et al., 2005; Hornung et al., 2006).

Determination of topography may be difficult. While elevation measurement is key to success, it is not just elevation measurement, but the development of topographic shapes that are important to soil development and soil water movement, and therefore crop productivity. Elevation can be measured with a transit, as has been done for hundreds of years. However, with GPS, particularly differential GPS, the height measurement of the GPS location is provided along with latitude and longitude (Department of the Army, 1998). A remote sensing approach was provided with the development of LiDAR (light detecting and ranging). LiDAR originated shortly following the invention of the laser in the early 1960s. It combined the

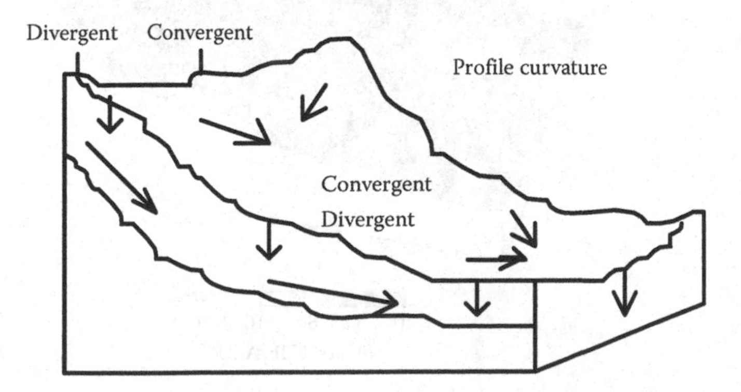

FIGURE 1.3 Water movement through and over the landscape. (Adapted from Ruhe, R.V. 1969. *Quaternary Landscapes in Iowa.* Ames, IA: Iowa State University Press.)

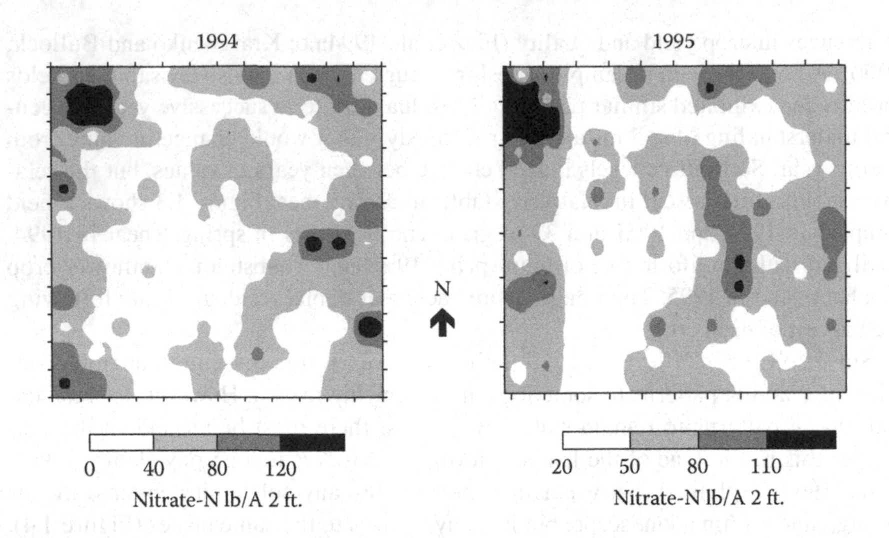

FIGURE 1.4 Residual soil nitrate in a 12.5-ha field near Valley City, North Dakota, after spring wheat (left) and the following year after sunflower.

narrow-focused properties of lasers with the distance calculating ability of radar (Carter et al., 2012). Its use in determining small-scale elevation differences and mapping them has been useful in many aspects of site-specific management, including soil conservation (Galzki et al., 2011).

In addition to topography, several other tools have been used to develop nutrient management zones. Satellite imagery has been used by numerous researchers to

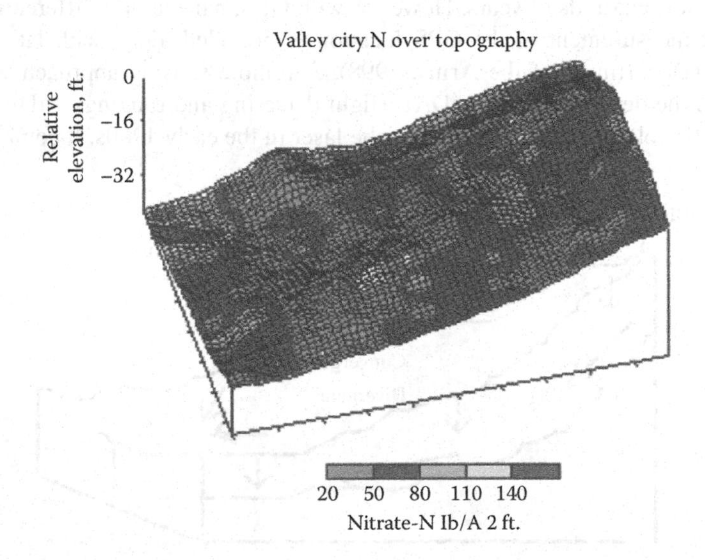

FIGURE 1.5 **(See color insert.)** Residual soil nitrate from Valley City, North Dakota, over the landscape.

delineate nutrient management zones, particularly for nitrogen. Some of the early research on the use of satellite imagery for precision nitrogen management include Anderson and Yang (1996), Bausch et al. (1994, 1995), Faleide and Rosek (1996), Henry and Nielsen (1998), Johannsen et al. (1998), Schepers et al. (1996). The earliest research on the use of satellite imagery for precision agriculture was hopeful of using it to direct in-season nitrogen availability status for crops. Elaborate procedures including sheets of plywood painted with specific colors from specific paint companies were erected in the corners of the research fields. However, problems with degree of light, light angle, and differences between crop cultivar colors made this objective unattainable. However, satellite imagery has been found to be useful in delineating nitrogen management zones for in-season nitrogen management using soil sampling and other methods (Franzen, 2004; Franzen et al., 2011).

The sensors that exploit soil electrical transmittance properties are also used in delineating nutrient management zones. The two most common sensors are the EM-38 (Geonics Ltd., Missasauga, Ontario, Canada) and the Veris EC detector (Veris Technologies, Salina, Kansas). Various models of the EM-38 have been available since 1980 (McNeil, 1980). The Veris electrical conductivity detector was commercialized in the late 1990s. Field zones can be delineated due to soil clay content (Doolittle et al., 1994; Kitchen et al., 1996; Banton, 1997). Electrical conductivity has also been related directly to soil nitrate levels in otherwise uniform soils (Eigenberg, 2002). In addition, EC and magnetism can be used to detect differences in water-holding capacity or soil water content, cation exchange capacity, porosity, salinity, and temperature gradients (Grisso et al., 2009). If fields have relatively uniform properties except for one measurable variable, as in the case of soil nitrate in some fields in Nebraska, or low salt coastal plain soils of the eastern United States, electrical conductivity can be used to directly relate to any of the variable factors that are singularly present in the field. However, in many fields, more than one factor varies independently of others. This is particularly the case in the northern Great Plains, where salinity may be present in soils with more or less clay content in different levels of landscape position due to internal water movement. In multivariable fields, the EC and magnetic flux sensors are pattern detectors with similar ability to delineate zones compared to other tools (Franzen, 2008a; Franzen et al., 2011).

Other common soil delineation tools are multiyear yield maps (Basnet et al., 2003; Franzen, 2008b; Franzen et al., 2011), aerial imagery (Blackmer and Schepers, 1996; Franzen et al., 2011), and grower information (Khosla et al., 2002). Although it may be sometimes possible to delineate nutrient management zones using just one tool, greater stability of the constructed zones is made possible with the use of more than two zone delineation tools (Franzen et al., 2011).

Soil sampling for nutrients may not be practical due to very small spatial variability. This was found in Oklahoma (Raun et al., 1998). Residual nitrate variability before top-dress timing of Bermuda grass was in the range of <1 m. Soil sampling for variability at this small scale is unreasonable to expect. The result of this preliminary grid sampling study resulted in the development of an active-optical sensor to detect differences in nitrogen status for winter wheat, Bermuda grass, and other crops, which was linked to a variable-rate real-time on-the-go nitrogen fertilizer application.

Directly after the microvariability of nitrogen status of Bermuda grass was realized, Raun was approached by Marvin Stone and John Solie, two agricultural engineers at Oklahoma State University with an idea for a normalized differential vegetative index (NDVI) detector and light emitter and asked if he thought there might be an application. Thus, the team of Raun, Stone, and Solie was formed, with the objective to address small-scale variability of nitrogen in regional crops (W.R. Raun, 1995, personal communication). Sensor development and the link between sensor readings and winter wheat nitrogen began very soon (Stone et al., 1996). This report included early sensor descriptions and the first correlation data of sensor readings with winter wheat yield. Estimating plant nitrogen status in Bermuda grass using sensors was reported in 1998 (Taylor et al., 1998). As work in Oklahoma progressed, the relationship between sensor reading and wheat yield prediction became very evident. The construction of algorithms with the goal of in-season nitrogen application is based on sensor prediction of yields in a nitrogen nonlimiting strip compared to other areas within the field (Raun et al., 2001). The result of the work was the commercialization of the GreenSeeker™ active-optical sensor (Solie et al., 2002). This applicator had the ability to operate at field sprayer speeds and apply N to each m^2 of crop independently due to its array of sensors and nozzle clusters arranged every 1 m of spray boom width.

The GreenSeeker sensor has since been tested and is used in many countries in wheat, corn, and other crops (http://www.nue.okstate.edu/). Most recently, a series of algorithms for North Dakota corn growers was published (Franzen et al., 2014). The procedure for use is to establish an N nonlimiting strip within field, within cultivar. At the time of top-dress/side-dress, the algorithm to be used is loaded into the rate-controller software, along with the growing degree days from the date of planting to the date of in-season application. When the field is entered with the in-season nitrogen applicator, the first activity is to operate the sensor over the N-rich strip. The reading is divided by the growing degree days to provide a value called INSEY, which the Raun group coined as an acronym for in-season-estimate-of-yield. Algorithms are produced for a specific crop growth stage; however, expecting a grower to arrive at the field at exactly the specific algorithm growth stage is unreasonable due to workload challenges combined with weather restrictions. Dividing the readings by the growing degree days results in a normalized value that makes the algorithm useful for plus–minus one or two growth stages before and after the stage for which the algorithm was developed. As the applicator moves through the field, the yield prediction between the N nonlimiting standard and the other parts of the field are developed. The difference in yield is multiplied by the controller software times the presumed N content of the grain to be produced and divided by an efficiency factor from the grower reflective of field conditions and the method of application. The result is a rate of N applied to a small area of the field on-the-go as the applicator moves through the field.

More recently, the Holland Scientific Crop Circle Sensor was introduced. Some of the basic principles of active-optical sensors developed by Oklahoma State were incorporated into the development of this sensor (Holland and Schepers, 2010); however, the instrument relies on an internal field standard rather than a separate N-rich strip for application (Holland and Schepers, 2013). When the applicator with the sensor enters the field, the greenest portion of the field is used as the standard. Using the sensor in this manner assumes that the greenest area of the field has as much nitrogen as required for the highest yield.

Active-optical sensors have clear advantages over passive light/radiation sensors from satellites or airplanes because they only detect the radiation that they emit. The light emitted by these sensors operates on the same principle as a remote control for a television or an automatic garage-door opener. The light emitted is released in rapid pulses of varying lengths of time, separated by gaps in light emission of varying lengths of time. The light emission is very much similar to the UPC code in effect found on many grocery item packages in the United States. For that reason, active-optical sensors can be used in any kind of light, at night or during the day whether clear or hazy or with intermittent cloud cover. The only conditions not conducive to gathering good data are rain or wet leaves, which would refract and disperse the light.

A number of additional sensors have been developed since the precision agriculture movement was initiated. One sensor was developed and marketed as the "Soil Doctor" during the early 1990s. Although the commercial operational procedure was not disclosed to the public, the device apparently grew out of an electrical conductivity apparatus developed from a Department of Energy grant (Colburn, 1986). The unit was sold through the early 1990s, but it is apparently not marketed at present. Other sensors were developed during the late 1980s into the early 1990s, but most did not come to market. These included a real-time organic matter sensor developed at Purdue (Shonk et al., 1991), an organic matter sensor using near-infrared spectroscopy (Sudduth and Hummel, 1993), and ion-selective electrodes (Birrell and Hummel, 1997). Successful commercialization of nutrient-related soil sensors includes the Veris Technologies soil pH sensor (Schirrmann et al., 2011). In one study, the use of the Veris pH sensor related mapped sensor pH better with actual field pH patterns compared with standard site-specific soil sampling techniques, with errors in the liming rate recommendations reduced by half. Veris also markets the Optic Mapper™, which uses a within-soil near-infrared detector to estimate soil organic matter content (Lund, 2011).

Analysis of grain protein during harvest was researched largely by Long first at Montana State in Havre, and then in Oregon (Engel et al., 1998; Long et al., 1998; Long and Rosenthal, 2005). Although several combine protein monitors were tested over the years, the last one referred to in 2005, the Zeltex model marketed as AccuHarvest™, is commercialized for use in wheat, barley, corn, and soybean (http://www.zeltex.com/products/grain, 2014).

Adoption of precision agriculture technologies by growers and their suppliers and practitioners has been slow for input control, but the adoption of machinery traffic control systems has been relatively rapid. The adoption of yield monitors by growers is provided in Table 1.1.

Variable-rate technology has been adopted by many farm supplier retailers, but its adoption by grower is relatively low, with 12% of corn growers adopting by 2005, and about 14% of wheat growers.

In contrast, adoption of tractor guidance systems by growers increased from about 5% in 2001 to 35% in 2009 (Schimmelpfennig and Ebel, 2011). In a 2010 survey of

TABLE 1.1
Yield Monitor Adoption for Corn and Soybeans by Region 1996–2006

	1996	1997	1998	1999	2000	2001	2002	2005	2006
					Percent of Planted Acres				
Corn									
Corn Belt		21	18	28	34	28		44	
Lake states		10	19	18	29	17		39	
Northern Plains		15	24	19	28	27		43	
Soybean									
Corn Belt	13	15	21	23	28		24		49
Lake states	16	11	16	15	30		29		48
Northern Plains		9	17	21	22		17		48

Source: Adapted from Schimmelpfennig, D. and R. Ebel. 2011. *On the Doorstep of the Information Age. Recent Adoption of Precision Agriculture*. Economic Information Bulletin No. 80, August, 2011. USDA, Washington, DC, http://www.ers.usda.gov/media/81195/eib80_1_.pdf.

corn growers, yield monitor use increased to over 72%, with about 34% of those using a yield monitor able to develop a map. Variable-rate fertilizer application was made by 19.3% of producers (USDA, 2010).

1.3 SUMMARY OF CURRENT STATUS

A 2013 survey of agricultural retailers in the United States (Erickson et al., 2013) found that some precision agriculture technologies increased while some remained at 2011 levels. Guidance with light bars with GPS of applicators was being used by 82% of retailers, which is about the same as in 2011. GPS-enabled sprayer booms (differential shut-offs) increased from 39% to 53% adoption from 2011 to 2013. For diagnosis of crop nutrient status and presumably to help develop nutrient management zones, satellite imagery and aerial imagery increased from 31% in 2011 to 40% in 2013. Active-optical sensor use increased from 4% in 2011 to 7% in 2013. Soil EC sensors remained about 12%, while use of other soil sensors remained about 3% from 2011 levels. Variable-rate single-nutrient fertilizer application is offered by more than 70% of retailers, while about 60% of retailers offer multinutrient variable-rate application. About 29% of retailers offer variable-rate pesticide application in 2013, but the survey indicates intentions to increase to 45% of retailers by 2016.

Site-specific weed control research has received limited research compared with crop nutrients. The focus of the research has been in mapping weed infestations (Koller and Lanini, 2005) and weed imaging discrimination (Lin, 2009). Of the two approaches, the weed imaging strategy has been the most difficult to implement. Studies on weed identification by instruments date to the 1990s, but have achieved little commercial success (Singher, 1999; Tellaeche et al., 2011).

Site-specific strategies for insects and disease are researched less than even weed control. However, one potentially useful tool to use is the ability of plants to emit unique volatile organic compounds when influenced by specific stressors (Spinelli et al., 2011; Niinemets et al., 2013). Specific compounds have been identified as a result of *Fusarium* infestation in winter wheat (Wenda-Piesik, 2011). With the proper nano-sensing or remote sensing instrument, one could imagine that in the future an array of sensors to a specific insect or weed in a field, and at an early infection, directed treatment could be provided.

REFERENCES

Allen, D.W., N. Ashby, and C.C. Hodge. 2004. *The Science of Timekeeping. Hewlett Packard Application Note 1289*, http://www.cs.cmu.edu/~sensing-sensors/readings/TheScienceOfTimekeeping.pdf, *With GPS History Appendix*, http://www.cs.cmu.edu/~sensing-sensors/readings/GPS_History-MR614.appb.pdf.

Anderson, G.L. and C. Yang. 1996. Multispectral videography and geographic information systems for site-specific farm management. In *Precision Agriculture Proceedings of the 3rd International Conference*, Minneapolis, MN, June 23–26, 1996, Madison, WI: ASA-CSSA-SSSA, pp. 681–692.

Banton, O., M.K. Seguin, and M.A. Cimon. 1997. Mapping field-scale physical properties of soil with electrical resistivity. *Soil Science Society of America Journal*, 61:1010–1017.

Basnet, B., R. Kelly, T. Jensen, W. Strong, A. Apan, and D. Butler. 2003. Delineation of management zones using multiple crop yield data. In *16th International Soil Tillage Research Organisation Conference*, Brisbane, Australia.

Bausch, W.C., H.R. Duke, and L.K. Porter. 1994. *Remote Sensing of Plant Nitrogen Status in Corn*. St. Joseph, MI: ASAE, Paper No. 94-2117.

Birrell, S.J. and J.W. Hummel. 1997. Multi-sensor ISFET system for soil analysis. In Stafford, J.V. (ed.), *Precision Agriculture '97*. Oxford, UK: BIOS Scientific Publishers, pp. 459–468.

Blackmer, T.M. and J.S. Schepers. 1996. Aerial photography to detect nitrogen stress in corn. *Journal of Plant Physiology*, 148:440–444.

Blackmer, T.M., J.S. Schepers, and G.E. Meyer. 1995. Remote sensing to detect nitrogen deficiency in corn. In *Proceedings of Site-Specific Management for Agricultural Systems. 2nd Annual Conference*, Minneapolis, MN, March 27–30, 1994, Madison, WI: ASA-CSSA-SSSA, pp. 505–512.

Blackmer, T.M., J.S. Schepers, and G.E. Varvel. 1994. Light reflectance compared to other nitrogen stress measurements in corn leaves. *Agronomy Journal*, 86:934–938.

Box, J.F. 1978. *R.A. Fisher: The Life of a Scientist*. New York: Wiley Publishers.

Brimeyer, J. 2005. Nothing runs like a precision farming system. *Progressive Engineer*. March, 2005.

Bullock, D. 2002. Soil sampling for conventional and precision agriculture. *Corn and Soybean Classic, 2002*. http://extension.cropsci.illinois.edu/fieldcrops/classics/2002/soilsampling.php.

Burgess, T.M., R. Webster, and A.B. McBratney. 1981. Optimal interpolation and isarithmic mapping of soil properties. IV Sampling strategies. *Journal of Soil Science*, 32:643–659.

Cato. 160 BC. *De Agri Cultura*. Latin text, G. Goetz. English translation, W.D. Hooper and H.B. Ash. Loeb Classical Library. 1934, http://penelope.uchicago.edu/Thayer/E/Roman/Texts/Cato/De_Agricultura/A*.html

Carr, P.M., G.R. Carlson, J.S. Jacobsen, G.A. Nielsen, and E.O. Skogley. 1991. Farming soils, not fields: A strategy for increasing fertilizer profitability. *Journal of Production Agriculture*, 4:57–61.

Carter, J., K. Schmid, K. Waters, L. Betzhold, B. Hadley, R. Mataosky, and J. Halleran. 2012. *Lidar 101: An Introduction to Lidar Technology, Data, and Applications*. Charleston, SC: National Oceanic and Atmospheric Administration (NOAA) Coastal Services Center.

Cline, M.G. 1944. Principles of soil sampling. *Soil Science*, 58:275–288.

Colburn, J. 1986. *R&D on a Fertilizer Sensor and Control System*. DOE ID/12518-1 (DE87014929). Springfield, VA: NTIS, US Department of Commerce.

Danchik, R.J. 1998. An overview of transit development. *John Hopkins APL Technical Digest*, 19:18–26.

Department of the Army. 1998. *Using Differential GPS Positioning for Elevation Determination*. Technical Letter No. 1110-1-183. http://www.dtic.mil/dtic/tr/fulltext/u2/a403475.pdf.

Doolittle, J.A., K.A. Sudduth, N.R. Kitchen, and S.J. Indorante. 1994. Estimating depths to claypans using electromagnetic induction methods. *Journal of Soil and Water Conservation*, 49(6):572–575.

Eigenberg, R.A., J.W. Doran, J.A. Nienaber, R.B. Ferguson, and B.L. Woodbury. 2002. Electrical conductivity monitoring of soil condition and available N with animal manure and a cover crop. *Agriculture, Ecosystems and Environment*, 88:183–193.

Engel, R.E., D. Long, and G. Carlson. 1998. Grain protein sensing for precision N management of wheat. I. Why? In *Proceedings, Intensive Wheat Management Conference*, Brookings, SD, March 4–5, 1998, Denver, CO: Potash & Phosphate Institute, pp. 41–46.

Erickson, B., D. Widmar, and J. Holland. 2013. *Survey: An Inside Look at Precision Agriculture in 2013. Crop Life*, May 29, 2013.

Faleide, L. and M.J. Rosek. 1996. Application of remote sensing for selecting soil sampling sites. In *Multispectral Imaging for Terrestrial Applications. SPIE Proceedings*. SPIE, Bellingham, WA: The International Society for Optical Engineering, Vol. 2818, pp. 59–60.

Fiez, T.E., B.C. Miller, and W.L. Pan. 1994a. Winter wheat yield and grain protein across varied landscape positions. *Agronomy Journal*, 86:1026–1032.

Fiez, T.E., B.C. Miller, and W.L. Pan. 1994b. Assessment of spatially variable nitrogen fertilizer management in winter wheat. *Journal of Production Agriculture*, 7:86–93.

Fisher, R.A. 1935. *The Design of Field Experiments*. Edinburgh, UK: Oliver and Boyd.

Fleming, K.L., D.G. Westfall, and W.C. Bausch. 2000. Evaluating management zone technology and grid soil sampling for variable rate nitrogen application. In Robert, P.C. et al. (eds.), *Precision Agriculture Proceedings of the International Conference, 5th*, Bloomington, MN, July 16–19, 2000, Madison, WI: ASA-CSSA-SSSA.

Franzen, D.W. 2004. Delineating nitrogen management zones in a sugarbeet rotation using remote sensing—A review. *Journal of Sugar Beet Research*, 41:47–60.

Franzen, D.W. 2007. *Illinois Grid Sampling Project Technical Bulletin*. http://www.ndsu.edu/fileadmin/soils/pdfs/Summary_of_Grid_Sampling_07.pdf. Electronic file of data available on request (david.franzen@ndsu.edu).

Franzen, D.W. 2008a. *Developing Zone Soil Sampling Maps*. NDSU Extension Circular 1176-2.

Franzen, D.W. 2008b. *Yield Mapping*. NDSU Extension Circular 1176-3.

Franzen, D.W., L.J. Cihacek, V.L. Hofman, and L.J. Swenson. 1998. Topography-based sampling compared with grid sampling in the Northern Great Plains. *Journal of Production Agriculture*, 11:364–370.

Franzen, D.W., D. Long, A. Sims, J. Lamb, F. Casey, J. Staricka, M. Halvorson, and V. Hofman. 2011. Evaluation of methods to determine residual soil nitrate zones across the northern Great Plains of the USA. *Precision Agriculture*, 12:594–606.

Franzen, D.W. and T.R. Peck. 1992. Spatial variability of soil pH, phosphorus and potassium levels. In Hoeft, R.G. (ed.), *1992 Proceedings of the Illinois Fertilizer and Chemical Association Conference*, Springfield, IL, pp. 103–111.

Franzen, D.W. and T.R. Peck. 1995. Field soil sampling density for variable rate fertilization. *Journal of Production Agriculture*, 8:568–574.

Franzen, D.W., L.K. Sharma, and H. Bu. 2014. *Active Optical Sensor Algorithms for Corn Yield Prediction and a Corn Side-Dress Nitrogen Rate Aid*. NDSU Extension Circular SF-1176-5.

Fridgen, J.J., N.R. Kitchen, and K.A. Sudduth. 2000. Variability of soil and landscape attributes within sub-field management zones. In Robert, P.C. et al. (eds.), *Precision Agriculture Proceedings of the 5th International Conference*, Bloomington, MN, July 16–19, 2000, Madison, WI: ASA-CSSA-SSSA.

Galzki, J.C., A.S. Birr, and D.J. Mulla. 2011. Identifying critical agricultural areas with three-meter LiDAR elevation data for precision conservation. *Journal of Soil and Water Conservation*, 66:423–430.

Gotway, C.A., R.B. Ferguson, G.W. Hergert, and T.A. Peterson. 1996. Comparison of kriging and inverse-distance methods for mapping soil parameters. *Soil Science Society of America Journal*, 60:1237–1247.

Graham, E.R. 1959. *An Explanation of the Theory and Methods of Soil Testing*. Missouri Agricultural Experiment Station Bulletin 734.

Grisso, R., M. Alley, W.G. Wysor, D. Holshouser, and W. Thomason. 2009. Precision farming tools: Soil electrical conductivity. *Virginia Cooperative Extension Circular*, pp. 442–508. http://pubs.ext.vt.edu/442/442-508/442-508.html.

Harris, J.A. 1920. Practical universality of field heterogeneity as a factor influencing plot yields. *Journal of Agricultural Research*, 19(7):279–314.

Hatch, R.A. 1967. *An Early View of the Land-Grant Colleges. Convention of Friends of Agricultural Education in 1871*. Urbana, IL: University of Illinois Press.

Hemingway, R.G. 1955. Soil sampling errors and advisory analyses. *Journal of Agricultural Science*, 46:1–8.

Henry, M.P. and G. Nielsen. 1998. Potentials for satellite remote sensing for nitrogen management of wheat. In Schlegel, A.J. (ed.), *1998 Great Plains Soil Fertility Conference Proceedings*, Brookings, SD, March 2–4, 1998, Denver, CO: Potash & Phosphate Institute, pp. 121–127.

Holland, K.H. and J.S. Schepers. 2010. Derivation of a variable rate nitrogen application model for in-season fertilization of corn. *Agronomy Journal*, 102:1415–1424.

Holland, K.H. and J.S. Schepers. 2013. Use of a virtual-reference concept to interpret active crop canopy sensor data. *Precision Agriculture*, 14:71–85.

Hornung, A., R. Khosla, R. Reich, D. Inman, and D.G. Westfall. 2006. Comparison of site-specific management zones: Soil-color-based and yield-based. *Agronomy Journal*, 98:407–415.

Inman, D.J., R. Khosla, and D.G. Westfall. 2005. Nitrogen uptake across site-specific management zones in irrigated corn production systems. *Agronomy Journal*, 97:169–176.

Isaaks, E.H. and R.M. Srivastava. 1989. *Introduction to Geostatistics*. New York: Oxford University Press.

James, D.W. and A.I. Dow. 1972. *Source and Degree of Soil Variation in the Field: The Problem of Sampling for Soil Tests and Estimating Soil Fertility Status*. Washington Agricultural Experiment Station Bulletin 749.

Jefferson, T. 1824. *Farm Book, 1774–1824*. Collections of the Massachusetts Historical Society, Boston, MA. Available online http://www.masshist.org/thomasjeffersonpapers/farm/

Jenny, H. 1941. *Factors of Soil Formation: A System of Quantitative Pedology*. New York: McGraw Hill.

Jensen, D. and J. Pesek. 1962. Inefficiency of fertilizer use resulting from nonuniform spatial distribution: Theory. *Proceedings of the Soil Science Society of America*, 26:170–173.

Johannsen, C.J., P.G. Carter, and P. Willis. 1998. Remote sensing of wheat in the Great Plains. In *Proceedings, Intensive Wheat Management Conference*, Brookings, SD, 4–5 March, Denver, CO: Phosphate & Potash Institute, pp. 25–29.

Khosla, R., K. Fleming, J.A. Delgado, T.M. Shaver, and D.G. Westfall. 2002. Use of site-specific management zones to improve nitrogen management for precision agriculture. *Journal of Soil and Water Conservation*, 57:513–518.

Kitchen, N.R., K.A. Sudduth, and S.T. Drummond. 1996. Mapping sand deposition from 1993 midwest floods using electromagnetic induction measurements. *Journal Soil and Water Conservation*, 51(4):336–340.

Knight, R.E. and D.W. James. 1985. Soil test phosphorus as a regionalized variable in leveled land. *Soil Science Society of America Journal*, 49:675–679.

Koller, M. and W.T. Lanini. 2005. Site-specific herbicide applications based on weed maps provide effective control. *California Agriculture*, 59:182–187.

Kravchenko, A.N. 2003. Influence of spatial structure on accuracy of interpolation methods. *Soil Science Society of America Journal*, 67:1564–1571.

Kravchenko, A. and D.G. Bullock. 1999. A comparative study of interpolation methods for mapping soil properties. *Agronomy Journal*, 91:393–400.

Kravchenko, A.N. and D.G. Bullock. 2000. Correlation of corn and soybean grain yield with topography and soil properties. *Agronomy Journal*, 92:75–83.

Krige, D.G. 1951. *A Statistical Approach to Some Mine Valuations and Allied Problems at the Witwatersrand*. MS Thesis, University of Witwatersrand. Union of South Africa.

Laslet, G.M. 1994. Kriging and splines: An empirical comparison of their predictive performance in some applications. *Journal of the American Statistics Association*, 89:391–409.

Laslett, G.M. and A.B. McBratney. 1990. Further comparison of spatial methods for predicting soil pH. *Soil Science Society of America Journal*, 54:1533–1558.

Laslett, G.M., A.B. McBratney, P.J. Pahl, and M.J. Hutchinson. 1987. Comparison of several spatial prediction methods for soil pH. *Journal of Soil Science*, 38:325–341.

Lin, C. 2009. *A Support Vector Machine Embedded Weed Identification System*. MS Thesis, University of Illinois, Urbana, IL.

Linsley, C.M. and F.C. Bauer. 1929. *Test Your Soils for Acidity*. University of Illinois Circular 346, Urbana, IL.

Long, D.S., R.E. Engel, and G.R. Carlson. 1998. Grain protein sensing for precision N management II. How? In *Proceedings, Intensive Wheat Management Conference*, Brookings, SD, March 4–5, 1998, Denver, CO: Potash & Phosphate Institute, pp. 56–61.

Long, D.S. and T. Rosenthal. 2005. Evaluation of an on-combine wheat protein analyzer on Montana hard red spring wheat. In Stafford, J. et al. (eds.), *Proceedings of the 5th European Conference on Precision Agriculture*, Uppsala, Sweden, 9–12 June.

Lund, E. 2011. Proximal sensing of soil organic matter using the Veris® OpticMapper™. In *Proceedings of The Second Global Workshop on Proximal Soil Sensing*, Montreal, 2011, pp. 76–79.

Malo, D.D. and B.K. Worcester. 1975. Soil fertility and crop responses at selected landscape positions. *Agronomy Journal*, 67:397–401.

Malo, D.D., B.K. Worcester, D.K. Cassel, and K.D. Matzdort. 1974. Soil-landscape relationships in a closed drainage system. *Soil Science Society of America Proceedings*, 38:813–818.

Matheron, G. 1963. Principles of geostatistics. *Economic Geology*, 58:1246–1266.

Mausbach, M.J., D.J. Lytle, and D.R. Nielsen. 1993. Application of soil survey information to soil specific farming. In Robert, P.C. et al. (eds.), *Proceedings, Soil Specific Crop Management*, Minneapolis, MN, April 14–16, 1992, Madison, WI: ASA-CSSA-SSSA, pp. 57–68.

McBratney, A.B. and R. Webster. 1983. How many observations are needed for regional estimation of soil properties. *Soil Science*, 135(3):177–183.

McIntyre, G.A. 1967. Soil sampling for soil testing. *Journal of Australian Institute of Agricultural Science*, 18:309–318.

McNeil, J.D. 1980. *Electromagnetic Terrain Conductivity Measurement at Low Induction Numbers*. Technical Note TN-6. Mississauga, Canada: Geonics Limited.

Melsted, S.W. 1967. The philosophy of soil testing. In *Soil Testing and Plant Analysis. Part 1. SSSA Special Publication Series No. 2*. Madison, WI: Soil Science Society of America, pp. 13–23.

Melsted, S.W. and T.R. Peck. 1973. Field sampling for soil testing. In *Soil Testing and Plant Analysis*, Revised edition, Madison, WI: Soil Science Society of America, pp. 67–75.

Mulla, D.J. 1991. Using geostatistics and GIS to manage spatial patterns in soil fertility. In Kranzler, G.W. et al. (eds.), *Automated Agriculture for the 21st Century*, St. Joseph, MI: American Society of Agricultural Engineers, pp. 336–345.

Mulla, D.J. and A.B. McBratney. 2002. Soil spatial variability. In Warrick, A.W. (ed.), *Soil Physics Companion*, Boca Raton, FL: CRC Press, LLC, pp. 343–373.

Niinemets, U., A. Kannaste, and L. Copolovici. 2013. Quantitative patterns between plant volatile emissions induced by biotic stresses and the degree of damage. *Frontiers of Plant Science*, 4:262.

Nolan, S.C., D.J. Heaney, T.W. Goddard, D.C. Penney, and R.C. McKenzie. 1995. Variation in fertilizer response across soil landscapes. In *Proceedings of Site-Specific Management for Agricultural Systems. 2nd Annual Conference*, Minneapolis, MN, March 27–30, 1994, Madison, WI: ASA-CSSA-SSSA, pp. 553–558.

Penney, D.C., R.C. McKenzie, S.C. Nolan, and T.W. Goddard. 1996. Use of crop yield and soil landscape attribute maps for variable rate fertilization. In Havlin, J. (ed.), *1996 Great Plains Soil Fertility Conference Proceedings*, Denver, CO, Kansas State University, March 5–6, 1996, Norcross, GA: Potash & Potash Institute, pp. 126–140.

Pennock, D.J., C. van Kessel, R.E. Farrell, and R.A. Sutherland. 1992. Landscape-scale variations in denitrification. *Soil Science Society of America Journal*, 56:770–776.

Pennock, D.J., B.J. Zebarth, and E. DeJong. 1987. Landform classification and soil distribution in hummocky terrain, Saskatchawan, Canada. *Geoderma*, 40:297–315.

Petersen, R.G. and L.G. Calvin. 1986. Sampling. In *Methods of Soil Analysis, Part 1. Physical and Mineralogical Methods*. Madison, WI: ASA-SSSA, pp. 33–50.

Raun, W.R., J.B. Solie, G.V. Johnson, M.L. Stone, E.V. Lukina, W.E. Thomason, and J.S. Schepers. 2001. In-season prediction of potential grain yield in winter wheat using canopy reflectance. *Agronomy Journal*, 93:131–138.

Raun, W.R., J.B. Solie, G.V. Johnson, M.L. Stone, R.W. Whitney, H.L. Lee, H. Sembiring, and S.B. Phillips. 1998. Microvariability in soil test, plant nutrient, and yield parameters in bermudagrass. *Soil Science Society of America Journal*, 62:683–690.

Reed, J.F. and J.A. Rigney. 1947. Soil sampling for fields of uniform and nonuniform appearance and soil types. *Journal American Society of Agronomy*, 39:26–40.

Ruhe, R.V. 1969. *Quaternary Landscapes in Iowa*. Ames, IA: Iowa State University Press.

Russo, D. 1984. Design of an optimal sampling network for estimating the variogram. *Soil Science Society of America Journal*, 48:708–716.

Sabbe, W.E. and D.B. Marx. 1987. Soil sampling: Spatial and temporal variability. In Brown, J. (ed.), *Soil Testing: Sampling, Correlation, Calibration, and Interpretation*. SSSA Special Publication No. 21. Madison, WI: SSSA, pp. 1–14.

Salter, R.M. 1938. *Methods of Applying Fertilizers*. 1938 Soils and Men, Yearbook of Agriculture. Washington, DC: USDA, pp. 546–562.

Schepers, J.S., T.M. Blackmer, T. Shah, and N. Christensen. 1996. Remote sensing tools for site-specific management. In *Precision Agriculture. Proceedings of the 3rd International Conference*, Minneapolis, MN, June 23–26, 1996, Madison, WI: ASA-CSSA-SSSA, pp. 315–320.

Schimmelpfennig, D. and R. Ebel. 2011. *On the Doorstep of the Information Age. Recent Adoption of Precision Agriculture*. Economic Information Bulletin No. 80, August, 2011. USDA, Washington, DC. http://www.ers.usda.gov/media/81195/eib80_1_.pdf.

Schirrmann, M., R. Gebbers, E. Kramer, and J. Seidel. 2011. Soil pH mapping with an on-the-go sensor. *Sensors*, 11:573–598.

Shonk, J.L., L.D. Gaultney, D.G. Schulze, and G.E. Van Scoyoc. 1991. Spectroscopic sensing of soil organic matter content. *Transactions of the ASAE*, 32(3):826–829.

Singher, L. 1999. Electro-optical-based machine vision for weed identification. In *Proceedings SPIE 3543, Precision Agriculture and Biological Quality, 319*, January 14, 1999.

Slavin, P. 2012. Church and food provisioning in Late-Medieval England, 1250–1450: Production costs, markets and the decline of direct demesne management. In Ammannati, F. (ed.), *Religion and Religious Institutions in the European Economy, 1000-1800*, Firenze, Italy: Firenze University Press, pp. 597–618.

Soil Science Society of America/NAPT. 2004. *QA/QC Model Plan for Soil Testing Laboratories*. SSSA and Oregon State University.

Solie, J.B., M.L. Stone, W.R. Raun, G.V. Johnson, K. Freeman, R. Mullen, D.E. Needham, S. Reed, and C.N. Washmon. 2002. Real-time sensing and N fertilization with a field scale GreenSeeker™ applicator. In *Proceedings of the 2002 International Conference on Precision Agriculture*, Minneapolis, MN.

Spinelli, F., A. Cellini, L. Marchetti, K.M. Nagesh, and C. Piovene. 2011. Emission and function of volatile organic compounds in response to abiotic stress. In Shanker, A. (ed.), *Abiotic Stress in Plants-Mechanisms and Adaptations*. ISBN: 978-953-307-394-1. InTech, DOI: 10.5772/24155. http://www.intechopen.com/books/abiotic-stress-in-plants-mechanisms-and-adaptations/emission-and-function-of-volatile-organic-compounds-in-response-to-abiotic-stress.

Spratt, E.D. and R.N. McIver. 1972. Effects of topographical positions, soil test values, and fertilizer use on yields of wheat in a complex of black chernozemic and gleysolic soils. *Canadian Journal of Soil Science*, 52:53–58.

Sprengel, C. 1839. *Die Lehre vom Dünger oder Beschreibung aller bei der Landwirthschaft gebräuchlicher vegetablilischer, animalischer und mineralischer Düngermaterialien, nebst Erklärung ihrer Wirkungsart*. Leipzig.

Stone, M.L., J.B. Solie, R.W. Whitney, and B.R. Raun. 1996. Sensors for detection of nitrogen in winter wheat. In *1996 SAE Symposium on Off-Highway and Engines Conference*, Indianapolis, IN, August 26, 1996.

Sudduth, K.A. and J.W. Hummel. 1993. Portable near infrared spectrophotometer for rapid soil analysis. *Transactions of the ASAE*, 36(1):187–195.

Taylor, S.L., W.R. Raun, J.B. Solie, G.V. Johnson, M.L. Stone, and R.W. Whitney. 1998. Use of spectral radiance for correcting nitrogen deficiencies and estimating soil test variability in an established bermudagrass pasture. *Journal of Plant Nutrition*, 21:2287–2302.

Tellaeche, A., G. Pajares, X.P. Burgos-Artizzu, and A. Ribeiro. 2011. *A Computer Vision Approach for Weeds Identification through Support Vector Machines*. http://oa.upm.es/13714/2/INVE_MEM_2011_115514.pdf.

Tillet, N.D. 1991. Automatic guidance sensors for agricultural field machines: A review. *Journal of Agricultural Engineering Research*, 50:167–181.

USDA. 2010. *Economic Research Service and National Agricultural Statistics Service, Agricultural Resource Management Survey*. Washington, DC.

Webster, R. and T.M. Burgess. 1984. Sampling and bulking strategies for estimating soil properties in small regions. *Journal of Soil Science*, 35:127–140.

Wenda-Piesik, A. 2011. Volatile organic compound emissions by winter wheat plants (*Triticum aestivum* L.) under *Fusarium* spp. infestation and various abiotic conditions. *Polish Journal of Environmental Studies*, 20:1335–1342.

Wibawa, W.D., D.L. Dludlu, L.J. Swenson, D.G. Hopkins, and W.C. Dahnke. 1993. Variable fertilizer application based on yield goal, soil fertility and soil map unit. *Journal of Production Agriculture*, 6:255–261.

Wittry, D.J. and A.P. Mallarino. 2004. Comparison of uniform-and variable-rate phosphorus fertilization for corn-soybean rotations. *Agronomy Journal*, 96:26–33.

Wollenhaupt, N.C. 1996. Sampling and testing for variable rate fertilization. In *Proceedings of the 1996 Information Agriculture Conference*, Urbana, IL, 30 July–1 August, Norcross, GA: Potash & Phosphate Institute, Vol. 1, pp. 33–34.

Wollenhaupt, N.C., R.P. Wolkowski, and M.K. Clayton. 1994. Mapping soil test phosphorus and potassium for variable-rate fertilizer application. *Journal of Production Agriculture*, 7:395–396.

Yates, F. 1948. Systematic sampling. *Philosophical Transactions of the Royal Society*, 241:345–378.

2 Sensing Technology for Precision Crop Farming

Marvin L. Stone and William R. Raun

CONTENTS

2.1 INTRODUCTION

Precision agriculture (PA) or site-specific crop management is a concept based on sensing or observing and responding with management actions to spatial and temporal variability in crops. The "sensing" component of the concept is a fundamental element of PA. Conventional PA technology is commonly associated with geolocation through global positioning system (GPS) or global navigation satellite system (GNSS) technology. A conventional PA application might employ yield monitoring where yield is sensed at GNSS-defined positions in a field. The sensed data are later used to manage the treatment of particular regions of the field based on the earlier yield at those locations. The process between sensing and management actions and the associated time delay can result in two fundamentally different PA techniques, real-time sense and apply (RTSA), and conventional georeferenced PA.

RTSA is based on sensing a parameter and immediately using that information to effect a management action. Geolocation is not required for RTSA, as the site of the measurement can be the same as the site where the management action is performed. In contrast, conventional georeferenced PA technology employs sensing and association of location with sensed data, a map of the sensed information. The management action is not performed immediately but at a later time and the location data of the map is used to allow the management action to be performed at the appropriate site. Both RTSA and conventional georeferenced PA technology employ sensing technology, but the latter required GPS technology to emerge for the technique to be viable. The inclusion of RTSA technology under the definition of PA broadens the definition of PA. RTSA technology emerged earlier than conventional PA and is not necessarily distinguished by its association with GPS technology.

The application of sensing to better manage crop production is a long-practiced technology. Crop irrigation is an example where irrigation is performed to mitigate crop water stress. The Egyptians and Mesopotamians irrigated agricultural crops before 2000 BC (Garbrecht, 1983). Sensing no doubt consists of observing wilt in the crops and irrigating the site to counter the observed crop water stress. Before the current widespread availability of electronics and sensing technology, nonelectronic technologies were developed for sensing soil water availability. Classical tensiometers, for example, required only mechanical technology to effect sensing and allowed irrigation to be performed at sites where water stress existed, perhaps an example of early PA.

The emergence of modern electronics has made broad sensing technologies available for management of crop production. Electronics and sensor systems were exploited for application in agricultural equipment early after the availability of emerging electronics technologies. Logic integrated circuits were introduced in 1964 by Texas Instruments, followed by monolithic amplifiers by Fairchild in 1965. These introductions were quickly followed by the introduction of a seed flow planter monitor by DICKEY-john in 1967 and a spray rate controller by ASCI in 1970 (Stone et al., 2008). Both of these controllers sensed operating parameters on an agricultural machine and in the first case allowed the operator to adjust the machine and in the second case automatically adjusted the machine. By the late 1980s, research was being conducted on conventional PA technologies. A yield measuring system was demonstrated with a local microwave positioning system by Searcy et al. in 1989.

The first satellite in the Navstar GPS system was launched in 1989 and the system became fully operational in 1995 (Hegarty and Chatre, 2008). One of the first combine yield monitors to utilize GPS location data was introduced by Ag Leader in 1982. Soon after, other yield monitoring systems became available, including the John Deere Greenstar system in 1985 (Stone et al., 2008). The sensing component of these systems was a grain flow sensor coupled with a grain water content sensor.

The emergence of yield sensing in a PA system focuses our attention to the matter of the need to effect a management action based on yield measurements. That is, if yield can be effectively measured, what beneficial management action should be taken? This relationship between the sensed parameters and the management action will be referred to here as the "control algorithm." The fundamental components of a PA system consist of sensor or sensors, the control algorithm, and management action. Figure 2.1 shows the relationship in a classical block diagram form and as a feedback system. In the case of yield monitoring, the "controlled variable" would be yield and the sensor system would provide the measured variable, measured yield. The measured variable would be compared to the "desired level of the measured

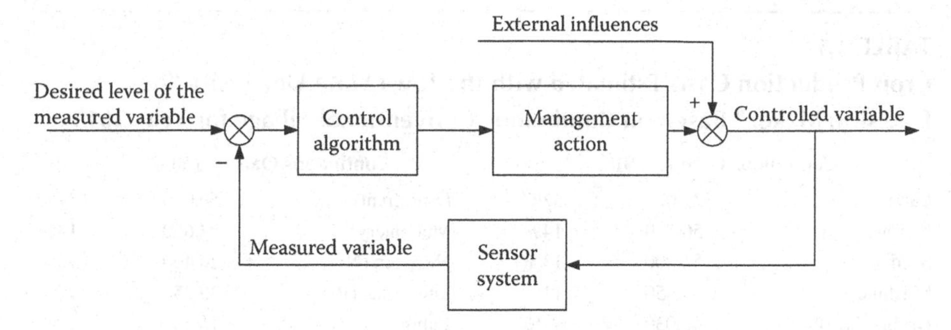

FIGURE 2.1 Block diagram of a precision farming control system.

variable," the yield, and the control algorithm would act upon this comparison to drive a management action. The action might be to increase or decrease fertilizer application rate. External influences, for example, the weather, affect the yield. The process for managing fertilizer application based on harvested yield may encompass many years of farming. In contrast, the timing in some RTSA systems, for example, using canopy reflectance to manage fertilizer application, may encompass only a few hundred microseconds between sensing and taking management action.

The ultimate controlled variable in a precision farming system may be profitability. Profitability should at least be a major factor as we consider development of sensing technologies. It is worthwhile to look at a typical farm budget to allow identification of promising sensing technologies. There is little sense in focusing significant efforts into developing technology if some reasonable return cannot be envisioned. The potential for significant increases in yield as well as significant decreases in production costs become the targets for potential sensor technology. Environmental issues, labor issues, and a myriad of other issues may affect profitability, and must play a role, but a simple examination of a typical farm budget provides some assistance in identifying important target areas for a focus on sensor technology.

Table 2.1 presents example farm production costs for continuous farming of corn and for small grains in Iowa. For both of the cases shown, the fixed land and machinery costs rank in the top four categories. The primary variable costs identified are for nitrogen, phosphorus, and seed. The potential for both seed technologies in corn and nitrogen fertilizer management technologies are very good. In the case of nitrogen fertilizer, nitrogen use efficiency (NUE) is 33% worldwide (Raun and Johnson, 1999), indicating that two-thirds of the fertilizer that is applied is not recovered in harvests. The environmental impact of nitrogen use inefficiency is significant and well understood (Matson et al., 1997; Whitson and Walster, 1912). Major portions of that inefficiency can be addressed through precision nitrogen management (Cassman et al., 2002; Roberts, 2007) and crop reflectance-based sensor technology has been demonstrated to be effective in that role (Li et al., 2009).

Iowa corn and small grains production costs cannot of course provide the broad perspective regarding the potential for sensor technology, but do provide a technique for identification of the potential with those crops. A very different perspective exists with irrigated crop production where water use is a major factor and with other crops

TABLE 2.1
Crop Production Costs Estimated with the Iowa State University "Ag Decision Maker" Based on Continuous Conventional Tillage for Year 2012

Continuous Corn (US $)			Continuous Oats (US $)		
Land (rent)	129,000	32%	Land (rent)	50,000	31%
Nitrogen (N)	56,700	14%	Machinery	22,650	14%
Seed	51,000	13%	Nitrogen (N)	20,400	13%
Machinery	43,650	11%	Phosphate (P)	20,250	13%
Drying, handling	30,030	7%	Labor	16,500	10%
Phosphate (P)	19,840	5%	Lime	14,500	9%
Labor	16,673	4%	Seed	10,000	6%
Potash (K)	13,750	3%	Potash (K)	7,200	4%
Insurance	11,250	3%	Herbicide	0	0%
Herbicide	10,000	2%	Insecticide	0	0%
Insecticide	9,200	2%	Insurance	0	0%
Interest	7,331	2%	Miscellaneous	0	0%
Lime	4,835	1%	Interest	0	0%
Miscellaneous	4,500	1%	Drying, handling	0	0%
Total	407,759	100%	Total	161,500	100%

where other potential applications exist. The sensor technologies reviewed below are not comprehensive, but address those technologies that appear to have good potential.

2.2 CONTROL ALGORITHMS

2.2.1 Sensor-Based Algorithms for Fertilizer Nitrogen

It is important to begin this section with work coming from the late Norman E. Borlaug. Borlaug (2000) stated that agricultural scientists have an obligation to inform others concerning the magnitude and seriousness of arable land, food, and population problems that lie ahead, even with breakthroughs in biotechnology. One of these problems has been the seriousness of how fertilizer nitrogen is used and the aftermath/consequences of its potential misuse for cereal grain production in the world. Algorithms that can objectively apply the right source, at the right rate, at the right time, and in the right place (IPNI, 2012) will be those that impact our world both today and well into the future.

In this light, this review hopes to highlight algorithms and methodologies that will make a difference in how nitrogen is managed. What exactly comprises these algorithms and how each of the components is used is delineated in the following discussion.

2.2.2 Algorithm Components

Yield goal: Grain yield goals have been used for many years to estimate preplant fertilizer N rates. Early work by Dahnke et al. (1988) noted that the yield goal was the "yield per acre you hope to grow." They further noted that what you hope to grow and what you end up with are two different things. Early work by Allison (1955) noted that accurate values for N removed in harvested crops were commonly available. However, leaching of N as NO_3–N and volatilization losses remained problematic in N balance studies. As a result, he noted that N balance sheets seldom added up (outputs minus inputs) and that helped to explain why recoveries of nitrogen in the crop were often only 50% of that added as fertilizer.

The yield goal concept at North Dakota State University recommends using the highest yield attained in the last 4–5 years and is usually 30%–33% higher than the average yield. Rehm and Schmitt (1989) noted that with favorable soil moisture at planting, it would be smart to aim for a 10%–20% increase over the recent average when selecting a grain yield goal. They also indicated that if soil moisture is limiting, the use of history and past maximums (used to generate the average) may not be the best method for setting a grain yield goal for the upcoming crop. In Nebraska, fertilizer N rate recommendations for corn and that use yield goals are 1.2 pounds N/bushel (0.02 kg N/kg grain), and that can include soil test and soil organic matter credits (Shapiro et al., 2008). For winter wheat, the yield goal-based N rate recommendation is 2 pounds of N for every bushel (0.03 kg N/kg grain) you hope to produce (Zhang et al., 2010).

The use of farm or county averages was not suggested for progressive farmers concerned with high farm profitability (Rehm and Schmitt, 1989). Black and Bauer (1988) reported that the grain yield goal should be based on how much water is available to the winter wheat crop from stored soil water to a depth of 1.5 m in the spring plus the anticipated amount of growing-season precipitation.

More recent sensor technologies, weather forecasting, and crop modeling have enabled the development of methods for predicting potential grain yields, and have allowed for in-season nutrient adjustments to reflect early crop development and growing conditions. Nonetheless, "yield goals" as we understand them are an incredibly useful term/concept because they embed intrinsic knowledge about the environment, climate, terrain, and a holistic understanding of the crop management system that is being practiced. Furthermore, these yield goals are understood to be unique for each and every producer, as he or she will know within certain boundaries, what their yield goal should be for any given year. This is predicated on years of experience, on that farm, and encompass the environmental conditions encountered over time.

Their "yield goal" embeds added information and experiences that include seasonally late or early planting, variety and/or hybrid, weed pressure, and even sensitivity to whether or not the ground is being rented, or that it was owned. All of this impacts the "yield goal" that a producer might establish at the very beginning of the season, with no knowledge whatsoever of what the environment/climate might deliver. As such, the yield goal can take on transitory properties that may or may not be the same, for the same producer, next year. It is also likely that for the exact same set of conditions, land being farmed, and crop being grown, two different producers would have completely different yield goals. This variability does not diminish the value and/or importance of yield goals, but rather highlights its intrinsic anthropological nature.

Yield potential (YP0): Because "yield goal" can take on so many different forms, some clearly numeric and linear, and others that are more composite and literal, it remains in fact, a highly desirable term. Sequential to this is thinking that researchers could indeed generate the parameters and inputs to predict yield goal. Is it even possible? If yield goals could indeed be predicted, they would make complete sense since known nutrient removal amounts are understood at a given level of yield (IPNI, 2014). Replenishing those nutrients at X-yield level using an expected efficiency for each nutrient in question would indeed be diligent and acceptable to any producer. Or, would it rather be possible to predict one of the important fragments of "yield goal" that is again understood as a more holistic all-encompassing quantity? An all-important fragment of "yield goal" is "yield potential" or the upper level yield boundary that is dictated by a host of factors. Nonetheless, can the menagerie of information and data available to a producer prior to planting be used to predict yield potential? At this juncture, the answer is likely no, again, for a preplant decision.

Raun et al. (2001) has focused on the task of predicting what that upper yield boundary might be midseason (rather than preplant), not what the actual yield would be, but what it could be. And this is based on midseason "progress" or the midseason "report card." Their work further showed that a normalized difference vegetative index (NDVI)-based formula accounted for 83% of the variability in actual grain yield, and ended up being the cornerstone of their entire YP0-RI concept paper (Raun et al., 2002). In this work, they went on to define the prediction of yield potential as INSEY or in-season estimated yield, and that continues to be used in the literature today. Their prediction of yield potential or INSEY was robust in that it accounted for the number of days from planting to sensing where growth was possible. This for winter wheat would be days where GDD >0 (GDD = $[T_{min} + T_{max}]/2–4.4°C$), where T_{min} and T_{max} represent daily ambient low and high temperatures. This index in turn represented growth rate, since total biomass can be estimated using NDVI, and when divided by the number of days where growth was possible was equivalent to biomass produced per day. This midseason estimate of growth rate was then found to highly correlate with actual yield. INSEY or yield potential could thus be estimated at any location where the planting date was known, and an accurate field value of NDVI was collected. And this was provided that sufficient data had been collected to generate the INSEY-measured grain yield equation (Lukina et al., 2001; Raun et al., 2001) and that is likely to be environment-specific as present research has shown (http://www.nue.okstate.edu/Yield_Potential.htm).

Basing fertilizer N rates on the amounts removed in forages and/or from cereal grains has also been an important practice. Work by Lukina et al. (2001) noted that INSEY could be used to compute the potential N that was removed from the grain. In-season N fertilization needs were then considered to be equal to the amount of predicted grain N uptake (potential yield times grain N) minus predicted early-season plant N uptake (at the time of sensing), divided by an efficiency factor. The use of INSEY could replace N fertilization rates determined using production history (yield goals), provided that the production system allows for in-season application of fertilizer N.

Response index (RI) or N responsiveness: Some of the earlier sensor work with algorithms was conducted by Mullen et al. (2003) who predicted N responsiveness or the likelihood of obtaining a response to applied fertilizer using NDVI data collected from an N-rich strip (area where preplant N had been applied at a rate where no N deficiency would be encountered during the growing season) and the farmer practice. Dividing the NDVI from the N-rich strip by the NDVI value from the farmer practice was termed the response index or RI_{NDVI}. The response index determined from these same two plots, but using grain yield at harvest, or $RI_{Harvest}$ (yield of the N-rich strip/yield of the farmer practice) was found to be highly correlated with RI_{NDVI}. Or in other words, the midseason RI_{NDVI} could be used to project what kind of final RI would be found at the end of the season, or $RI_{Harvest}$. Being able to predict this environmentally dependent statistic empowered those developing sensor-based applications for nutrient management.

What was interesting from this work was the value of the N-rich strip by itself. Independent of any methodology for predicting midseason yield potential, knowledge of what the N-rich strip or RI was could dictate whether or not a producer applied N fertilizer. This was simply a yes or no question and that was in no way bound to a specific rate. If producers could not see the N-rich strip in the middle of the season, it was then highly likely that they would not see a response to added fertilizer N and the decision to not apply any more N was embraced. Alternatively, if visible differences in biomass and intensity of green color were clearly different, the demand for added N was evident. Equally important concerning work with the N-rich strip was that the responsiveness or RI changed radically from year to year and at the same location (Arnall et al., 2013). This was noted for both corn and wheat throughout the Great Plains region (Raun et al., 2011).

Combined use of yield potential (YP0) and RI or (YP0-RI): Algorithms and N fertilization methods have been present in the sensing world for some time. Raun et al. (2004) noted that basing midseason N fertilizer N rates on predicted yield potential and a response index could increase NUE by over 15% in winter wheat when compared to conventional methods. Their work further noted that using a sensor-based algorithm employing both yield potential and N responsiveness could increase yields and decrease environmental contamination due to excessive N fertilization.

Fundamental to this work were pointed findings in the entire Great Plains region documenting that yield potential (yield level) and nitrogen responsiveness were independent of one another (Raun et al., 2011; Arnall et al., 2013). Their independence and knowledge that both impact crop demand for N necessitates the inclusion of both in any reasonable algorithm expected to determine accurate midseason fertilizer N rate recommendations. This was facilitated by early NDVI sensor work that made midseason fertilizer N rate recommendations possible (Stone et al., 1996).

A variant of the YP0-RI approach has two added findings that clearly apply to sensor-based N rate algorithms. The first is the knowledge that not only did the YP0-RI approach work to improve fertilizer NUE over conventional methods, but that the spatial scale at which sensor-based systems should operate is 0.4 m^2 (Raun et al., 2005b). Furthermore, and that is tied specifically to the GreenSeeker™ NDVI sensor, is determining the spatial variability within each 0.4 m^2 area using the coefficient of variation (CV) from NDVI readings. When CVs were higher (within each 0.4 m^2), N rates were lowered due to the expressed variability within that area. When

CVs were lower, N rates were increased due to the improved homogeneity of the plant stand and that reflected a higher yield potential.

Sufficiency: Roger Bray's original nutrient mobility concept helped to clarify why sufficiency could only be used for *immobile* nutrients (Bray, 1954). In this work, he delineated the root system sorption zone for mobile nutrients and the root surface sorption zone for immobile nutrients.

In soils, plants respond to the total amount present for mobile nutrients and to the actual concentration for immobile nutrients. Or, yield is directly related (proportional) to the total amount of a mobile nutrient present in the soil. As such, nutrient depletion of the root system sorption zone is dependent on the environment, or growing conditions. Response of crops to mobile nutrients is considered to be linear (mobile nutrient availability is unaffected by reaction with the soil). However, yield response to an immobile nutrient is not related to the total amount present in the soil, but instead is a function of the concentration at, or near, the root surface. Because of this, nutrient depletion in the root surface sorption zone is considered to be independent of the environment.

Varvel et al. (1997) used chlorophyll meter readings to calculate a sufficiency index ([as-needed treatment/well-fertilized treatment] × 100). This was then used to make in-season N fertilizer applications when index values were below 95%. The rate used was 30 lb N/ac, checked every 7 days and N was applied (if needed) all the way to the R3 corn growth stage.

Why is this so important? If sufficiency is dependent on the environment and in-season growing conditions, midseason fertilizer rates should be tied to yield level. This is precisely why this issue has been raised because those using the sufficiency approach have not included yield level or an expected yield potential to refine the final recommended rate.

Maximum return to nitrogen (MRTN): MRTN provides N rate recommendations based directly from analysis of N response trials and return to N (corn nitrogen rate calculator; http://extension.agron.iastate.edu/soilfertility/nrate.aspx). This approach does not employ the use of sensors, and remains relatively popular. Fertilizer N rates are determined from yield increases to applied N and current grain and fertilizer prices; but not to yield level (Sawyer et al., 2006). Further clarification by Larry Bundy noted that their results provided no clear indication of a change in N rates over time. Reasons for similar optimum N rates where yields have increased substantially include more efficient utilization of available N by the crop and increased soil N supplying capability.

Their summary indicates that the flat net return surrounding the N rate at MRTN reflects small yield changes near optimum N thus indicating that choosing an exact N rate is not critical to maximize profit. In light of the known environmental problems coming from excess N in agriculture, particularly corn, this laissez faire approach is highly disconcerting.

2.2.3 Algorithm Applications and Concepts

Work by Roberts et al. (2009) noted that because sensor information can be processed into an N rate that approximates optimum N, sensor-based N applications can also have environmentally beneficial effects. Added studies in Missouri (Scharf et al., 2011) showed that sensor-based N management reduced N applied by 25% beyond what was removed in the grain, thus reducing unused N that can move to water or air. They further noted that a midseason sensor-based approach could choose N rates for corn that perform better than rates chosen by producers. Comprehensive field work by Kitchen et al. (2010) showed that crop-canopy reflectance sensing delivered improved N management over conventional single-rate applications. More recent studies by Torino et al. (2014) showed that early season estimates of crop N status and yield potential may be more accurate if red-edge vegetation indices were used.

What is apparent in the literature is that sensor-based N management has taken hold in the developed world. It is a technology that embeds sound agronomic principles within engineering applications that can be delivered at whatever scale producers seek.

2.3 YIELD MONITORING

Yield monitoring is an important source of information in PA systems. Comparative yield information may be used to assess the performance of farming practices. Absolute yield levels may also be used directly in control algorithms. The obvious example is in managing fertilization where removal of nutrients through harvesting can be determined as a fraction of yield and an algorithm may be used to compute the necessary nutrient replacement. In both cases, site-specific yield information may be required at a spatial resolution and accuracy suitable for the application.

Georeferenced yield monitoring emerged as a commercial product in the early 1990s and was the first conventional PA technology to become widely used. Fundamental types of yield monitor sensing systems have evolved and are reviewed by Demmel (2013). Table 2.2 presents a classification of yield sensing systems and identifies typical crop applications where the technology has been used.

Yield is normally expressed as a volume per unit area and might be better referred to as area-specific yield. Expressed this way, yield comparisons may be made without regard to area and total harvest mass or volume (total yield) can be readily computed from average yield and area. The expression of yield as an area-specific quantity requires monitoring area harvested as an input to provide yield data. Harvested area

TABLE 2.2
Yield Sensor Types

Type	Typical Crop Application
Momentum difference	Grain
Mass flow detection	Grain
Load cell based	Grain
Radiation based	Grain
Optical density	Cotton, sugarcane, peanuts
Volumetric metering	Grain
Batch weighing systems	Potatoes, specialty crops

TABLE 2.3
Variables in Yield Monitoring Equations

Variable	Definition	Unit (L—Length, M—Mass, T—Time)
Q	Volumetric rate	L^3/T
A	Cross-sectional area	L^2
V	Average velocity of the material	L/T
v_t	Travel velocity of the machine	
$\dot{m}$	Mass flow rate	M/T
M	Mass measurement	M
ρ	Density	M/L^3
ρ_b	Crop bulk density	M/L^3
V	Volume of the control volume	L^3
L	Control volume length	L
F	Reaction force	ML/T^2
Y	Yield	L^3/L^2
W	Effective harvesting width	L

is usually computed as effective harvesting width w multiplied by the length of travel over which the measurement of harvested mass or volume is made. Yield may be expressed as the volumetric flow rate divided by the product of travel velocity and effective harvesting width as shown in Equation 2.1 (the variables are defined further in Table 2.3). Error in yield monitoring systems arises from not only the crop flow measuring system but also from the speed and effective width inputs.

$$Y = \frac{Q}{v_t w} \tag{2.1}$$

The first principle equations governing flow measuring techniques are reviewed below. The purpose of examining the governing equations is to gain a better understanding of the operating principles and identification of some of the potential interferences affecting these sensors.

2.3.1 Mass Flow-Based Yield Measuring Systems

Figure 2.2 depicts a continuous weighing system. This type of flow measuring device may be modeled as a control volume of fixed length, l, and a mass detection means denoted in the figure by m. The harvested crop flows through the control volume at an average velocity, v.

Flow through the system may be calculated as cross-sectional area multiplied by average material velocity as shown in Equation 2.2. This equation may be expressed in terms of mass flow rate by multiplying Equation 2.2 by density, ρ, as show in Equation 2.3.

$$Q = Av \tag{2.2}$$

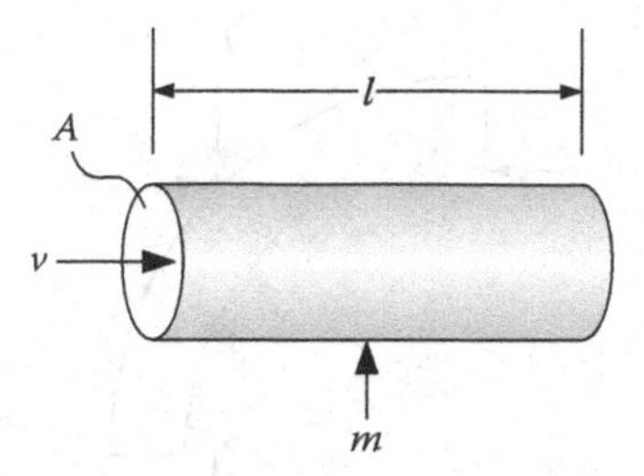

FIGURE 2.2 Weighing conveyor (mass flow detecting) sensor.

$$\dot{m} = \rho Q = \rho A v \tag{2.3}$$

The density of the crop flowing through the sensor can be expressed as mass divided by the control volume or cross-sectional area multiplied by the length as shown in Equation 2.4. Substituting this density into Equation 2.3 as shown in Equation 2.5 provides a mass flow equation:

$$\rho = \frac{m}{V} = \frac{m}{Al} \tag{2.4}$$

$$\dot{m} = \rho A v = \frac{m}{Al} A v = \frac{mv}{l} \tag{2.5}$$

The mass flow Equation 2.5 no longer includes density of the crop in the control volume and issues of partial fill in the control volume are eliminated. Equation 2.5 may be converted to a volumetric flow equation as shown in Equation 2.6 by dividing both sides of the equation by bulk density of the harvested crop. Equation 2.1 may then be used to compute yield from the mass measurement by substituting Q from Equation 2.6 into Equation 2.1.

$$Q = \frac{\dot{m}}{\rho_b} = \frac{mv}{\rho_b l} \tag{2.6}$$

$$Y = \frac{(mv/\rho_b l)}{v_t w} = \frac{mv}{\rho_b l v_t w} \tag{2.7}$$

Equation 2.7 demonstrates that the variables that contribute to yield sensing with this type of flow meter include mass detection, m, as well as crop bulk density, travel velocity, and effective harvesting width. The control volume length and material velocity would normally be fixed and probably included in a calibration coefficient.

Figure 2.3 depicts a momentum difference-based mass flow meter. The momentum equation from fluid mechanics can be applied in this concept to determine the effect of the momentum change of the material flowing through the sensor. Equation

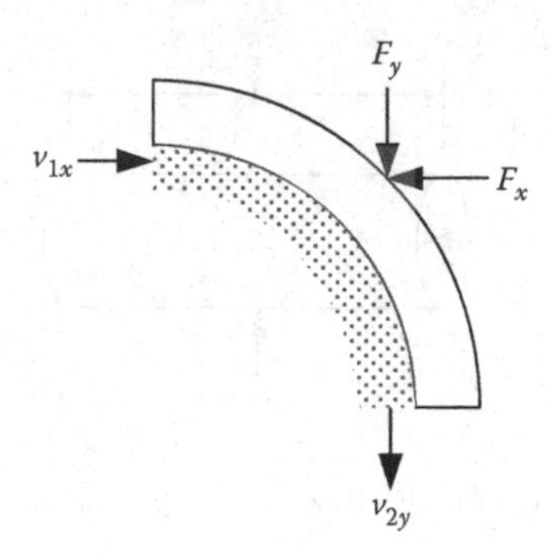

FIGURE 2.3 Deflection plate-based (momentum difference) sensor.

2.8 presents the classical fluid mechanics equation for conservation of momentum in one direction (x). Based on that equation, we can write the equations for reaction forces in F_x and F_y due to the change in momentum of the material flowing through the sensor, Equations 2.9 and 2.10. The volumetric flow rate based on F_y can be found by rearranging Equation 2.10 as shown in Equation 2.11 and an equation for yield found by substituting Equation 2.11 into 2.1 giving Equation 2.12. Equation 2.12 demonstrates that momentum-based flow meters are susceptible to errors in bulk density, travel speed, and effective harvesting width and require that the material velocity through the meter remain constant or be accounted for

$$\sum F_x = \rho Q(v_{x2} - v_{x1}) \tag{2.8}$$

$$F_x = -\rho Q v_1 \tag{2.9}$$

$$F_y = \rho Q v_2 \tag{2.10}$$

$$Q = \frac{F_y}{\rho v_2} \tag{2.11}$$

$$Y = \frac{F_y}{\rho_b w v_2 v_t} \tag{2.12}$$

Equation 2.12 does not account for friction of the material flowing through the meter on the force measurements. That effect is important and can be visualized with a geometry where there is no curvature through the meter. In this case, the incoming and outgoing velocity would be the same. In this case too, the friction of the material would tend to force the deflector toward the discharge. This frictional effect is managed by minimization where a deflector friction coefficient with the flowing material is kept low and through calibration.

Crop flow meters that utilize force or mass measurement are susceptible to errors due to motion of the harvesting machine. Gravity forces act on the devices and though they may be removed by calibration for a machine on the level, tilting of the machine produces errors. In addition, accelerations of the machine also act on the meters. Taylor et al. (2011) investigated these effects in grain flow yield monitors with corn in field tests and found typical errors of $-1.1 \pm 3.6\%$ (where the random error is expressed as ± 2 standard deviations here), which were confirmed in bench testing by Demmel (2013) with errors of $-1.9 \pm 3.4\%$. Yield monitors with these errors have been successful in the market. Micromachined silicon accelerometers and gyroscopes are now available that would allow measurements of accelerations and removal of gravitational and motion effects.

Force-based crop flow meters have also been applied in sugarcane and root crops. Root crop systems have been reviewed by Demmel (2013) and research regarding sugarcane systems has been reviewed by Price et al. (2011). These root crop systems were found to have errors similar to those for grain. The sugarbeet systems had significantly higher error and Price et al. (2011) developed an alternative optical system.

2.3.2 Optical and Radiation-Based Yield Measuring Systems

Optical and radiation-based crop flow meters have been reported and eliminate the inertial and body force effects of the force measuring devices. Optical-based systems typically utilize an optical density measurement where light transmission through the flowing crop is related to the crop mass (Thomasson et al., 1999; Wilkerson et al., 2001). These systems have been applied in cotton and are the basis of commercial cotton yield monitors (Vellidis et al., 2003). Optically based flow meters have been applied in sugarcane where optical fiber was used to effect a volume measurement of conveyor sections (Price et al., 2011). Thomasson (2006) described an optical method used in peanut yield monitoring and this technology is now available from several commercial manufacturers. Porter et al. (2013) described field testing of Ag leader cotton yield monitors applied in peanuts with under 10% error. Persson et al. (2004) demonstrated an optical yield measurement with potatoes and the sensor worked well and is a potential option for tuber yield monitoring with errors in the 1% range.

Radiation-based mass detection using a radioactive sealed source has been employed in commercial grain yield monitors in Europe. These devices delivered accuracy comparable to force-based designs (Demmel, 2013) but face heavy regulatory limitations in some countries and decommissioning the devices is expensive and regulated (Government of the UK, 2008). X-ray systems have been considered for forage applications (Kormann, 2004; Wild et al., 2014) with promising results.

Most combine yield monitors also incorporate grain moisture measurement. This measurement can contribute to applications in precision farming. With regard to nutrient replacement control algorithms, the nutrient contents in the grain are calculated on a dry weight basis and correction for moisture content of yields should be done.

2.3.3 Yield Maps

Data from combine yield maps are normally recorded with geographical coordinates associated with each measurement. The data are typically placed into commercial farm GIS packages where the data are placed into map form. Interpolation is used to allow a spatially continuous representation of yield to be made. Noack et al. (2006) discuss the issues regarding the effect of interpolation methods on the representation of the data.

2.4 CROP CANOPY REFLECTANCE SENSING

2.4.1 Spectral Indices and Their Relationship to Crop Management

The green color of plants is largely due the scarcity of red light from their reflected energy. The strong absorption of red light by chlorophyll is associated with this effect and provides a convenient remote sensing method for the assessment of chlorophyll in plant canopies. Canopy reflectance sensing technology is based on measurements of reflected energy in different portions of the spectrum and is generally reliant on ratios of one portion of the spectra to another. This ratio technique allows spectral differences to be assessed and yet the ratio is largely invariant to the variation of total spectral energy.

Rouse et al. (1974) identified several "vegetative indices" and developed the NDVI. NDVI was used to relate green biomass development during spring "green-up." NDVI is a spectral index relating the difference between reflected energy in the near-infrared (NIR) portion of the spectrum and reflected energy in the red portion of the spectrum. The difference was normalized into a ratio by dividing this difference by the sum of both the NIR and red portions of the spectral energy, as shown in Equation 2.13.

$$\text{NDVI} = \frac{\rho_{\text{NIR}} - \rho_{\text{red}}}{\rho_{\text{NIR}} + \rho_{\text{red}}} \tag{2.13}$$

NDVI is effective in distinguishing chlorophyll-containing plants against a soil background as seen in Figure 2.4. Spectra for wheat at Feekes 5 (a short grass stage) is compared to bare soil spectra. The NIR and red regions of the spectra are marked by the arrows. The NIR reflectance of the wheat is much greater than the NIR reflectance of the soil while the red reflectance of the wheat plant is much less than the red reflectance of the soil. The resulting NDVI for the wheat is much greater than that for the soil.

The potential interference due to variations in soil background is a potential issue. Figure 2.5 shows the spectral reflectance of soils taken from across Oklahoma. Soils in this region vary widely in color and texture. NDVI for each of the soils shown are tabulated on the right side of the figure. The resulting variation in NDVI is small enough so as not to be a significant factor in the measurement of crop NDVI against a soil background.

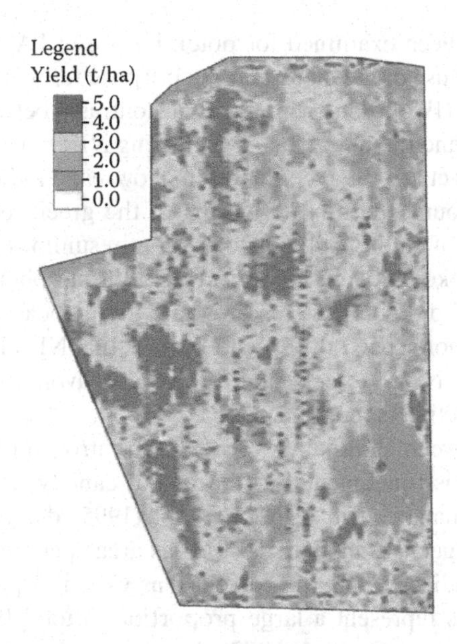

FIGURE 2.4 (**See color insert.**) Typical combine-derived yield map.

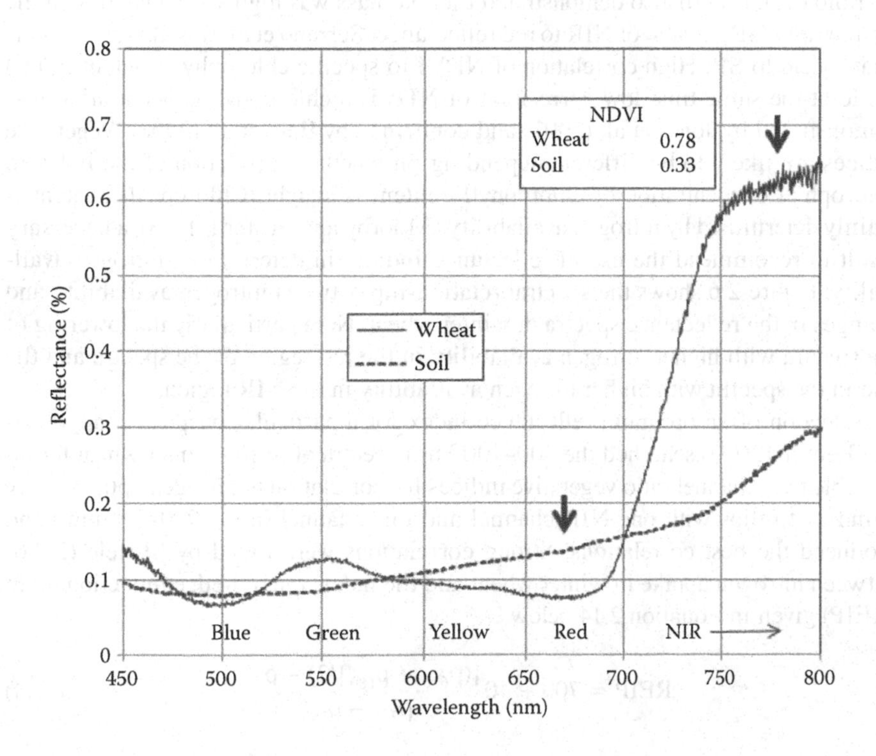

FIGURE 2.5 Spectral reflectance of wheat and soil.

Many indices have been examined for potential use in PA. Care must be taken regarding the proposed use of the index and the purpose for which the index is being selected. Tucker et al. (1979) demonstrated that contrasts between the red chlorophyll absorbing bands and the internal leaf-scattering NIR bands could be related to crop biomass. Peñuelas et al. (1994) compared narrow-band indices and NDVI taken on single leaves and found the first derivative of the green reflectance at 525 nm (Dg) a better indicator of chlorophyll concentration in sunflower leaves than NDVI. Narrow-band indices like Dg were found to track diurnal photosynthetic light use efficiency where NDVI did not. This result would first appear to be a disadvantage for application of the more spectrally spread indices like NDVI, but lack of diurnal variation and ability of NDVI to correlate to biomass favors the use of wider band indices if biomass detection is required.

The connection between chlorophyll content and nitrogen uptake by the vegetative portions of plants is fundamental to the use of canopy reflectance indices for nitrogen fertilization management. Gamon et al. (1995) demonstrated that NDVI is well correlated to vegetative biomass as well as area-specific chlorophyll content (g/m^2). Their results indicated that the correlations were independent of plant type. Photosynthetic proteins represent a large proportion to total leaf nitrogen content (Evans, 1983). Thomas and Oerther (1972) demonstrated that nitrogen content in plant vegetative biomass could be estimated with 550 nm reflectance measurements. Serrano et al. (2000) also demonstrated that biomass was highly correlated with the simple ratio (SR), a ratio of NIR to red reflectance. Serrano et al. was also able to correlate yield to SR. High correlation of NDVI to specific chlorophyll content (g/m^2) while at the same time low correlation of NDVI to chlorophyll concentration was demonstrated by Jones et al. (2007) and confirmed by Eitel et al. (2008). Vegetative indices are likely to be different depending on whether correlation of the index to chlorophyll concentration or chlorophyll content is sought. Chlorophyll content is mainly determined by nitrogen availability (Moorby and Besford, 1983), a necessary result to recommend the use of reflectance indices in determining nitrogen availability. Figure 2.6 shows the spectral relationship between nitrogen availability and changes in the reflectance spectra of winter wheat. Note particularly the lowering of the spectra with higher nitrogen availability in the red region of the spectra and the rise in the spectra with higher nitrogen availability in the NIR region.

Selection of an optimum reflectance index for a particular purpose is appropriate. Reusch (2005) searched the 400–1000 nm spectra of *in situ* winter wheat for all possible two-channel ratio vegetative indices for correlation to nitrogen uptake. They found that ratios with one NIR channel and one channel in the 730–750 nm band produced the best correlations. Strong correlations were found by Mistele (2006) between nitrogen uptake in winter wheat and the index, the red-edge inflection point (REIP) given in Equation 2.14 below.

$$\text{REIP} = 700 + 40\frac{[(\rho_{670} + \rho_{680}]/2) - \rho_{700}}{\rho_{740} - \rho_{700}} \tag{2.14}$$

REIP was developed by Guyot et al. (1988) and has been associated with better sensitivity to biomass where canopy coverage is high. Horler et al. (1983) attributed

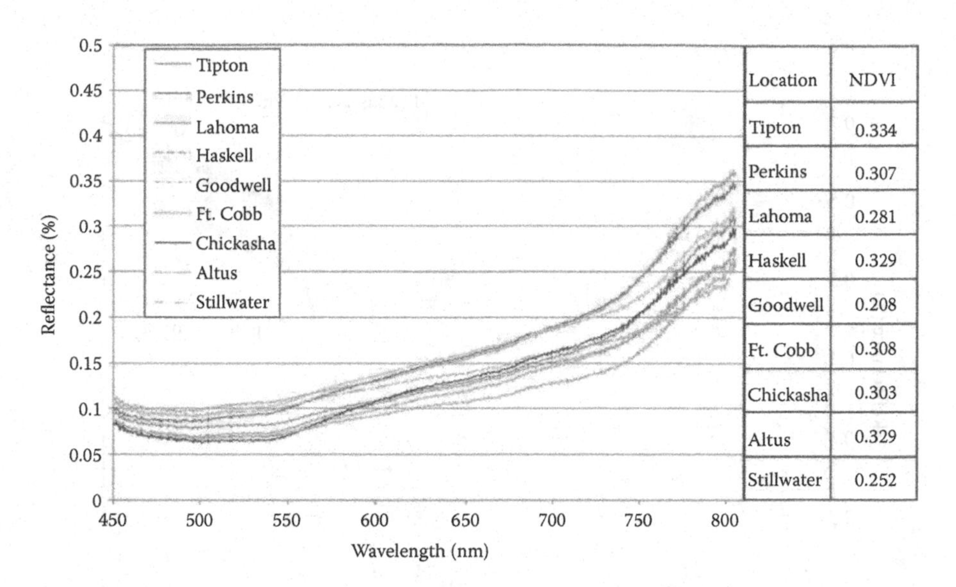

Location	NDVI
Tipton	0.334
Perkins	0.307
Lahoma	0.281
Haskell	0.329
Goodwell	0.208
Ft. Cobb	0.308
Chickasha	0.303
Altus	0.329
Stillwater	0.252

FIGURE 2.6 Spectral variation in Oklahoma soils (bare soils taken from the surface). Left: Spectral reflectance. Right: Associated NDVI.

the shift of the inflection point toward the NIR for higher chlorophyll concentrations to higher internal scattering within the leaf tissue. Figure 2.7 illustrates the shift where the horizontal grey arrow is placed near the inflection points. The inflection moved to the right (toward the NIR) with the higher nitrogen availability.

Normalized difference red edge index (NDRE) has been successfully used to describe nitrogen stress in wheat by Rodriguez et al. (2006). Barnes et al. (2000) developed the NDRE to detect crop nitrogen and water stress as shown in Equation 2.15. The spectral ranges are shown in the subscripts for reflectance data with this index. Barnes et al. (2000) used NDRE with NDVI to compute a canopy chlorophyll content index (CCCI), which they found well correlated with nitrogen uptake.

$$\text{NDRE} = \frac{\rho_{760-850} - \rho_{690-730}}{\rho_{760-850} + \rho_{690-730}} \tag{2.15}$$

2.4.2 Application of Canopy Reflectance Indices

Canopy reflectance technology has been developed for use in nitrogen management for most major field crops. Nitrogen uptake, the nitrogen content of the vegetative biomass, has been used in control algorithms as an input parameter in the algorithms for top-dressing fertilization (fertilizer applied after the crop has emerged) (Raun et al., 2002; Liang et al., 2005). In winter wheat, for example, nitrogen uptake by the plant provides a measure of nitrogen availability. The availability of nitrogen in the soil is heavily dependent on the climatic conditions and canopy reflectance is used to assess the need for additional nitrogen fertilizer (Johnson and Raun, 2003).

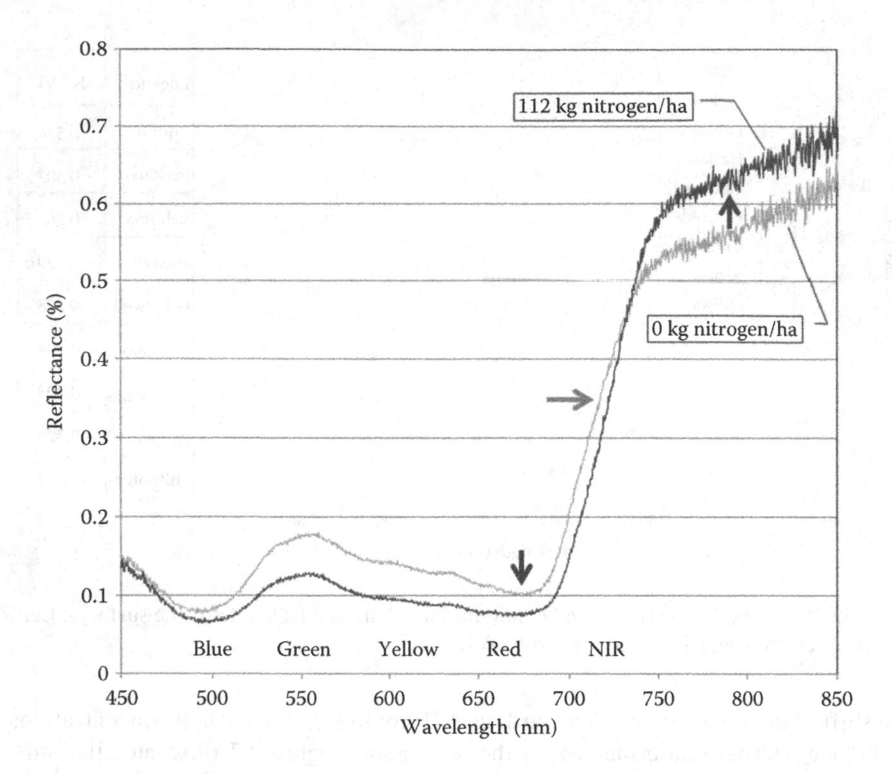

FIGURE 2.7 Spectral reflectance response to higher levels of nitrogen fertilization in wheat.

Raun et al. (2002) proposed an algorithm for use in wheat that uses canopy reflectance to predict yield potential as well as assessing soil nitrogen availability. Figure 2.8 illustrates their yield potential prediction method. Yield potential is the maximum yield that would be expected for a particular soil condition with ideal weather conditions. The actual yield as seen in the figure is distributed along the yield axis and is normally reduced from the yield potential due to weather, insects, disease, and other stressors. The yield potential is represented by some maximum envelope and a standard deviation above the average was used by Raun et al. (2002) to estimate yield potential. The reflectance index used in their method, INSEY, was NDVI modified by a measure of growing days since planting. NUE averaged over all of their sites was improved by 15% over conventional nitrogen fertilization practice. Ortez-Monasterio et al. (2014) reported improved profitability through nitrogen savings of greater than $60 per ha over 432 field trials covering more than 6000 ha in farmers' fields in Mexico using a similar system.

Mistele and Schmidhalter (2010) validated the use of REIP for application in biomass and nitrogen uptake measurement in wheat with a tractor mounted system in a 3-year field study. They concluded that the system could be used for nitrogen management in heterogeneous fields.

Wright et al. (2004) examined the use of canopy reflectance as a means for managing grain protein in wheat. They examined various spectral indices and

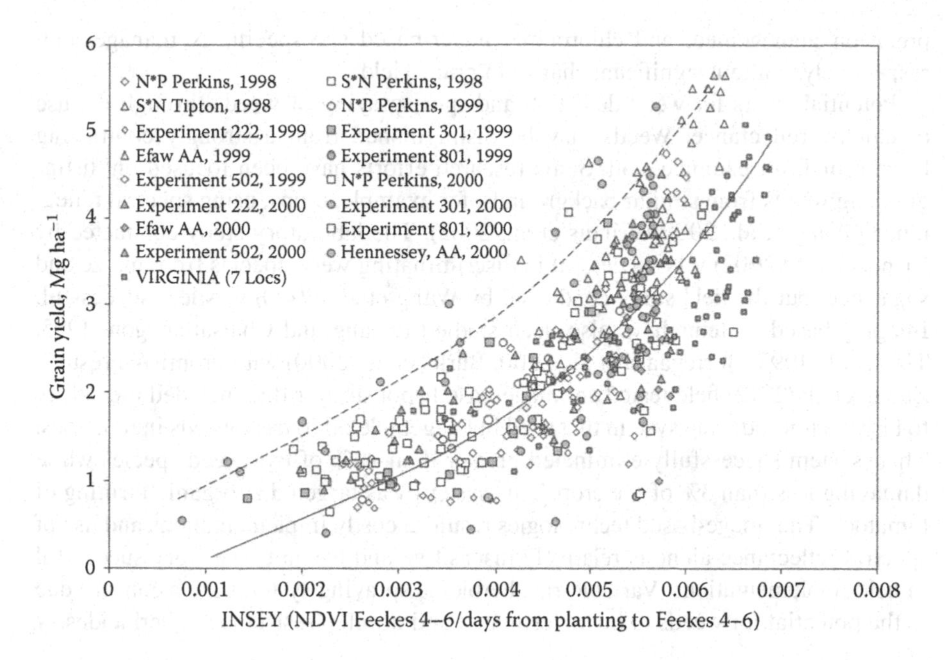

FIGURE 2.8 Relationship between the NDVI-based INSEY index and yield. (After Raun, W.R. et al. 2005a. *Communications in Soil Science and Plant Analysis*, 36:2759–2781.)

concluded that NDVI was as effective as other indices and recommended that canopy reflectance could be effective in using nitrogen fertilization to manage grain protein.

Kitchen et al. (2010) used a transformed NDVI index, that is, the inverse simple ratio (ISR) of a nitrogen-sufficient strip created in maize fields to allow computation of a sufficiency index, SI, ISR sufficient/ISR target crop using a Holland Scientific ACS-210-based system. They conducted 16 field-scale experiments in maize fields over four seasons in three different soil areas. They developed and applied a maximum profitability algorithm and achieved a $25–$50 per ha profit improvement using the technique.

Raper et al. (2013) evaluated three commercial sensor systems for application in management of nitrogen in cotton. The Yara N-Sensor® (Yara International ASA, Oslo, Norway), GreenSeeker® Model 505 Optical Sensor Unit (NTech Industries, Inc., Ukiah, CA), and Crop Circle® Model ACS-210 (Holland Scientific, Inc., Lincoln, NE) systems were evaluated. Plant height relationships with NDVIs were strong but sensor readings did not consistently predict cotton leaf N status before early flowering.

In rice culture in China, Xue et al. (2004) found the relationships between leaf N accumulation and reflectance in the green band and NIR to green ratio index were consistent across the whole growth cycle. The ratio of NIR to green (R810/R560) was linearly related to total leaf N accumulation, independent of N level and growth stage with correlation in the 0.96 range. Yao et al. (2012) demonstrated an increased partial factor productivity of rice farmers by 48% and 65% with a GreenSeeker-based precision management and chlorophyll meter-based site-specific N management, respectively, without significant change in grain yield.

Potential exists for weed detection and spot spraying of weeds through the use of canopy reflectance. Weeds may be distinguished from a strongly contrasting background, for example, soil. Some research efforts have been focused on distinguishing weeds from similar backgrounds, for example, crops, using spectral reflectance (Wang et al., 2000; Vrindts et al., 2002). The laboratory study conducted by Vrindts et al. (2002) was successful in discriminating weed species from maize and sugar beet but the field study conducted by Wang et al. (2000) was less successful. Imaging-based systems have also been studied (Zhang and Chaisattapagon, 1995; Tian et al., 1997; Borregaard et al., 2000; Burks et al., 2000) with promising results. Zhang et al. (2012) field tested an image-based spot sprayer that included the ability to target a microdosing system that used hot vegetable oil to treat weeds in tomatoes. Their system successfully eliminated greater than 90% of two weed species while damaging less than 3% of the crop. This system was targeted at organic farming of tomatoes. The image-based technologies result in costly implementations and use of spectral reflectance alone is relatively insensitive and has not been very successful in field implementations. Variable-rate herbicide spraying systems are a concern due to the potential for weeds to be undertreated and develop resistance to herbicides.

2.4.3 Implementations of Canopy Sensing for Field Applications

Field canopy reflectance measuring systems are generally based on multispectral measuring systems. Various implementations of canopy reflectance-based systems exist, including those targeted for fertilizer management and one targeted for weed management.

Commercial systems using canopy reflectance for weed management employ simple ratio (ρ_{NIR}/ρ_{red}) to discriminate between weed and soil background. An early commercial system patented by McCloy and Felton (1992) used natural illumination. The system was effective in discriminating between weeds and soil backgrounds (Felton et al., 1991). The commercial system, DetectSpray®, based on the patent was marketed and appears to no longer be available. A later system patented by Beck and Vyse (1995) is available in the market today. This system is known as the Weedseeker® and is marketed by Trimble Navigation Ltd. This system uses a ratio of NIR to red to distinguish green weeds from background material and uses a threshold comparison to trigger a spray nozzle integrated with the sensor (Beck, 1995).

At least four commercial field-machine-based canopy reflectance systems designed for fertility management exist. Table 2.4 summarizes the characteristics of those systems. A vegetative index that can be used with the system is given, though most systems are capable of providing alternate indices. The Claas system uses two derived indices to present to users, the IRMI vegetation index and the IBI biomass index. The Claas system uses a yield potential map combined with sensing in their algorithm to vary fertilization.

The Claas, AgLeader, and Trimble units are designed for boom mounting and view the plant canopy from directly above. The Claas system provides the boom for mounting as shown in Figure 2.9a. The Trimble and AgLeader systems are designed

TABLE 2.4
Commercial Field Machine-Based Canopy Reflectance Systems

	Claas	AgLeader	Trimble	Topcon
Model number	ISARIA™	OptRx©	GreenSeeker™	CropSpec™
Visible wavelength (nm)	670	670	660	735
NIR wavelengths (nm)	700, 740, and 780	730, 780	770	808
Sensor geometry	NADIR	NADIR	NADIR	Oblique 45–55°
Sensor to crop distance (m)	0.4–1.0	0.25–2.1	0.6–1.6	2–4
Sensing footprint	–	32° × 6°	0.61 × 0.015 m (invariant with height)	2–4 m
Vegetative index	REIP	NDRE	NDVI	SR
Resource	Haas (2014)	Suddeth et al. (2011), AgLeader Technology, 2014. OptRx Crop Sensors. http://www.agleader.com/products/directcommand/optrx-crop-sensors/, Holland (2008)	Suddeth et al. (2011), Stone et al. (2003)	Tevis (2012), Reusch (2010), Kumagai and Shugo (2011)

to be mounted on the booms of sprayers. Figure 2.10 shows the sensed area for these sensors, which are typically operated at 1 m above the crop canopy. The Topcon system is designed to be mounted on a tractor cab and views the crop at an oblique angle as shown in Figure 2.9b.

Satellite- and aircraft-borne sensing systems may be used to develop variable-rate application maps using the same algorithms as those for ground-based systems. Some differences exist between sensor systems. Satellite data must be corrected for atmospheric interference and illumination effects (Mather, 2004). It is common that NDVI and other indices are computed directly from the corrected data. Some differences between systems exist due to the variations in bandwidth of the particular spectral channels of the measuring system. Figure 2.11 shows the relative spectral response (RSR) of sensors on the Quickbird satellite (Forestier et al., 2011). In contrast, Figure 2.12 shows a typical spectral transmittance for interference filters that might be used in a polychromatic LED-based sensor design (Holland, 2008). Sensors based on either technique may be used to generate an NDVI but the response may be slightly different due to the bandwidths used in the sensor.

A third sensor design, one that depends on monochromatic LEDs to provide the bandwidth control (Stone et al., 2003), provides again a slightly different spectral sensitivity. Figure 2.13 provides a typical LED relative spectral power output. The

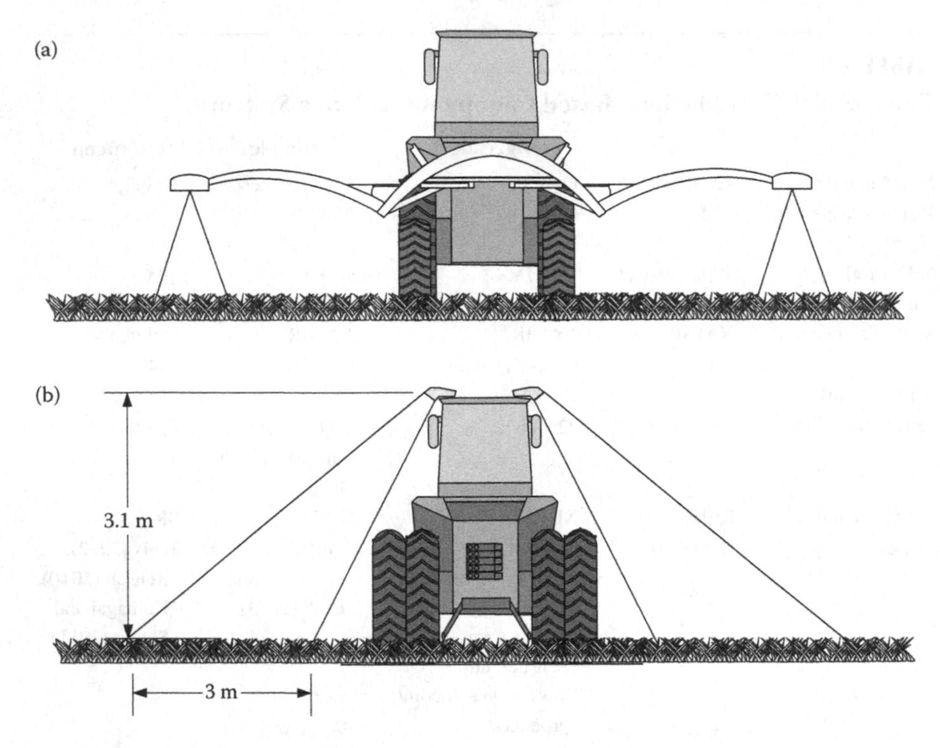

FIGURE 2.9 Sensor mounting geometries for boom-mounted reflectance sensors, Claas Isario™ (a) and the Topcon CropSpec™ (b).

spectral response is similar to the satellite bands, though narrower in the NIR band and NDVIs calculated from this sensor do not match exactly those calculated from Quickbird data.

Without standardization of bandwidths, band shape, as well as center wavelengths, we should expect that algorithms using a particular sensor will have to be adjusted for the particular sensor type.

Another issue that must sometimes be addressed in utilizing data from image-based systems is the need to convert camera image data into reflectance images. This task normally requires that a white plate (100% reflectance object) be included in the image or that the camera capture a white plate image for use in calibration. This process is difficult but necessary if the data are to be used in an algorithm that uses canopy reflectance. Ideally, each pixel of the camera would be calibrated in the same way a reflectance is calibrated. The typical calibration equation is given in Equation 2.16 where the DN values are the raw response for each pixel. Dark current, the DN value for the case where the camera is viewing a completely dark field must be subtracted from any camera measurement. Reflectance is then the result of the ratio of the dark corrected measurement to the dark corrected white plate measurement.

$$\rho_\lambda = \frac{DN_\lambda - DN_{\lambda,\text{dark\_current}}}{DN_{\lambda,\text{white\_plate}} - DN_{\lambda,\text{dark\_current}}} \tag{2.16}$$

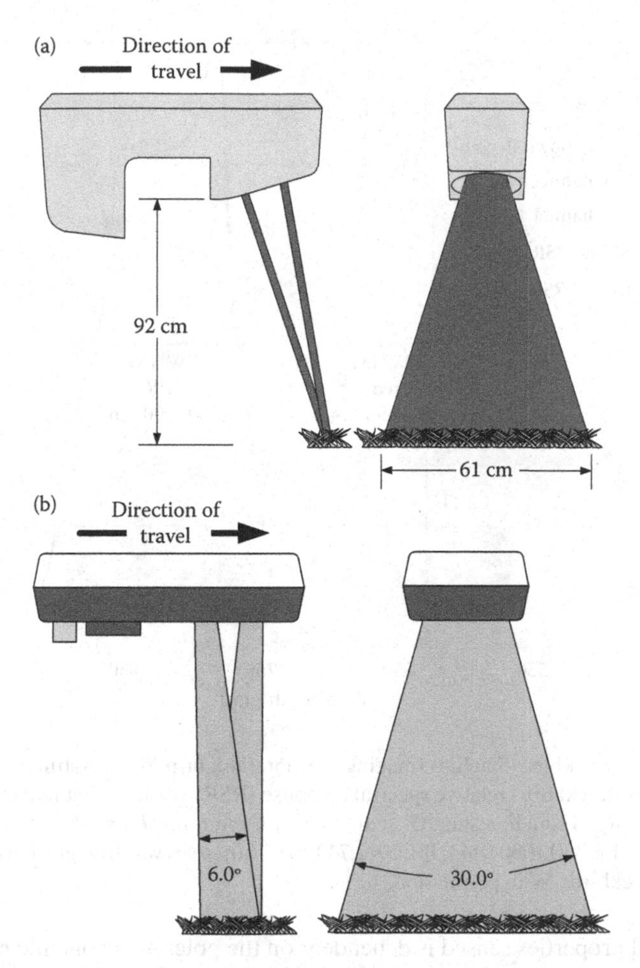

FIGURE 2.10 Optical geometry for the Trimble GreenSeeker reflectance sensor (a), and the AgLeader OptRx (b).

Once the camera has undergone correction for dark current and white plate response, the camera is susceptible to changes in scene illumination between those under which the camera was calibrated and when the images were captured. These changes may be due to sun-angle changes or changes in clouds or haze. It would be preferable to have a white plate object within the captured image to compensate for these changes. With proper calibration, imaging-based reflectance measurements calibrate well with sensor-based measurements (Jones, 2007).

2.5 SOIL PROPERTY SENSING

Crop performance is predicated on an adequate crop environment, including soils in which the crop grows. Management to improve crop performance may include decisions based on soil characteristics and dictate sensing of soil properties. The

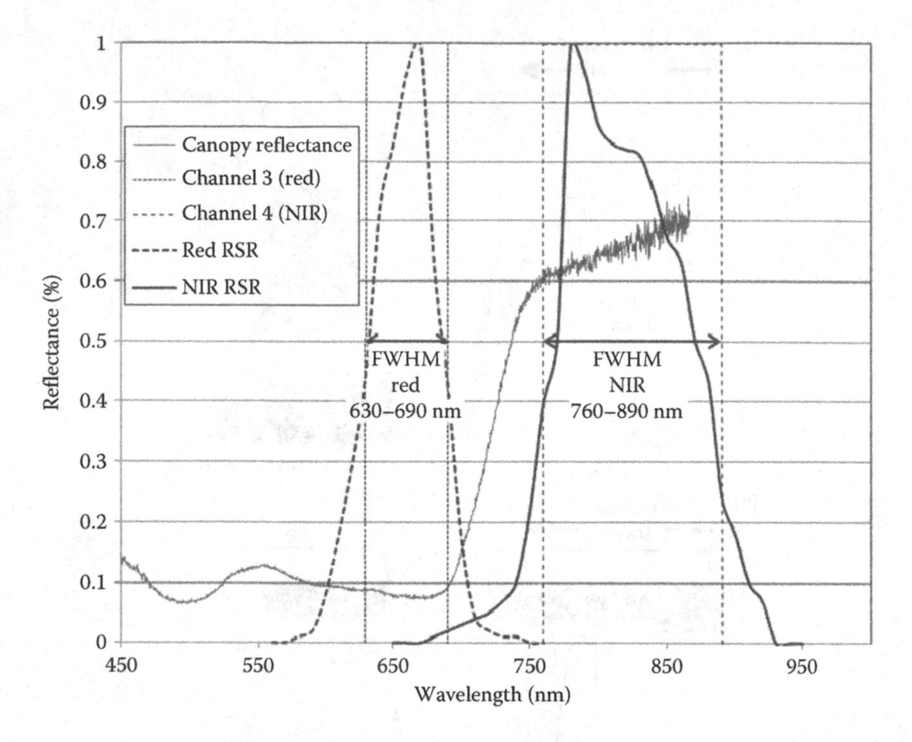

FIGURE 2.11 Quickbird (Satellite Imaging Corporation, http://www.satimagingcorp.com/satellite-sensors/quickbird/) relative spectral response (RSR) versus wheat reflectance (based on estimated RSR). (From Forestier, G. et al. 2011. *International Journal of Remote Sensing*, 34:2327–2349, doi: 10.1080/01431161.2012.744488, http://www.satimagingcorp.com/satellite-sensors/quickbird/. With permission.)

particular soil properties sensed is dependent on the potential economic return from management actions based on the sensed properties and the difficulty in integrating the sensing technology into management practices for the crop. The type of sensing necessary for precision farming is also dependent on the spatial as well as temporal variation in properties.

Temporal variability of soil nutrients is well understood and the major soil nutrients are classified with regard to mobility (Bray, 1954). Nitrogen is a mobile nutrient and its availability varies from season to season (Johnson and Raun, 2003). Phosphorus (P) and potassium (K) are classified as immobile and uptake changes depending on crop demand. Organic carbon (OC) changes in soils where the total content is largely controlled by climate. Carbon supply to plants actually comes via CO_2 assimilation in the atmosphere via photosynthesis. The impact on sensing is that parameters that are associated with nitrogen need sampling before, during, and after the growing season whereas sensing parameters associated with P, K, and OC may be sampled once per season.

Spatial variability of soil nutrients impacts the requirements for soil sampling. Solie et al. (1996) reported semivariance of soil N, P, K, OC, and pH, where soils were sampled at a 0.3 m resolution for two sites. They reported soil property

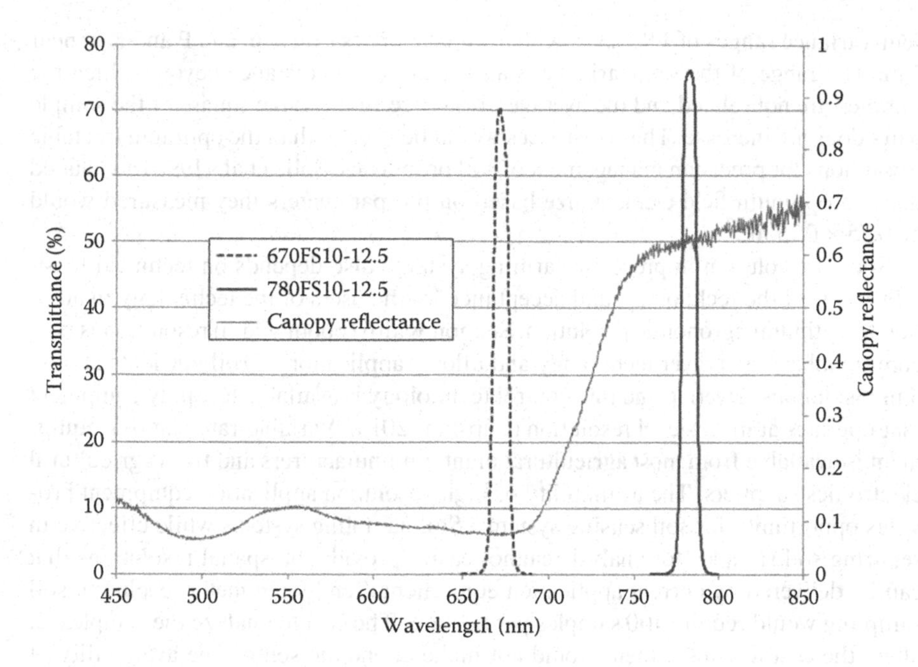

FIGURE 2.12 Interference filter bandwidths (Edmund Optics Inc., Barrington, NJ) and shapes for interference filter-polychromatic LED-based sensor designs.

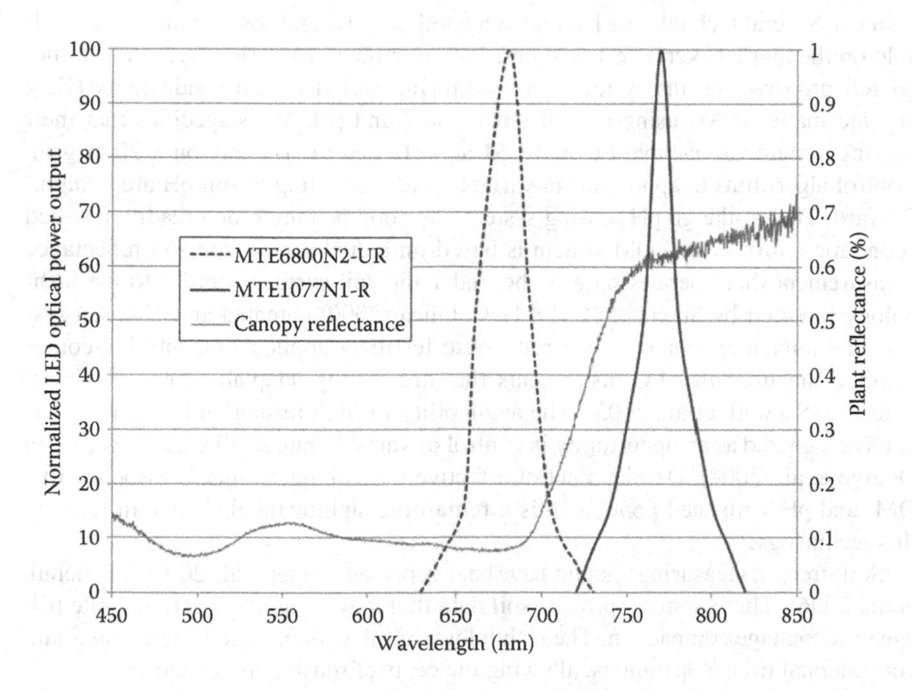

FIGURE 2.13 LED transmitted bandwidths for monochromatic LED-based canopy reflectance sensor designs.

semivariance ranges of 1.9–5.3 m with most properties near 4 m and P and K at near 2 m. The range of the semivariance is a measure of the distance beyond which the samples are not related and the average difference in the value square of the sample pairs does not increase. These distances would be greater than the optimum treatable resolutions for precision management of soil properties. Solie et al. (1996) concluded that the optimum field element size based on the parameters they measured would be 0.75 × 0.75 m.

Spatial resolution in precision farming systems also depends on technical feasibility, cost of the technology, and acceptance by the users of the technology in addition to optimum agronomic resolution. Section width (boom width) resolution is now commonplace in sprayer technology and allows application of soil nutrients at near 3 m resolutions. Overlap and rate control technology is available for spray equipment that operates at meter-level resolution (Capstan, 2013). Variable-rate planting equipment is available from most agricultural planter manufacturers and from agricultural electronics suppliers. The availability of high-resolution application equipment provides opportunity for soil sensing systems. Soil sampling systems while effective in securing soil samples for analysis cannot easily provide the spatial resolutions that can be delivered by current application equipment. Ten by ten meter resolution soil sampling would require 100 samples per hectare. The cost to analyze the samples let alone the cost to handle them would not make economic sense. The availability of high-resolution application equipment obviates the need for on-the-go soil sensing systems that can deliver cost-effective soil property information.

Adamchuk et al. (2004) and Heege (2013) have reviewed on-the-go soil sensing systems. Several technologies have shown good promise and some of those are available on the market. Veris Technologies, Inc. manufactures various types of on-the-go soil property sensing systems for monitoring soil electrical conductivity (EC), organic matter (OM) using optical reflectance, and pH. Veris specifies that their sensing system can sample EC and OM at 1 Hz and pH roughly on a 20-m grid. Control algorithms to apply lime at variable rates according to soil pH are straightforward. The on-the-go pH sensing system has good potential for a readily justified economic return. Veris' OM system is based on a dual-wavelength soil reflectance measurement that operates on a probe under the soil surface, similar to the technology reported by Shonk et al. (1991). Colburn (2000) patented an EC-based system that implements on-the-go variable-rate fertilizer application control. Geonics Limited manufactures EC instruments that are readily adaptable for agricultural field use (Sudduth et al., 2005). The availability of high-resolution EC sensing has been recognized as an opportunity in control of variable-rate seeding (Doerge, 1999; Doerge et al., 2006). Development of effective control algorithms to associate EC, OM, and pH with seed population is a remaining significant challenge in refining this technology.

Soil strength measuring systems have been reported (Raper et al., 2003; Adamchuk et al., 2006). These systems provide soil data that may be used for variable-rate tillage or to manage compaction. The technology has the potential to be integrated into conventional tillage operations, allowing the costs of sensing to be reduced.

Direct sensing of soil nutrient concentrations with on-the-go sensors could dramatically improve the efficiency of the production of agricultural crops. These technologies would have the potential to provide low-cost sensed data and could be integrated into precision farming systems. Technology for direct sensing of soil nutrients is currently an active part of research programs but not yet commercialized (Kim et al., 2009). Success of on-the-go pH sensing may be a precursor to practical soil nutrient sensing. Some promising technologies have been investigated, including solid-state ion-selective membrane technology (Birrell and Hummel, 2001) but further development is needed in this important area.

2.6 ISOBUS SUPPORT FOR SENSOR SYSTEMS

Much of the technology being developed for application in precision farming is based on integrated ISOBUS support. ISOBUS is the standardized network communications system for agricultural equipment. ISO 11783, the standard on which ISOBUS is based, was developed with the support of precision farming as a requirement. The capabilities for transferring application maps to mobile implement control systems (MICS), that is, field systems and to transfer back to farm management information systems (FMIS) as applied maps is integrated into ISOBUS (ISO/TC23/SC19, 2014). Not surprisingly, support for sensor-based control systems is also integrated into ISOBUS through ISO 11783 Part 10.

ISOBUS support for sensor-based rate control systems allows standardized communication between sensor systems and rate control systems. This capability allows sensor systems manufactured by one manufacturer to communicate with rate control systems made by another manufacturer as well as between components made by the same manufacturer. The provisions in the part 10 document specify that the task controller, the component of ISOBUS systems that sends prescription-based commands to implements, manages the connections between sensor systems and rate control systems. ISOBUS support for sensor systems also includes the capability to allow map-based prescriptions information to be provided to sensor systems and allow sensor systems to use algorithms that combine map-based information with sensed information to command rate controllers. An expected ISOBUS supported function of the rate control system is that it would supply "as applied" information back to the task controller, which would allow the task controller to supply as applied maps back to the FMIS.

REFERENCES

Adamchuk, V.I., J.W. Hummel, M.T. Morgan, and S.K. Upadhyaya. 2004. On-the-go soil sensors for precision agriculture. *Computers and Electronics in Agriculture*, 44(1):71–91.

Adamchuk, V.I., K.A. Sudduth, T.J. Ingram, and S. Chung. 2006. *Comparison of Two Alternative Methods to Map Soil Mechanical Resistance On-The-Go*. St. Joseph, MI: ASABE, ASABE Paper No. 61057.

Allison, F.E. 1955. The enigma of soil nitrogen balance sheets. *Advances in Agronomy*, 7:213–250.

Arnall, D.B., A.P. Mallarino, M.D. Ruark, G.E. Varvel, J.B. Solie, M.L. Stone, J.L. Mullock, R.K. Taylor, and W.R. Raun. 2013. Relationship between grain crop yield potential and nitrogen response. *Agronomy Journal*, 105:1335–1344, doi:10.2134/agronj2013.0034.

Barnes, E.M., T.R. Clarke, S.E. Richards, P.D. Colaizzi, J. Haberland, M. Kostrzewski, P. Waller et al. 2000. Coincident detection of crop water stress, nitrogen status and canopy density using ground-based multispectral data. *Proceedings of the Fifth International Conference on Precision Agriculture*, July 16–19, 2000, Bloomington, MN.

Beck, J. and T. Vyse. 1995. Structure and method usable for differentiating a plant from soil in a field. US Patent No. 5389781.

Birrell, S.J. and J.W. Hummel. 2001. Real-time multi ISFET/FIA soil analysis system with automatic sample extraction. *Computers and Electronics in Agriculture*, 32(1):45–67.

Black, A.L. and A. Bauer. 1988. Setting winter wheat yield goals. In Havlin, J.L. (ed.), *Central Great Plains Profitable Wheat Management Workshop Proceedings*, Wichita, KS, August 17–20, 1988, Atlanta, GA: Potash & Phosphate Institute.

Borlaug, N.E. 2000. Ending world hunger. The promise of biotechnology and the threat of antiscience zealotry. *Plant Physiology*, 124(2):487–490, doi: 10.1104/pp.124.2.487.

Borregaard, T., H. Nielsen, L. Nørgaard, and H. Have. 2000. Crop-weed discrimination by line imaging spectroscopy. *Journal of Agricultural Engineering Research*, 75:389–400.

Bray, R. 1954. A nutrient mobility concept of soil–plant relationships. *Soil Science*, 78:9–22.

Burks, T.F., S.A. Shearer, R.S. Gates, and K.D. Donohue. 2000. Backpropagation neural network design and evaluation for classifying weed species using color image texture. *Transactions of the ASAE*, 43(4):1029–1037.

Capstan Ag Systems, Inc. 2013. Nozzle by Nozzle: Individual Nozzle Control. Capstan Ag Systems, Inc. Specifications brochure 56652 7/13.

Cassman, K.G., A. Doberman, and D.T. Walters. 2002. Agroecosystems, nitrogen-use efficiency, and nitrogen management. *AMBIO*, 31(2):132–40.

Colburn Jr., J.W. 2000. Soil constituent sensor and precision agrichemical delivery system and method. US Patent No. 6138590.

Dahnke, W.C., L.J. Swenson, R.J. Goos, and A.G. Leholm. 1988. Choosing a crop yield goal. SF-822. North Dakota State Ext. Serv. Fargo.

Demmel, M. 2013. Site specific recording of yields. In Heege, H.J. (ed.), *Precision Crop Farming: Site Specific Concepts and Sensing Methods: Applications and Results.* Dordrecht: Springer Science + Business Media.

Doerge, T. 1999. Soil electrical conductivity mapping. *Crop Insights*, 9(19), Johnston, IA: Pioneer Hi-Bred.

Doerge, T., K. O'Bryan, and D. Gardner. 2006. Putting variable-rate seeding to work on your farm. *Field Facts*, 6(12), Johnston, IA: Pioneer Hi-Bred.

Eitel, J.U.H., D.S. Long, P.E. Gessler, and E.R. Hunt. 2008. Combined spectral index to improve ground-based estimates of nitrogen status in dryland wheat. *ASA Agronomy Journal*, 100(6):1694–1702.

Evans, J.R. 1983. Nitrogen and photosynthesis in the flag leaf of wheat (*Triticum aestivum* L.). *Plant Physiology*, 72:297–302.

Felton, W.L., A.F. Doss, P.G. Nash, and K.R. McCloy. 1991. A microprocessor controlled technology to selectively spot spray weeds. In *Proceedings of Automated Agriculture for the 21st Century Symposium*, Chicago, IL, St. Joseph, MI: ASAE, pp. 427–432.

Forestier, G., J. Inglada, C. Wemmert, and P. Gancarski. 2011. Comparing optical sensors spectral resolution using spectral libraries. *International Journal of Remote Sensing*, 34:2327–2349, doi: 10.1080/01431161.2012.744488.

Gamon, J.A., C.B. Filed, M.L. Goulden, K.L. Griffin, A.E. Hartley, G. Joel, J. Peñuelas, and R. Valentini. 1995. Relationship between NDVI, canopy structure, and photosynthesis in three Californian vegetation types. *Ecological Applications*, 5(1):28–42.

Garbrecht, 1983. Ancient water works—Lessons from history. In Richardson, J. (ed.), *Managing Our Fresh-Water Resources, Impact of Science on Society*, 1st Edition, Paris: UNESCO.

Government of the UK. 2008. Radioactive Substances Act Guidance (RASAG): Chapter 1 Registrations, Operational Instruction 371_04. Environment Agency, Government of the UK.

Guyot, G., F. Baret, and D.J. Major. 1988. High spectral resolution: Determination of spectral shifts between the red and the near infrared. *International Archives of Photogrammetry and Remote Sensing*, 11:750–760.

Haas, T. 2014. Measuring device for determining a vegetation index value of plants. US Patent No. 8823945.

Heege, J.H. 2013. Precision in crop farming: Site specific concepts and sensing methods: Applications and results. In Heege, J.H. (ed.) *Chapter 5: Sensing of Natural Soil Properties*. Springer International Publishing AG. Heidelburg, New York, and London. ISBN: 978-94-007-6759-1.

Hegarty, C.J. and E. Chatre. 2008. Evolution of the Global Navigation Satellite System (GNSS). *Proceedings of the IEEE*, 96(12):1902–1917, doi: 10.1109/JPROC.2008.2006090.

Holland, K. 2008. Light sensor with modulated radiant polychromatic source. US Patent No. 7408145.

Horler, D.N.H., M. Dockray, and J. Barber. 1983. The red edge of plant leaf reflectance. *International Journal of Remote Sensing*, 4(2):273–288.

ISO/TC23/SC19. 2014. *ISO 11783 Part 10: Task Controller and Management Information System Data Interchange, Tractors and Machinery for Agriculture and Forestry—Serial Control and Communications Data Network*, Geneva: ISO.

International Plant Nutrition Institute (IPNI). 2012. 4R Plant nutrition: A manual for improving the management of plant nutrition. 3500 Parkway Lane, Suite 550, Peachtree Corners, Georgia, 30092-2844.

International Plant Nutrition Institute (IPNI). 2014. Crop nutrient removal calculator. 3500 Parkway Lane, Suite 550, Peachtree Corners, Georgia, 30092-2844.

Johnson, G.V. and W.R. Raun. 2003. Nitrogen response index as a guide to fertilizer management. *Journal of Plant Nutrition*, 26:249–262.

Jones, C.L., P.R. Weckler, N.O. Maness, R. Jayasekara, M.L. Stone, and D. Chrz. 2007. Remote sensing to estimate chlorophyll concentration in spinach using multi-spectral plant reflectance. *Transactions of the ASABE*, 50(6):2267–2273.

Kim, H., K.A. Sudduth, and J.W. Hummel. 2009. Soil macronutrient sensing for precision agriculture. *Journal of Environmental Monitoring*, 2009(11):1810–1824.

Kitchen, N.R., K.A. Sudduth, S.T. Drummond, P.C. Scharf, H.L. Palm, D.F. Roberts, and E.D. Vories. 2010. Ground-based canopy reflectance sensing for variable-rate nitrogen corn fertilization. *Agronomy Journal*, 102:71–84, doi: 10.2134/agronj2009.0114.

Kormann, G. 2004. Mass flow measurement based on X-ray absorption. In: *Agricultural Engineering Conference 2004*, Leuven. Paper No. 034.

Kumagai, K. and A. Shugo. 2011. Plant sensor. US Patent No. 7910876.

Li, F., Y. Miao, F. Zhang, Z. Cui, R. Li, X. Chen, H. Zhang, J. Schroder, W.R. Raun, and L. Jia. 2009. In-season optical sensing improves nitrogen-use efficiency for winter wheat. *Soil Science Society of America Journal*, 73:1566–1574, doi: 10.2136/sssaj2008.0150.

Liang, H., C. Zhao, W. Huang, L. Liu, J. Wang, and Y. Ma. 2005. Variable-rate nitrogen application algorithm based on canopy reflected spectrum and its influence on wheat. *Proceedings of SPIE 5655, Multispectral and Hyperspectral Remote Sensing Instruments and Applications II*, Honolulu, HI, January 20, 2005, doi: 10.1117/12.582987.

Lukina, E.V., K.W. Freeman, K.J. Wynn, W.E. Thomason, R.W. Mullen, A.R. Klatt, G.V. Johnson et al. 2001. Nitrogen fertilization optimization algorithm based on in-season estimates of yield and plant nitrogen uptake. *Journal of Plant Nutrition*, 24:885–898.

Mather, P.M. 2004. *Computer Processing of Remotely Sensed Images: An Introduction*, Chichester, UK: John Wiley & Sons Ltd.

Matson, P.A., W.J. Parton, A.G. Power, and M. Swift. 1997. Agricultural intensification and ecosystem properties. *Science*, 277:504–509.

McCloy, K. and W. Felton. 1992. Controller for agricultural sprays. US Patent No. 5144767.

Mistele, B. 2006. *Tractor Based Spectral Reflectance Measurements Using an Oligo View Optic to Detect Biomass, Nitrogen Content and Nitrogen Uptake*. PhD Dissertation, Technische Universität München.

Mistele, B. and U. Schmidhalter. 2010. Tractor-based quadrilateral spectral reflectance measurements to detect biomass and total aerial nitrogen in winter wheat. *Agronomy Journal*, 102:499–506.

Moorby, J. and R.T. Besford. 1983. Mineral nutrition and growth. In Lauchi, A. and Bieleski, R.L. (eds.), *Encyclopedia of Plant Physiology New Series*, Berlin, Germany: Springer-Verlag, pp. 481–529.

Mullen, R.W., K.W. Freeman, W.R. Raun, G.V. Johnson, M.L. Stone, and J.B. Solie. 2003. Identifying an in-season response index and the potential to increase wheat yield with nitrogen. *Agronomy Journal*, 95:347–351.

Noack, P.H., T. Muhr, and M. Demmel. 2006. Effect of interpolation methods and filtering on the quality of yieldmaps. In *Precision Agriculture '05. Papers Presented at the 5th European Conference on Precision Agriculture*, Uppsala, Sweden, 2005, Wageningen, Netherlands: Wageningen Academic Publishers, pp. 701–706, 1005 p.

Ortez-Monasterio, U.S., M.E. Cardenas, and A. Mendoza. 2014. Sensor use in Mexico. *12th Annual NUE Conference*, Sioux Falls, SD, August 4–6, 2014.

Peñuelas, J., J.A. Gamon, A.L. Fredeen, J. Merino, and C.B. Field. 1994. Reflectance indices associated with physiological changes in nitrogen-and water limited sunflower leaves. *Remote Sensing of Environment*, 48:135–146.

Persson, D.A., L. Eklundh, and P.A. Algerbo. 2004. Evaluation of an optical sensor for tuber yield monitoring. *Transactions of the ASAE*, 47(5):1851–1856, doi: 10.13031/2013.17602.

Porter, W.M., J. Ward, K.R. Kirk, R.K. Taylor, J.B. Fravel, and C.B. Godsey. 2013. *Application of an Agleader® Cotton Yield Monitor for Measuring Peanut Yield: A Multi-State Investigation*. Kansas City, Missouri: ASABE Paper Number 131596295, July 21–24, 2013, doi: 10.13031/aim.20131596295.

Price, R.R., R.M. Johnson, R.P. Viator, J. Larsen, and A. Peters. 2011. Fiber optic yield monitor for a sugarcane harvester. *Transactions of the ASABE*, 54(1):31–39, doi: 10.13031/2013.36250.

Raper, R.L. and E.H. Hall. 2003. Soil strength measurement for site-specific agriculture. US Patent No. 6647799.

Raper, T.B., J.J. Varco, and K.J. Hubbard. 2013. Canopy-based normalized difference vegetation index sensors for monitoring cotton nitrogen status. *Agronomy Journal*, 105(5):1345–1354.

Raun, W.R. and G.V. Johnson. 1999. Improving nitrogen use efficiency for cereal production. *Agronomy Journal*, 91:357–363.

Raun, W.R., G.V. Johnson, M.L. Stone, J.B. Solie, E.V. Lukina, W.E. Thomason, and J.S. Sehepers. 2001. In-season prediction of potential grain yield in winter wheat using canopy reflectance. *Agronomy Journal*, 93:131–138.

Raun, W.R., J.B. Solie, G.V. Johnson, M.L. Stone, R.W. Mullen, K.W. Freeman, W.E. Thomason, and E.V. Lukina. 2002. Improving nitrogen use efficiency in cereal grain production with optical sensing and variable rate application. *Agronomy Journal*, 94: 815–820.

Raun, W.R., J.B. Solie, K.L. Martin, K.W. Freeman, M.L. Stone, K.L. Martin, G.V. Johnson, and R.W. Mullen. 2005b. Growth stage, development, and spatial variability in corn evaluated using optical sensor readings. *Journal of Plant Nutrition*, 28:173–182.

Raun, W., J. Solie, J. May, H. Zhang, J. Kelly, R. Taylor, B. Arnall, and I. Ortiz-Monasterio. 2010. Nitrogen rich strips for wheat, corn and other crops. Oklahoma Cooperative Extension Service, E-1022, Stillwater, OK, 3 pp.

Raun, W.R., J.B. Solie, and M.L. Stone. 2011. Independence of yield potential and crop nitrogen response. *Precision Agriculture*, 12:508–518.

Raun, W.R., J.B. Solie, M.L. Stone, K.L. Martin, K.W. Freeman, R.W. Mullen, H. Zhang, J.S. Schepers, and G.V. Johnson. 2005a. Optical sensor based algorithm for crop nitrogen fertilization. *Communications in Soil Science and Plant Analysis*, 36:2759–2781.

Rehm, G. and M. Schmitt. 1989. Setting realistic crop yield goals. Minnesota Extension Service, AG-FS-3873. University of Minnesota.

Reusch, S. 2005. Optimum waveband selection for determining the nitrogen uptake in winter wheat by active remote sensing. *Papers Presented at the 5th European Conference on Precision Agriculture*, Uppsala, Sweden, 9–12th June 2005, pp. 261–266.

Reusch, S., J. Jasper, and A. Link. 2010. Estimating crop biomass and nitrogen uptake using CropSpec™, a newly developed active crop-canopy reflectance sensor. *10th International Conference on Precision Agriculture*, July 18–21, Denver, Colorado.

Rodriguez, G., J. Fitzgerald, R. Belford, and L.K. Christensen. 2006. Detection of nitrogen deficiency in wheat from spectral reflectance indices and basic crop eco-physiological concepts. *Australian Journal of Agricultural Research*, 57(7):781–789.

Rouse, J.W., R.H. Haas, J.A. Schell, and D.W. Deering. 1974. Monitoring vegetation systems in the Great Plains with ERTS. In Fraden S.C., Marcanti, E.P., and Becker, M.A. (eds.), *Third ERTS-1 Symposium*, Washington DC: NASA SP-351, December 10–14, 1973, pp. 309–317.

Roberts, D.F., N.R. Kitchen, P.C. Scharf, and K.A. Sudduth. 2009. Will variable-rate nitrogen fertilization using corn canopy reflectance sensing deliver environmental benefits? *Agronomy Journal*, 102:85–95, doi: 10.2134/agronj2009.0115.

Roberts, T.L. 2007. Right product, right rate, right time, and right place…the foundation of BMPs for fertilizer. *International Fertilizer Industry Association (IFA) Workshop on Fertilizer Best Management Practices*, Brussels, Belgium, March 7–9, 2007.

Sawyer, J., E. Nafziger, G. Randall, L. Bundy, G. Rehm, and B. Joern. 2006. *Concepts and Rationale for Regional Nitrogen Rate Guidelines for Corn*. PM 2015. Ames: Iowa State Univ. Extension.

Scharf, P.C., D.K. Shannon, H.L. Palm, K.A. Sudduth, S.T. Drummond, N.R. Kitchen, L.J. Mueller, V.C. Hubbard, and L.F. Oliveira. 2011. Sensor-based nitrogen applications out-performed producer-chosen rates for corn in on-farm demonstrations. *Agronomy Journal*, 103:1683–1691, doi: 10.2134/agronj2011.0164.

Searcy, S.W., J.K. Schueller, Y.H. Bae, S.C. Borgelt, and B.A. Stout. 1989. Mapping of spatially variable yield during grain combining. *Transactions of the ASABE*, 32(3):0826–0829, doi: 10.13031/2013.31077.

Serrano, L., I. Filella, and J. Peñuelas. 2000. Remote sensing of biomass and yield of winter wheat under different nitrogen supplies. *Crop Science*, 40:731–736.

Shapiro, C.A., R.B. Ferguson, G.W. Hergert, and C.S. Wortmann. 2008. *Fertilizer Suggestions for Corn*, Lincoln, NE: University of Nebraska Extension, EC117.

Shonk, J.L., L.D. Gaultney, D.G. Schulze, and G.E. Van Scoyoc. 1991. Spectroscopic sensing of soil organic matter content. *Transactions of the ASABE*, 34(5):1978–1984, doi: 10.13031/2013.31826.

Solie, J.B., W.R. Raun, R.W. Whitney, M.L. Stone, and J.D. Ringer. 1996. Optical sensor based field element size and sensing strategy for nitrogen application. *Transactions of ASAE*, 39:1983–1992.

Stone, M.L., D. Needham, J.B. Solie, W.R. Raun, and G.V. Johnson. 2003. US Patent No. 6596996.

Stone, M.L., J.B. Solie, W.R. Raun, R.W. Whitney, S.L. Taylor, and J.D. Ringer. 1996. Use of spectral radiance for correcting in-season fertilizer nitrogen deficiencies in winter wheat. *Transactions of ASAE*, 39(5):1623–1631.

Stone, M.L., R.K. Benneweis, and J. Van Bergeijk. 2008. Evolution of electronics for mobile agricultural equipment. *Transactions of the ASABE*, 51(2):385–390, doi: 10.13031/2013.24374.

Suddeth, K.A., N.R. Kitchen, and S.T. Drummond. 2011. *Nadir and Oblique Canopy Reflectance Sensing for N Application in Corn*. Louisville, KY: ASABE Paper No. 1111261.

Sudduth, K.A., N.R. Kitchen, W.J. Wiebold, W.D. Batchelor, G.A. Bollero, D.G. Bullock, D.E. Clay et al. 2005. Relating apparent electrical conductivity to soil properties across the north-central USA. *Computers and Electronics in Agriculture*, 46:263–283.

Taylor, R., J. Fulton, M. Darr, L. Haag, S. Staggenborg, D. Mullenix, and R. McNaull. 2011. *Using Yield Monitors to Assess On-Farm Rest Plots*. Louisville, KY: ASABE Paper No. 1110690, August 7–10, 2011, doi: 10.13031/2013.37280.

Tevis, J.W. 2012. *CropSpec™ Update. 2012 NUE Conference*, Fargo, ND, August 6–8 .

Thomas, J.R. and G.F. Oerther. 1972. Estimating nitrogen content of sweet pepper leaves by reflectance measurements. *ASA Agronomy Journal*, 64:11–13.

Thomasson, J.A., D.A. Pennington, H.C. Pringle, E.P. Columbus, S.J. Thomson, and R.K. Byler. 1999. Cotton mass flow measurement: experiments with two optical devices. *Applied Engineering in Agriculture*, 15(1):11–17.

Thomasson, J.A., R. Sui, G.C. Wright, and A.J. Robson. 2006. Optical peanut yield monitor: development and testing. *Applied Engineering in Agriculture*, 22(6):809–818, doi: 10.13031/2013.22249.

Tian, L., D.C. Slaughter, and R.F. Norris. 1997. Outdoor field machine vision identification of tomato seedlings for automated weed control. *Transactions of the ASAE*, 40(6):1761–1768.

Torino, M.S., B.V. Ortiz, J.P. Fulton, K.S. Balkcom, and W.C. Wood. 2014. Evaluation of vegetation indices for early assessment of corn status and yield potential in the southeastern united states. *Agronomy Journal*, 106:1389–1401, doi: 10.2134/agronj13.0578.

Tucker, C.J., J.H. Elgin Jr., and J.E. McMurtrey III. 1979. Monitoring corn and soybean crop development with hand-held radiometer spectral data. *Remote Sensing of Environment*, 8:237–248.

Varvel, G.E., J.S. Schepers, and D.D. Francis. 1997. Ability for in-season correction of nitrogen deficiency in corn using chlorophyll meters. *Soil Science Society of America Journal*, 61:1233–1239.

Vellidis, G., C.D. Perry, G.C. Rains, D.L. Thomas, N. Wells, and C.K. Kvien. 2003. Simultaneous assessment of cotton yield monitors. *Applied Engineering in Agriculture*, 19(3):259–272, doi: 10.13031/2013.13658.

Vrindts, E., J. De Baerdemaeker, and H. Ramon. 2002. Weed detection using canopy reflection. *Precision Agriculture*, 3(1):63–80.

Wang, N., N. Zhang, D. Peterson, and F. Dowell. 2000. Design of an optical weed sensor using plant spectral characteristics. In DeShazer, J.A. and Meyers, G.E. (eds.), *Proc. SPIE 4203, Biological Quality and Precision Agriculture II*, 63 (December 29, 2000); doi: 10.1117/12.411740.

Wright, D.L., V.P. Rasmussen, R.D. Ramsey, D.J. Baker, and J.W. Ellsworth. 2004. Canopy reflectance estimation of wheat nitrogen content for grain protein management. *GI Science and Remote Sensing*, 41(4):287–300.

Whitson, A.R. and H.L. Walster. 1912. *Soils and Soil Fertility*. St. Paul, MN: Webb Publishing Co.

Wilkerson, J.B., F.H. Moody, W.E. Hart, and P.A. Funk. 2001. Design and evaluation of a cotton flow rate sensor. *Transactions of the ASAE*, 44(6):1415–1420.

Wild, K., F. Martin, T. Schmiedel, E. Hirschberg, and J. Müller. 2014. Yield measurements in a self-propelled forage harvester by means of X-rays. *Proceedings International Conference of Agricultural Engineering, AgEng 2014*, July 6–10, 2014, Zurich.

Xue, L., W.X. Cao, W.H. Luo, T.B. Dai, and Y. Zhu. 2004. Monitoring leaf nitrogen status in rice with canopy spectral reflectance. *Agronomy Journal*, 96(1):135–142.

Yao, Y., Y. Miao, S. Huang, L. Gao, X. Ma, G. Zhao, R. Jiang et al. 2012. Active canopy sensor-based precision N management strategy for rice. *Agronomy for Sustainable Development* 32(4):925–933.

Zhang, H., W. Raun, and B. Arnall. 2010. *OSU Soil Test Interpretations*. Stillwater, OK: Oklahoma Cooperative Extension Service. PSS-2225.

Zhang, N. and C. Chaisattapagon. 1995. Effective criteria for weed identification in wheat fields using machine vision. *Transactions of the ASAE*, 38(3):965–974.

Zhang, Y., E.S. Staab, D.C. Slaughter, D.K. Giles, and D. Downey. 2012. Automated weed control in organic row crops using hyperspectral species identification and thermal micro-dosing. *Crop Protection*, 41(0):96–105.

3 Data Processing and Utilization in Precision Agriculture

Chunjiang Zhao, Liping Chen, Guijun Yang, and Xiaoyu Song

CONTENTS

3.1 INTRODUCTION

To achieve the goal of demand-oriented input and variable-rate fertilization for different crops and farmland environments, a high-density, high-speed, and low-cost supply of spatial information on crops, soil, and environmental conditions is necessary. Such information can guide management decisions, including variable-rate fertilization, variable-rate pesticide application, and irrigation to form a prescription map for agricultural production.

Conventional techniques for acquiring information on crop nutrition, crop growth, yield, and soil nutrition rely mainly on surveys, field sampling, and laboratory analysis. For production managers in precision agriculture, the antecedent data, real-time data, point data, and relevant materials can be acquired and used as a reference for decision making in variable-rate operations. So far, the greatest barrier to the implementation of precision agriculture lies in the rapid and cost-efficient acquisition of spatial information about farmlands. In addressing the requirements for decision making in precision agriculture, this section presents methods for acquiring spatial information about farmlands using remote sensing techniques.

3.1.1 Acquisition of Crop Information Based on Remote Sensing Techniques

Crop growth status can be delineated through leaf area, leaf color, leaf inclination angle, plant height, and stalk diameter during the growth period. Other characterizations use factors closely related to crop growth such as crop nitrogen, leaf area index, and biomass. The monitoring and diagnosis of crop nutrition is the core content of regulation and management of crop growth. Compared with healthy vegetation in the growth period, vegetation restricted by nitrogen deficiency may undergo a series of changes in physiological status, biochemical composition, and canopy structure. The vegetation may suffer from small leaves and low biomass if there is a reduction in nitrogen and chlorophyll and decreased synthesis of organic nitrogen-containing compounds such as proteins, nucleic acids, and lipids. Water stress is one of the most common limitations to photosynthesis and plant primary productivity, and its measurement is important for irrigation practices and in drought assessments of natural communities (Penuelas et al., 1993). In response to changes in these parameters, the plants' spectra would exhibit a certain change, which serves as the physical basis for remote sensing detection of nitrogen content or water content of crops.

3.1.1.1 Chlorophyll in Crops

Chlorophyll is the crucial compound for photosynthesis in crops. It also serves as an indicator of the growth status of crops. Standard methods for the measurement of chlorophyll content include the chemometrics method (McKinney, 1941) and nondestructive measurement with SPAD (Minolta Camera Co. Ltd., 1989).

Chlorophyll shows obvious absorption characteristics in the visible band, which is strongly correlated with nitrogen in plants, and it usually increases in the upper leaves at the expense of the lower leaves when fertilizer is deficient. Hinzman et al. (1986) reported that canopy chlorophyll density (CCD), the total amount of chlorophyll present in the canopy per unit of ground area, was a sensitive indicator of N deficiencies in wheat. During the past few decades, various types of spectral indices have been used to estimate chlorophyll content (Datt, 1994; Carter, 1994; Brantley et al., 2002; Huang et al., 2011; Hunt et al., 2013; Li et al., 2013). Liao et al. (2013) used a continuous wavelet transform (CWT) to estimate the chlorophyll content of maize leaves in different layers from their visible to near-infrared (NIR) (400–1000 nm) spectra.

3.1.1.2 Nitrogen in Crops

Nitrogen (N) is a very important nutrient element for crop growth. Timely and optimal N fertilizer supply can increase wheat production, minimize environment pollution, and increase N use efficiency (NUE). Laboratory-based methods, such as preplanting (or pre-sidedress) soil NO_3–N (nitrate N) (or NH_4^+–N, ammonium N) tests, and plant tissue (sap or petiole) tests, are effective ways for making N fertilizer recommendations (Fox et al., 1989; Wu et al., 2007).

Previous studies have shown that leaf chlorophyll concentration in plants was closely correlated to leaf N concentration (LNC) (Shadchina et al., 1998; Serrano et al., 2000). Thus, crop N status can be determined from the measurement of leaf or canopy spectral reflectance. Generally, assessment of crop N status is based on the relationships between LNC of single leaf or whole canopy and spectral parameters. Since the spectra are always acquired based on canopy level, the N status should also be based on canopy level, when establishing estimation models for N status assessment.

Canopy N density (CND) is a sensitive indicator to detect N deficiency in wheat (Zhao et al., 2011). CND defined as the total leaf nitrogen per unit land area can be calculated by the following formula:

$$CND = LNC \times SLW \times LAI \tag{3.1}$$

where LNC is leaf nitrogen content; SLW is specific leaf weight; and LAI is leaf area index.

Under nitrogen stress, the nitrogen in old leaves migrates to new leaves. As a result, the lower leaves turn yellow under nitrogen stress, and then such phenomenon propagates to the upper leaves (Lu et al., 1994). Thus, considering the vertical distribution of nitrogen and the corresponding spectral response is of practical significance. Few field studies have concentrated on the challenging issue of capturing leaf N distribution in the crop canopy using remote sensing technology. The existing studies can be grouped into three classes according to the hyperspectral data used. One class estimated the leaf N content of different vertical layers using spectral data obtained from top-view observations (Wang, 2004; Wang et al., 2007). Another class employed multiangle canopy reflectance data (Zhao et al., 2006), and the third investigated the spectral reflectance and fluorescence characteristics of different vertical leaf layers and their relationships to corresponding leaf N or chlorophyll content (Wang et al., 2004). These studies have made important progress in detecting leaf N distribution by means of remote sensing.

3.1.1.3 Water Content of Crops

Water is an important component of plants since it participates in photosynthesis and transpiration. It is also a critical parameter in agricultural irrigation. Tissue water content is an indicator of the physiological status of plants, which is usually measured by the weighting method (Woods et al., 1982; Zhang et al., 2012).

Spectral characteristics of leaves are determined by the light absorption and scattering characteristics of leaf water, pigments, and dry matter. Leaf water contributes primarily to the leaf spectrum by absorbing incident light at the doubling frequency or combination frequency of water molecule vibration (e.g., 975, 1200, 1450, and 1950 nm).

Remote sensing of liquid water in vegetation has important applications in agriculture and forestry (Jackson and Ezra, 1985; Gao and Goetz, 1995). Water stress is one of the most common factors limiting photosynthesis and plant primary productivity, and its measurement is important in irrigation practices and in drought assessments of natural communities (Penuelas et al., 1993). The primary effect of water content on leaf spectral reflectance is its absorption of radiation. The reflectance spectra of green vegetation in the 1300–2500 nm region are dominated by liquid-water absorption, and are weakly affected by absorption due to other biochemical components, such as protein, lignin, and cellulose (Carter, 1991; Gao and Goetz, 1995). The spectral reflectance in NIR bands is determined by the leaf's internal structure, its dry matter content (mainly protein, lignin, and cellulose) and two minor water-related absorption bands at 975 and 1200 nm (Jacquemoud et al., 1996; Penuelas et al., 1997). In addition, there are secondary effects of water content on reflectance that cannot be explained solely by the radiative properties of water. Some of the secondary effects of water content on leaf reflectance are influenced by the transmissive properties rather than the absorptive properties of water. When leaf water content (LWC) decreases, the internal structure (e.g., the fraction of air spaces in the spongy mesophyll) may also change, thereby inducing variations in NIR reflectance (Carter, 1991; Filella and Penuelas, 1994; Liu et al., 2004).

It may be possible to use crop reflectance to estimate LWC. LWC was calculated as

$$LWC = \frac{LFWC - LDWC}{LFWC} \times 100\% \quad (3.2)$$

where LFWC is the sample fresh leaf mass (kg) and LDWC is the sample dry leaf mass (kg).

3.1.1.4 Crop LAI

LAI is an important indicator of the growth status of crops and an important basis for variable-rate fertilization. It can also be used as a reference for variable-rate irrigation. LAI was estimated by multiplying the plant population by the leaf area per plant as described in Kar et al. (2006). Direct or semidirect methods involve a measurement of leaf area, using either a leaf area meter or a specific relationship of dimension to area via a shape coefficient (McKee, 1964; Marshall, 1968; Manivel and Weaver, 1974). LAI can also be measured using the LAI-2000 instrument (LI-COR, USA).

The method for LAI inversion based on remote sensing technology utilizes the variation in spectra of crops under LAI measurement data. Timely, accurately, and dynamically obtaining crop LAI is beneficial for suitable field management strategies in agricultural production. Until now, the methods for LAI estimation mainly include statistical algorithms (Broge et al., 2001; Wang et al., 2011), nonparametric algorithms (Smith et al., 1991; Fang et al., 2003; Kalacska et al., 2005), physical models (Qin et al., 2009; Xiao et al., 2009; Richter et al., 2011; Dorigo, 2012), and data assimilation algorithms (Dente et al., 2008; Sabater et al., 2008; Thorp et al., 2010; Wang et al., 2010).

3.1.2 Acquisition of Farmland Data by Machine

Acquisition of farmland information by machine refers to the acquisition of farmland information using calculators and sensors carried by tractors and reapers. In this section, a detailed description is provided for rapid acquisition of crop yield and soil nutrition information by machine-borne equipment (Zhao, 2009).

3.1.2.1 Acquisition of Crop Yield Information by Combine Harvester

Acquisition of crop yield data in the plot and plotting the spatial distribution diagram are the starting points of precision agriculture. They are also the basis for achieving scientific regulation of input and making decisions about crop production. The commercialized yield estimation systems carried by a combine harvester mainly include the AFS system (CASE, USA), the FieldStar system (Massey Ferguson, UK), the GreenStar system (John Deere, USA), and the PF system (Ag Leader, USA).

The yield estimation system mounted on axial flow–type combine harvesters (CASE) includes a DGPS device, an intelligent terminal, a wheel rotation speed sensor, a grain elevator rotation speed sensor, a header height potentiometer, a grain flow sensor, a grain water content sensor, a memory card, and graphic software (Figure 3.1). The grain flow sensor is located on top of the grain elevator. After the grain enters the top of the elevator, it hits the impact plate of the sensor under the guidance of the deflector. The impact signals are converted to electric signals as output. The output of the signal is proportional to the grain flow. The Hall sensor is used to measure grain elevator rotation speed. The output signals of the sensor are used to correct the output signals of the flow sensor and to restrict the working status. The signal processing unit is responsible for integration and processing of output from the wheel speed sensor, the elevator rotation speed sensor, the grain water content sensor, the header height potentiometer, and the grain flow sensor. The distance traveled by the machine, working area, transient grain water content, and transient grain flow are measured. Differential GPS (DGPS) provides position information on these signals. After these signals are transmitted to the intelligent terminal, the measurement errors are effectively reduced by software through *in situ* calibration. Then the data are recorded on the data card, and the crop yield of each plot at each spatial location is obtained. The data card is carried back to the office, and special data processing software is used to generate the spatial distribution diagram of yield. The result is utilized by yield analysis and serves as the basis for implementation of variable-rate farming.

Instant Yield Map, a software program for generating yield diagrams by CASE, can not only generate point diagrams, grid maps, smooth grid maps, and other line graphs of yield based on the original data stored in the data card, but also enables the management of yield data by harvest time and place. The yield data are classified at equal intervals or at those defined by the user. Each category is represented by different colors, allowing the field conditions to be rapidly visualized and low-yield areas to be marked out.

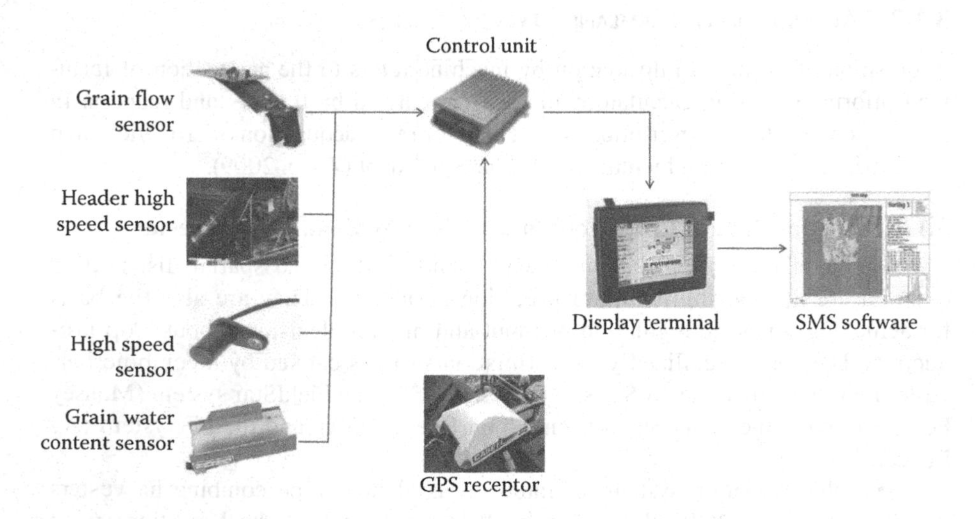

FIGURE 3.1 Yield estimation system on an axial flow–type combine harvester.

3.1.2.2 Acquisition of Soil Nutrition Data

Spatial distribution of soil fertility is a very important factor that affects decision making in precision agriculture. Determination of the spatial variation of soil fertility and the application of relevant data to mechanized variable-rate fertilization are major concerns in the field of precision agriculture. The Beijing Research Center for Agricultural Information Technology has developed an automatic soil sample collection system consisting of a sampling device and a recording device. This system can arrange the sampling points and design the sampling paths with geographic information system (GIS). Soil samples are positioned by GPS and the sampling is done automatically. The coordinates of the sampling points and the results of soil analysis are managed by this information technique.

1. Recording device
   The recording device is the auxiliary system for soil sampling. It not only acquires the geographical coordinates and other information on the sampling points, but also provides a platform for sampling design and management. A complete soil information sampling system consists of five basic functions: sampling design, sampling navigation, sample positioning, data analysis, and detection.
   Structurally, the soil information sampling system is composed of peripherals and the software. Peripnherals include the GPS receiver and various types of sensors. The software includes application programs and databases. Users are connected with each functional module via the man–machine interface. The structure of the system is shown in Figure 3.2.
2. Collection device
   The collection device has 12 components (Figure 3.3): sampling tube, a hydraulic cylinder of guider, a locator on sampling head, a slide rail for

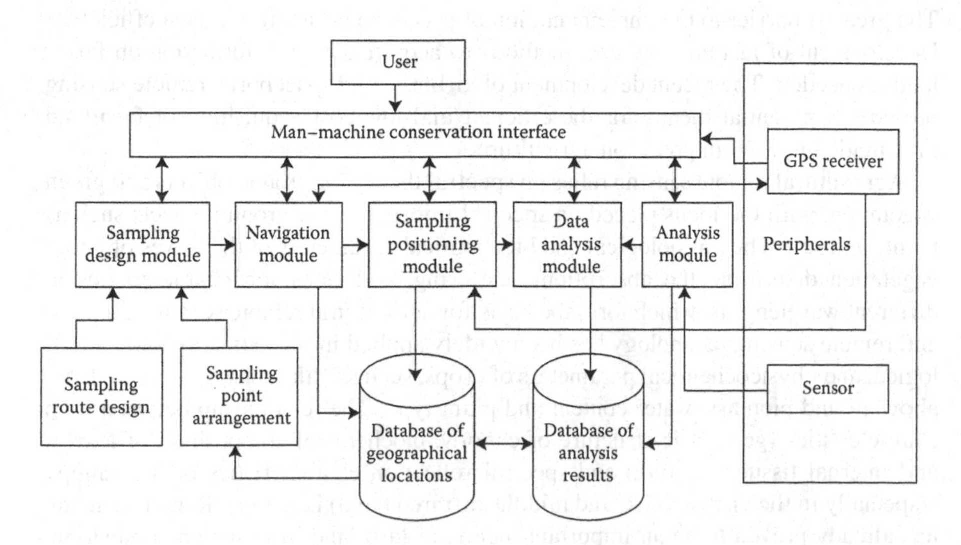

FIGURE 3.2 Schematic of the soil information sampling and recording system.

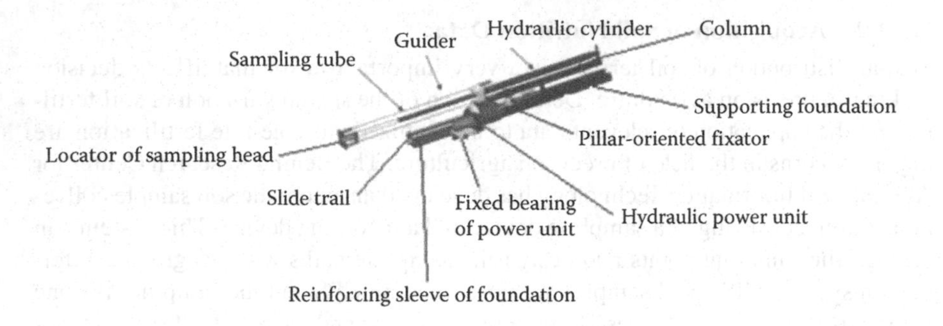

FIGURE 3.3 The automatic soil collection system. (From NERCITA, China.)

the supporting frame of the hydraulic cylinder, a reinforcing sleeve of the seat, a hydraulic power unit, a fixed bearing of the power unit, a supporting base of the pillar-oriented fixator, a pillar, and a GPS recording system. The hydraulic power unit is driven by a direct-current motor. Through the pressurization of hydraulic oil in the hydraulic oil tank, high-pressure hydraulic power is supplied to the system.

The automatic soil collection device is fixed on the top of the lateral trough of a pickup. The soil sampling device slides outward via the slide trail of the fixed frame of the hydraulic cylinder. The sampling depth is adjusted by the operator. Once the soil sampling depth is confirmed, the hydraulic module performs the work.

3.2 INTERPRETATION OF CROP INFORMATION FROM REMOTE SENSING

The greatest barrier to the implementation of precision agriculture is cost efficiency. Development of fast and low-cost methods to acquire spatial information on farmlands is needed. The recent development of airborne and spaceborne remote sensing provides a potential means for the efficient and low-cost acquisition of farmland information needed in precision agriculture.

Agricultural remote sensing relies on spectral theory for ground objects and green vegetation, with the focus placed on spectral information of ground objects such as plants and soil. The physiological and biochemical parameters of the leaves of green vegetation determine the absorption, scattering, and reflection characteristics at different wavelengths, which form the basis for agricultural remote sensing. At present, remote sensing technology has been widely applied in the extraction of key biological and physicochemical parameters of crops, such as chlorophyll, nitrogen, LAI, aboveground biomass, water content and plant type. The relationship between crop characteristics (geometric structure of canopy, biochemical composition of leaves, and internal tissue structure) and spectral reflection characteristics of the canopy (especially in the visible, NIR, and middle-infrared band) is critical. Remote sensing has already proven to be an important means of farmland information acquisition.

However, remote sensing data cannot be directly applied in decision making in precision agriculture. Information interpretation is needed to establish models representing the correlation between remote sensing information and the growth status of crops. A method for the inversion of agricultural parameter space and map plotting techniques should be established to provide support for farmland production management.

3.2.1 Crop Chlorophyll and Nitrogen Content Inversion through Remote Sensing

In this section, LNC and CND were inverted through remote sensing spectral parameters (Zhao et al., 2012). Then, using maize as the subject material, the vertical distribution of chlorophyll is retrieved with sensitive parameters and wavelet transform techniques (Liao et al., 2013).

3.2.1.1 Wheat Leaf Nitrogen Concentration and Canopy Nitrogen Density Estimation

3.2.1.1.1 Experiment and Data Collection

The experiment was conducted at the China National Experimental Station for Precision Agriculture. Twenty-five winter wheat cultivars, including Xiaoyan 54, Lumai 21, Laizhou 3279, P7, Gaocheng 8901,76-2, Jing 411, Jingwang 10, Nongda 3214, Nongda 3291, I-93, Lunxuan 201, Chaoyou 66, Chaoyou 69, CA 9722, 95128, 9428, Jingdong 8, Zhongyou 9507, Baili 981, Zhongmai 9, Zhongmai 16, 95021, and Linkang 2 and 6211, were investigated in the experiment. A randomized complete block design with three replications was used. The plot size was 5 × 3 m. Each plot received 180 kg ha^{-1} area, 225 kg ha^{-1} $(NH_4)_2HPO_4$ (diammonium phosphate), and 150 kg ha^{-1} K_2SO_4 before sowing. Topdressing N with 280 kg ha^{-1} urea was applied with two splits, 50% at Feekes 3.0 (March 25, 2003) and 50% at Feekes 7.0 (April 16, 2003). The Feekes scale is a system used by agronomists to identify the growth and development of cereal crops.

In each plot, a 1-m^2 area of wheat canopy was selected for canopy spectral reflectance measurements, and physiological and biochemical analyses. Measurements were performed eight times at Feekes 4.0 (April 4, 2003), Feekes 5.0 (April 12, 2003), Feekes 8.0 (April 21, 2003), Feekes 10.0 (April 29, 2003), Feekes 10.5.1 (May 8, 2003), Feekes 10.5.3 (May 16, 2003), Feekes 10.5.4 (May 24, 2003), and Feekes 11.1 (June 1, 2003). Feekes 4.0 is the beginning of erection of the pseudostem. At Feekes 5.0, the wheat plants become strongly erect. When approaching Feekes 8.0, the flag leaf is visible. Feekes 10.0 is the booting stage. At Feekes 10.5.1, the wheat is flowering. Flowering is complete at the base of the spike at Feekes 10.5.3. Upon reaching Feekes 10.5.4, wheat flowering is complete and the kernels are watery ripe. Feekes 11.1 is the milky ripe stage. Feekes 4.0, Feekes 5.0, Feekes 8.0, and Feekes 10.0 are growth stages in which vegetative growth develops synchronously with reproductive growth. These stages determine the spike number and grain number per spike. Field management is always applied at these stages. Feekes 10.5.1, Feekes

10.5.3, Feekes 10.5.4, and Feekes 11.1 are reproductive growth stages and the wheat 1000 kernels weight and grain quality are significantly affected by N availability at these stages.

Canopy spectral measurements were taken from a height of 1.3 m above ground (the height of the wheat was 90 ± 5 cm at maturity), under clear sky conditions between 10:00 and 14:00 hrs, using an ASD FieldSpec Pro spectrometer (Analytical Spectral Devices, Boulder, CO, USA) fitted with a 25° field of view (FOV) fiber optics, operating in the 350–2500 nm spectral region with a sampling interval of 1.4 nm between 350 and 1050 nm, and 2 nm between 1050 and 2500 nm, and with spectral resolution of 3 nm at 700 nm, and 10 nm at 1400 nm. After canopy spectral measurements were completed, a flag was placed in the FOV of the ASD to mark the location of measurement. Samples for LAI, SLW (g m^{-2}), and LNC determination were collected on the same day as canopy spectral reflectance measurements. All plants in the FOV of the ASD were cut at ground level with scissors immediately after spectral measurements, placed in a plastic bag, and transported to the laboratory for subsequent analysis. For each sample, all green leaves were separated from stems. LAI was determined by a dry weight method. LNC (%) was determined by the Kjeldahl method (Bremner, 1965) with a B-339 Distillation Unit.

3.2.1.1.2 Results and Conclusions

Thirteen narrow-band spectral indices (difference vegetation index [DVI], normalized difference vegetation index [NDVI], soil-adjusted vegetation index [SAVI], red-edge position [REP], photochemical reflectance index [PRI], structure insensitive pigment index [SIPI], green normalized difference vegetation index [GNDVI], optimized soil-adjusted vegetation index [OSAVI], normalized difference water index [NDWI], water band index [WBI], transformed chlorophyll absorption in reflectance index [TCARI], nitrogen reflectance index [NRI], and TCARI/OSAVI), three spectral features parameters associated with absorption bands centered at 670 and 980 nm, and another three related to reflectance maximum values located at 560, 920, 1690, and 2230 nm were calculated. It was demonstrated that REP is a good indicator for winter wheat LNC estimation. For CND, the largest R^2 among the growth stages was observed for GNDVI ($R^2 = 0.83^{**}$) at Feekes 10.5.3 (Table 3.1 and Figure 3.4). The LWC ranged from 61% to 84% from Feekes wheat cultivars and decreased with plant development. The strongest relationships of LNC and CND with spectral parameters at later growth stages were also attributed to the lower LWC. Plant water status provided information that can be used to assess crop growth under drought conditions. The two narrow-band spectral indices involving water absorption bands NDWI and WBI were well correlated with CND through the growth stages (Table 3.1). To show the relationships of LNC and CND to a given spectral parameter in two opposite growth stages, the NDVI and ABD normalized to the area of absorption feature (NBD) at 670 nm (NBD670) were plotted against LNC and CND (Table 3.2). When taking only these two spectral parameters, NDVI and NBD670, into consideration, the differences in correlation results for LNC with both NDVI and NBD670 between Feekes 4.0 and Feekes 11.1 were greater than those for CND (Figures 3.4 through 3.6).

TABLE 3.1
Pearson's Correlation Coefficients (*r*) for LNC and CND, and the 13 Narrow-Band Spectral Indices

	Growth Stages															
	Feekes 4.0 (Average LAI = 0.28)		Feekes 5.0 (Average LAI = 0.47)		Feekes 8.0 (Average LAI = 1.12)		Feekes 10.0 (Average LAI = 2.22)		Feekes 10.5.1 (Average LAI = 2.09)		Feekes 10.5.3 (Average LAI = 2.00)		Feekes 10.5.4 (Average LAI = 1.52)		Feekes 11.1 (Average LAI = 1.12)	
	LNC (%)	CND ($g\ m^{-2}$)	LNC (%)	CND ($g\ m^{-2}$)	LNC (%)	CND ($g\ m^{-2}$)	LNC (%)	CND ($g\ m^{-2}$)	LNC (%)	CND ($g\ m^{-2}$)	LNC (%)	CND ($g\ m^{-2}$)	LNC (%)	CND ($g\ m^{-2}$)	LNC (%)	CND ($g\ m^{-2}$)
DVI	0.293	0.791**	−0.066	0.522**	0.330	0.836**	0.048	0.470*	0.225	0.470*	0.340	0.743**	0.234	0.608**	0.598**	0.599**
NDVI	0.210	0.796**	−0.045	0.480*	0.145	0.705**	0.299	0.615**	0.342	0.462*	0.526**	0.890**	0.220	0.786**	0.741**	0.739**
SAVI	0.211	0.797**	−0.045	0.480*	0.149	0.710**	0.293	0.614**	0.342	0.466*	0.525**	0.892**	0.220	0.785**	0.740**	0.737**
REP	0.189	0.625**	0.122	0.353	0.457*	0.759**	0.352	0.222	0.443*	0.307	0.453*	0.846**	0.304	0.827**	0.857**	0.865**
PRI	0.357	0.634**	−0.091	0.387	0.107	0.620**	0.199	0.603**	0.340	0.516**	0.566**	0.769**	0.224	0.710**	0.749**	0.786**
SIPI	−0.209	−0.826**	0.113	−0.560**	−0.066	−0.650**	−0.230	−0.581**	−0.373	−0.528**	−0.596**	−0.798**	−0.169	−0.730**	−0.731**	−0.691**
GNDVI	0.185	0.790**	−0.029	0.494*	0.213	0.754**	0.378	0.614**	0.359	0.408*	0.474*	0.910**	0.236	0.822**	0.784**	0.789**
OSAVI	0.208	0.808**	−0.050	0.478*	0.142	0.706**	0.291	0.616**	0.339	0.466*	0.525**	0.888**	0.219	0.785**	0.740**	0.738**
NDWI	0.143	0.777**	−0.109	0.382	0.131	0.753**	0.296	0.679**	0.298	0.524*	0.424*	0.780**	0.257	0.812**	0.701**	0.702**
WBI	0.147	0.807**	−0.106	0.405*	0.205	0.796**	0.294	0.624**	0.242	0.525**	0.413*	0.765**	0.289	0.856**	0.673**	0.714**
TCARI	0.120	−0.361	0.001	−0.373	0.128	−0.196	−0.511**	−0.569**	−0.327	−0.277	−0.387	−0.569**	−0.200	−0.807**	−0.764**	−0.764**
NRI	0.219	0.820**	−0.082	0.405*	0.058	0.589**	0.222	0.640**	0.229	0.464*	0.488*	0.778**	0.212	0.719**	0.585**	0.605**
TCARI/OSAVI	−0.108	−0.752**	0.083	−0.546**	0.022	−0.465*	−0.419*	−0.583**	−0.349	−0.362	−0.457*	−0.706**	−0.177	−0.767**	−0.725**	−0.683**

Source: Zhao, C.J. et al. 2012. *International Journal of Remote Sensing*, 33(11):3472–3491. With permission.

Note: LNC is leaf N concentration and CND is canopy N density. DVI, NDVI, SAVI, REP, PRI, SIPI, GNDVI, OSAVI, NDWI, WBI, TCARI, NRI, and TCARI/OSAVI are the 13 narrow-band spectral indices. * and ** indicate significance at 0.05 and 0.01 levels, respectively. The number of samples is 25 except for the data at Feekes 10.5.1, where 24 samples were involved.

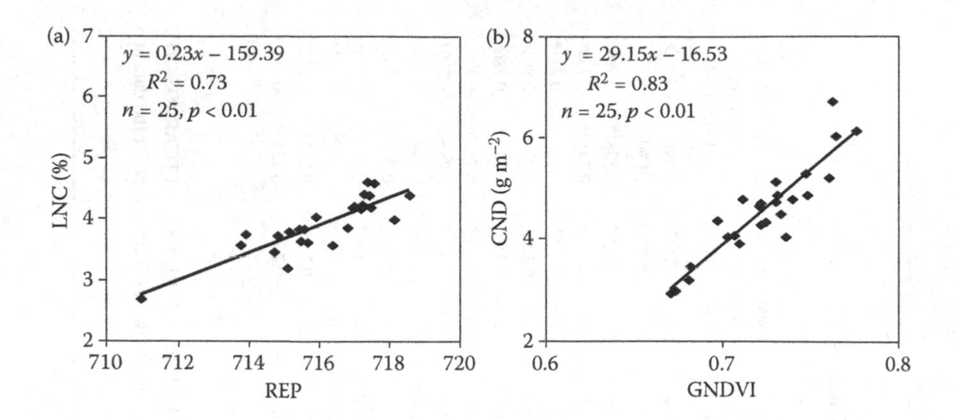

FIGURE 3.4 The best correlations between LNC (a) (at growth stage Feekes 11.1) or CND (b) (at growth stage Feekes 10.5.3) and the narrow-band spectral indices. Note: n is the sample number and p is the statistical significance. (From Zhao, C.J. et al. 2012. *International Journal of Remote Sensing*, 33(11):3472–3491. With permission.)

Pearson correlation analysis indicated that significant correlations between 31 spectral parameters and LNC existed at Feekes 8.0, Feekes 10.0, Feekes 10.5.1, Feekes 10.5.3, and Feekes 11.1 (Table 3.3). In contrast, relationships between the 31 spectral parameters and CND were consistently significant. Meanwhile, the correlation coefficient (r) values between the 31 spectral parameters and CND were generally higher than those of LNC. Thus, CND was more sensitive to winter wheat canopy spectral variation than was LNC. The spectra used in this study were canopy-level parameters, but LNC was a leaf-level parameter, which is a possible reason why LNC is less sensitive than CND to winter wheat canopy spectra.

For the 31 spectral parameters, REP showed the best relationship with LNC at Feekes 11.1 ($R^2 = 73$**, **significant at 0.01 level), followed by SIPI at Feekes 10.5.3 ($R^2 = 36$**) and TCAR at Feekes 10.0 ($R^2 = 26$**). REP was also significantly correlated with LNC at Feekes 8.0 ($R^2 = 21$*, *significant at 0.05 level) and Feekes 10.5.3 ($R^2 = 21$) (data not shown). Danson and Plummer (1995) stated that the red edge responded more linearly to LAI and chlorophyll compared to the classical NDVI, which often suffers from saturation problems, even at relatively low LAI values (<3.0). Our research demonstrated that REP is a good indicator for winter wheat LNC estimation. For CND, the largest R^2 values among the growth stages were observed for GNDVI ($R^2 = 0.83$**) at Feekes 10.5.3. A_Area670, SIPI, DVI, NDWI, R_Area1690, WBI, and REP were significantly correlated with CND at Feekes 4.0 ($R^2 = 0.68$**), Feekes 5.0 ($R^2 = 0.31$**), Feekes 8.0 ($R^2 = 0.69$**), Feekes 10.0 ($R^2 = 0.46$**), Feekes 10.5.1 ($R^2 = 0.29$**), Feekes 10.5.4 ($R^2 = 0.73$**), and Feekes 11.1 ($R^2 = 0.75$**), respectively, as well.

The most encouraging findings in this section were that REP was a good indicator for winter wheat LNC estimation and the absorption features derived from the wavelength centered at 670 nm, especially NBD670, proved to be reliable indicators for assessing wheat canopy N status. Therefore, winter wheat canopy N status can be assessed with both CND and spectral features parameters. This information is useful

TABLE 3.2
Pearson's Correlation Coefficients (*r*) for LNC and CND, the Spectral Features Parameters Associated with Absorption Features Located at 670 and 980 nm

	Growth Stages															
	Feekes 4.0 (Average LAI = 0.28)		Feekes 5.0 (Average LAI = 0.47)		Feekes 8.0 (Average LAI = 1.12)		Feekes 10.0 (Average LAI = 2.22)		Feekes 10.5.1 (Average LAI = 2.09)		Feekes 10.5.3 (Average LAI = 2.00)		Feekes 10.5.4 (Average LAI = 1.52)		Feekes 11.1 (Average LAI = 1.12)	
	LNC (%)	CND (g m^{-2})	LNC (%)	CND (g m^{-2})	LNC (%)	CND (g m^{-2})	LNC (%)	CND (g m^{-2})	LNC (%)	CND (g m^{-2})	LNC (%)	CND (g m^{-2})	LNC (%)	CND (g m^{-2})	LNC (%)	CND (g m^{-2})
ABD670	0.212	0.820**	–0.068	0.489*	0.120	0.692**	0.279	0.609**	0.338	0.467*	0.529**	0.879**	0.210	0.778**	0.734**	0.725**
A_Area670	0.216	0.826**	–0.055	0.475*	0.163	0.727**	0.304	0.619**	0.349	0.462*	0.523**	0.902**	0.232	0.804**	0.763**	0.767**
NBD670	–0.177	–0.805**	0.078	–0.552**	–0.292	–0.826**	–0.367	–0.620**	–0.372	–0.437*	–0.498*	–0.908**	–0.259	–0.835**	–0.802**	–0.789**
ABD980	0.199	0.634**	–0.081	0.397*	0.198	0.825**	0.394	0.611**	0.328	0.469*	0.326	0.712**	0.244	0.778**	0.756**	0.735**
A_Area980	0.180	0.717**	–0.063	0.391	0.209	***0.829*** **	0.354	0.589**	0.308	0.500*	0.361	0.749**	0.286	0.819**	0.738**	0.739**
NBD980	0.092	0.600**	–0.120	0.409*	0.065	0.539**	0.341	0.184	0.213	–0.072	–0.098	–0.013	–0.140	0.148	0.128	–0.022

Source: Zhao, C.J. et al. 2012. *International Journal of Remote Sensing*, 33(11):3472–3491. With permission.

Note: LNC is leaf N concentration and CND is canopy N density. ABD is the absorption band depth, A_Area is area of the absorption feature, and NBD is the absorption band depth normalized to the area of the absorption feature. * and ** indicate significance at 0.05 and 0.01 levels, respectively. The number of samples is 25 except for the data at Feekes 10.5.1, where 24 samples were involved.

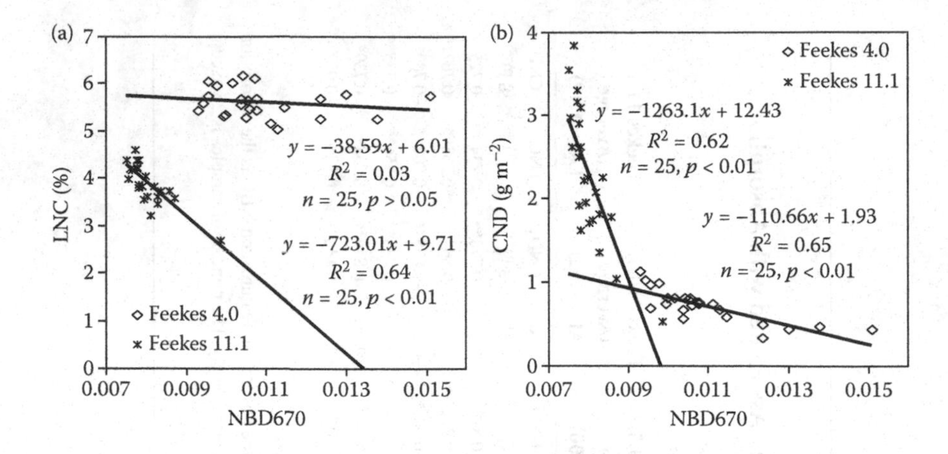

FIGURE 3.5 Relationship of LNC (a) and CND (b) with NBD670 at Feekes 4.0 and Feekes 11.1. Note: n is the sample number and p is the statistical significance. (Adapted from Zhao, C.J. et al. 2012. *International Journal of Remote Sensing*, 33(11):3472–3491.)

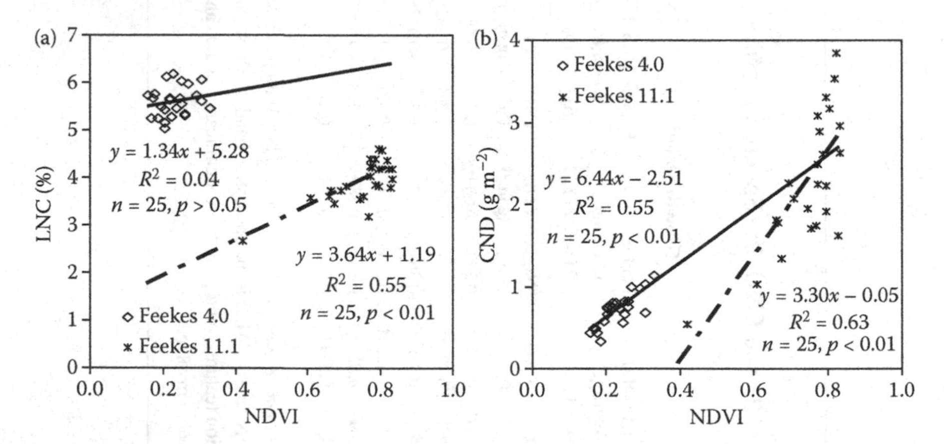

FIGURE 3.6 Relationship of LNC (a) and CND (b) with NDVI at Feekes 4.0 and Feekes 11.1. Note: n is the sample number and p is the statistical significance. (From Zhao, C.J. et al. 2012. *International Journal of Remote Sensing*, 33(11):3472–3491. With permission.)

for developing nondestructive monitoring techniques for spatial variation in N status in wheat with ground-based hyperspectral data or airborne and satellite imagery.

3.2.1.2 Estimation of Crop Vertical Chlorophyll Content and Nitrogen Content Distribution with Remote Sensing

3.2.1.2.1 Experiment and Data Collection

The experiment was conducted at the China National Experimental Station for Precision Agriculture during 2011–2012. The crop was summer maize, including Nongda 108 and Jinghua 8, a semicompact and a compact maize with respect to canopy morphology. Three nitrogen treatments with 0 kg (N0), 337 kg (N1), and 675 kg

TABLE 3.3
Pearson's Correlation Coefficients (*r*) between LNC and CND, and the Spectral Features Parameters Associated with Reflectance Features Located at 560, 920, 1690, and 2230 nm

	Growth Stages															
	Feekes 4.0 (Average LAI = 0.28)		Feekes 5.0 (Average LAI = 0.47)		Feekes 8.0 (Average LAI = 1.12)		Feekes 10.0 (Average LAI = 2.22)		Feekes 10.5.1 (Average LAI = 2.09)		Feekes 10.5.3 (Average LAI = 2.00)		Feekes 10.5.4 (Average LAI = 1.52)		Feekes 11.1 (Average LAI = 1.12)	
	LNC (%)	CND ($g\ m^{-2}$)	LNC (%)	CND ($g\ m^{-2}$)	LNC (%)	CND ($g\ m^{-2}$)	LNC (%)	CND ($g\ m^{-2}$)	LNC (%)	CND ($g\ m^{-2}$)	LNC (%)	CND ($g\ m^{-2}$)	LNC (%)	CND ($g\ m^{-2}$)	LNC (%)	CND ($g\ m^{-2}$)
RBH560	0.211	0.825**	−0.054	0.435	0.072	0.637**	0.244	0.626**	0.194	0.408*	0.431*	0.782**	0.215	0.714**	0.551**	0.555**
R_Area560	0.225	0.821**	−0.055	0.431*	0.068	0.639**	0.242	0.626**	0.177	0.388	0.415*	0.766**	0.215	0.712**	0.531**	0.534**
NBH560	−0.364	−0.007	0.118	−0.226	0.067	−0.556**	−0.163	−0.507*	−0.010	−0.083	−0.193	−0.429*	−0.177	−0.495*	−0.287	−0.238
RBH920	0.195	0.338	−0.055	0.388	0.186	0.758**	0.476*	0.582**	0.373	0.409*	0.287	0.659**	0.215	0.650**	0.737**	0.657**
R_Area920	−0.176	−0.218	0.083	−0.369	0.412*	0.785**	0.497*	0.548**	0.363	0.382	0.281	0.684**	0.254	0.704**	0.690**	0.634**
NBH920	0.254	0.414*	0.007	0.339	−0.064	0.577**	0.196	0.571**	0.306	0.461*	0.178	0.101	0.028	0.380	0.697**	0.556**
RBH1690	0.121	0.779**	−0.090	0.412*	−0.006	0.426*	0.375	0.604**	0.352	0.528**	0.417*	0.732**	0.250	0.811**	0.641**	0.703**
R_Area1690	0.121	0.784**	−0.083	0.410*	−0.045	0.357	0.379	0.613**	0.358	0.536**	0.442*	0.731**	0.250	0.801**	0.656**	0.715**
NBH1690	−0.135	−0.779**	0.100	−0.517**	0.357	0.240	−0.381	−0.588**	−0.382	−0.515**	−0.415*	−0.325	−0.200	−0.516**	−0.608**	−0.580**
RBH2230	−0.236	0.457*	0.140	0.277	0.028	0.339	0.335	0.565**	0.180	0.210	0.475*	0.374	0.240	0.689**	0.405*	0.410*
R_Area2230	−0.227	0.516**	0.123	0.281	−0.023	0.291	0.355	0.589**	0.222	0.264	0.499*	0.425*	0.250	0.709**	0.457*	0.472*
NBH2230	−0.201	−0.316	0.316	−0.085	0.367	0.463*	−0.139	−0.138	−0.273	−0.332	0.346	0.021	0.052	0.382	−0.134	−0.222

Source: Zhao, C.J. et al. 2012. *International Journal of Remote Sensing*, 33(11):3472–3491. With permission.

Note: LNC is leaf N concentration and CND is canopy N density. RBH is the reflectance band height, R_Area is the area of reflectance feature, and NBH is the reflectance band height normalized to the area of reflectance feature. * and ** indicate significance at 0.05 and 0.01 levels, respectively. The number of samples is 25 except for the data at Feekes 10.5.1, where 24 samples were involved.

(N2) per ha were designed to produce different chlorophyll contents. The maize was divided into two layers at the jointing stage and three layers at the other stages; leaf samples were collected from the different layers at each stage. Hyperspectral reflectance was measured using the ASD Fieldspec FR spectroradiometer (Analytical Spectral Devices; Boulder, CO, USA). After the spectra were measured, leaf samples were collected using a hole punch. The leaf samples were extracted with 80% acetone for 24 h in the dark at 22°C. Chlorophyll a, chlorophyll b, and total chlorophyll content were analyzed in lab (Liao et al., 2013).

To investigate the performance of chlorophyll spectral indices, 10 different spectral indices designed to estimate chlorophyll content were selected (Table 3.4). The determination coefficient (R^2) and root mean square error (RMSE) were used to evaluate precision of the fit of the linear regression model to experimental data. A CWT was used to extract accurate spectral information from the hyperspectral reflectance, using a mother wavelet function to convert the hyperspectral reflectance into several wavelet coefficients at specific scales, then generating the correlation scalogram between the wavelet coefficients and chlorophyll content. As the reflection peak of chlorophyll at 550 nm is similar to the Mexican hat wavelet, it was used as the basis for the mother wavelet. The CWT was conducted at dyadic scales

TABLE 3.4
Correlation Analysis between Spectral Indices and Chlorophyll Content Derived from the Calibration Dataset

Spectral Indices	Upper Layer		Middle Layer		Lower Layer		All Data	
	R^2 (%)	RMSE	R^2 (%)	RMSE	R^2 (%)	RMSE	R^2 (%)	RMSE
SR705: R_{750}/R_{705}	80.85[a]	5.55	82.40[a]	7.95	81.49[a]	9.68	78.87[a]	7.99
ND705: $(R_{750}-R_{705})/(R_{750}+R_{705})$	81.04[a]	5.52	79.81[a]	8.52	81.03[a]	9.67	78.13[a]	8.13
mSR705: $(R_{750}-R_{445})/(R_{705}-R_{445})$	81.20[a]	5.54	83.23[a]	7.87	78.03[a]	10.25	78.06[a]	8.21
mND705: $(R_{750}-R_{705})/(R_{750}+R_{705}-2R_{445})$	82.92[a]	5.32	83.46[a]	7.79	80.53[a]	9.59	79.70[a]	7.89
DD: $(R_{750}-R_{720})-(R_{700}-R_{670})$	82.50[a]	5.31	83.83[a]	7.63	83.24[a]	8.97	81.81[a]	7.43
BmSR: $(BR_{750}-BR_{445})/(BR_{705}-BR_{445})$	83.30[a]	5.35	83.31[a]	7.72	80.84[a]	9.78	81.85[a]	7.23
Green model: $(R_{800}/R_{550})-1$ $(R_{800}/R_{550})-1$	79.66[a]	5.73	82.87[a]	7.85	82.74[a]	9.09	81.05[a]	7.57
Red edge model: $(R_{800}/R_{700})-1$	75.72[a]	6.25	76.82[a]	9.15	78.08[a]	10.24	73.52[a]	8.94
Chlgreen: $(R_{760-800})/(R_{540-560})-1$	79.37[a]	5.75	82.59[a]	7.93	82.47[a]	9.16	80.72[a]	7.64
Chlred edge: $(R_{760-800})/(R_{690-720})-1$	80.14[a]	5.67	83.21[a]	7.78	81.87[a]	9.31	80.02[a]	7.78

Source: Liao, Q.H. et al. 2013. *Intelligent Automation & Soft Computing*, 19(3):295–304. With permission.

Note: [a]$p < 0.001$, [b]$p < 0.01$, [c]$p < 0.05$.

2^1, 2^2, 2^3,..., 2^{10}. The scales were defined as 1, 2, 3,..., 10, discarding scales greater than 2^{10} as they carried little spectral information. This CWT was conducted using MATLAB® 7.1 (Natick, MA, USA).

The maize dataset consisted of 65 upper layer leaves, 69 middle layer leaves, and 46 lower layer leaves. The dataset for each layer was randomly divided into calibration (60%) and validation (40%) subsets. The calibration dataset was used to build estimation models between the wavelet coefficients and chlorophyll content by simple linear regression, whereas the validation dataset was used to test the regression model. The R^2 and RMSE were used to assess the predictive performance of the estimation models. All statistical analyses were conducted using SPSS 16.0.

3.2.1.2.2 Results and Discussion

It is known that variation in chlorophyll content can be induced by different N treatments. For this reason, three nitrogen gradients were designed. Since the chlorophyll content differs little between Nongda 108 and Jinghua 8, in order to investigate the variation of chlorophyll content in different maize leaf layers, means and standard deviations were calculated. The mean values of chlorophyll content in the middle layer can reach 49.14 μg cm^{-2}, which is greater than that in the upper layer (40.02 μg cm^{-2}), but the differences among the three layers later vanished as nutrient (such as N, P, and K) were transported to the middle and upper layers in the trumpet stage. During the anthesis–silking and maturation stages, the chlorophyll content of the lower layer was obviously reduced as a result of nutrient accumulation in the earleaf for grain-filling, and as the lower and upper leaves aged, the chlorophyll content decreased from 38.29 to 25.30 μg cm^{-2}. Leaves with lower chlorophyll content had the highest reflectance in the visible range and the lowest reflectance in the NIR waveband, similar to other crops. The most obvious change in leaf spectral reflectance in the visible region was near 550 nm, especially at the maturation stage; the spectral reflectance ranged from 11.1% to 17.8% between the middle and lower layers. This waveband is usually not used directly to construct the spectral indices because of the effects of other pigments; the red-edge region (670–800 nm) is usually used to investigate the variation of chlorophyll content.

To compare the estimation capacity of the spectral indices with that of the wavelet transform, we used 10 leaf-scale spectral indices of chlorophyll content. These spectral indices mainly included simple ratios of reflectance (R_x/R_y), normalized ratios of differences of reflectance ($[R_x - R_y][R_x + R_y]$), and reflectance derivatives (dR_x/dR_y).

As spectral indices, first derivative reflectance of modified simple ratio index (BmSR) and double difference index (DD), which is defined in Table 3.4, considered plant functional type, leaf structure, leaf developmental stage, specular reflection, and the variation range of chlorophyll content. These indices exhibited good results for estimating chlorophyll content.

The hyperspectral reflectance of the different layers and all datasets were transformed using a CWT. Figure 3.7a, c, e, and g shows the correlation between the wavelet coefficients and chlorophyll content, with the vertical axis and horizontal axis representing the decomposition scale and wavelength, respectively. The highlighting represents the wavelet regions with high R^2, whereas the dark portion represents regions that are less sensitive to chlorophyll content. Figure 3.7b, d, f, and h

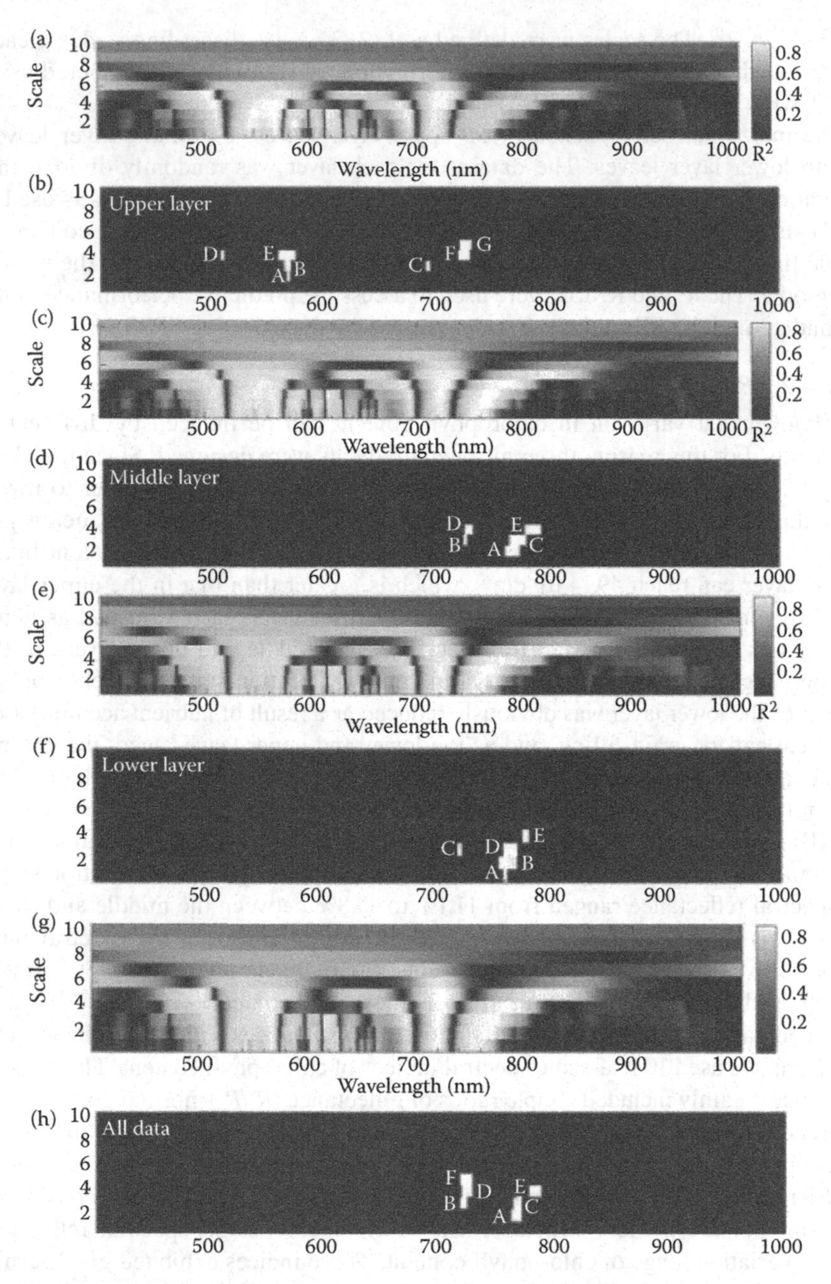

FIGURE 3.7 Hyperspectral reflectance of maize leaves in different layers and at different growth stages. (a, c, e, g) are correlation scalograms between the wavelet coefficients and chlorophyll content, with the vertical axis and horizontal axis representing the decomposition scales and wavelength, respectively, (b, d, f, h) are selected wavelet features sensitive to chlorophyll content. A, B, C, D, E, F in (b, d, f, h) indicate the wavelet feature regions that were found to be sensitive to the chlorophyll content for the different maize layers. (From Liao, Q.H. et al. 2013. *Intelligent Automation & Soft Computing*, 19(3):295–304. With permission.)

indicates the extraction of wavelet features that are sensitive to chlorophyll content. These features were chosen as follows. First, the R^2 values for chlorophyll content and the wavelet coefficients were computed using the CWT. Second, all significant features ($p < 0.0001$) were ranked in descending order based on R^2, and the top 1% was retained. The highest R^2 and corresponding scales and wavelengths for the significant features were then identified. Figure 3.7b shows that seven sensitive wavelet features were found for the upper layers, and the wavelengths of all the wavelet features were in the 550 or 700 nm region, which is sensitive to chlorophyll content. This result is similar to the selection of wavebands used to design the chlorophyll spectral indices. The highest R^2 of this layer was 92.25% (Table 3.5), which is higher than those of the spectral indices for this layer. It should be noted that the most sensitive wavelet region was located at 569 nm, which is close to the strong reflection peak of chlorophyll. Figure 3.7d demonstrates that five sensitive wavelet features were extracted from the correlation scalogram (Figure 3.7c). The most sensitive feature was located at 766 nm and scale 2; the corresponding R^2 was 91.46% (Table 3.5), which was also an improvement over the spectral indices. The chlorophyll content of this layer is much higher than that of the upper layer, such that all wavelet features moved to the red-edge region (680–800 nm), in accordance with other studies. Figure 3.7f shows that five wavelet features were sensitive to chlorophyll content in the lower layer; the highest R^2 (94.85%) was obtained at 760 nm and scale 1. These wavelet features were also evident in the red-edge region. Many studies have found that when chlorophyll content decreases, the red edge shifts toward a shorter wavelength.

Figure 3.7a, c, e, and g shows the correlation scalogram between the wavelet coefficients and the chlorophyll content; Figure 3.7b, d, f, h indicates the extraction of wavelet features that are sensitive to the chlorophyll content.

TABLE 3.5
Correlation Analysis between the Wavelet Features and Chlorophyll Content Derived from the Calibration Dataset

	Upper Layer		Middle Layer		Lower Layer		All Data	
Code	Feature Location and Scale	R^2 (%)	Feature Location and Scale	R^2 (%)	Feature Location and Scale	R^2 (%)	Feature Location and Scale	R^2 (%)
A	567, 2	91.62[a]	766, 2	91.46[a]	760, 1	94.85[a]	768, 2	89.91[a]
B	568, 3	91.99[a]	722, 3	90.31[a]	759, 2	94.18[a]	721, 3	89.89[a]
C	692, 3	92.13[a]	770, 3	91.06[a]	721, 3	93.54[a]	770, 3	89.90[a]
D	509, 4	91.77[a]	726, 4	90.76[a]	765, 3	93.75[a]	724, 4	90.50[a]
E	569, 4	92.25[a]	782, 4	90.43[a]	778, 4	93.41[a]	782, 4	89.64[a]
F	724, 4	91.96[a]	–	–	–	–	723, 5	89.80[a]
G	725, 5	91.87[a]	–	–	–	–	–	–

Source: Liao, Q.H. et al. 2013. *Intelligent Automation & Soft Computing*, 19(3):295–304. With permission.

Note: [a]$p < 0.001$, [b]$p < 0.01$, [c]$p < 0.05$.

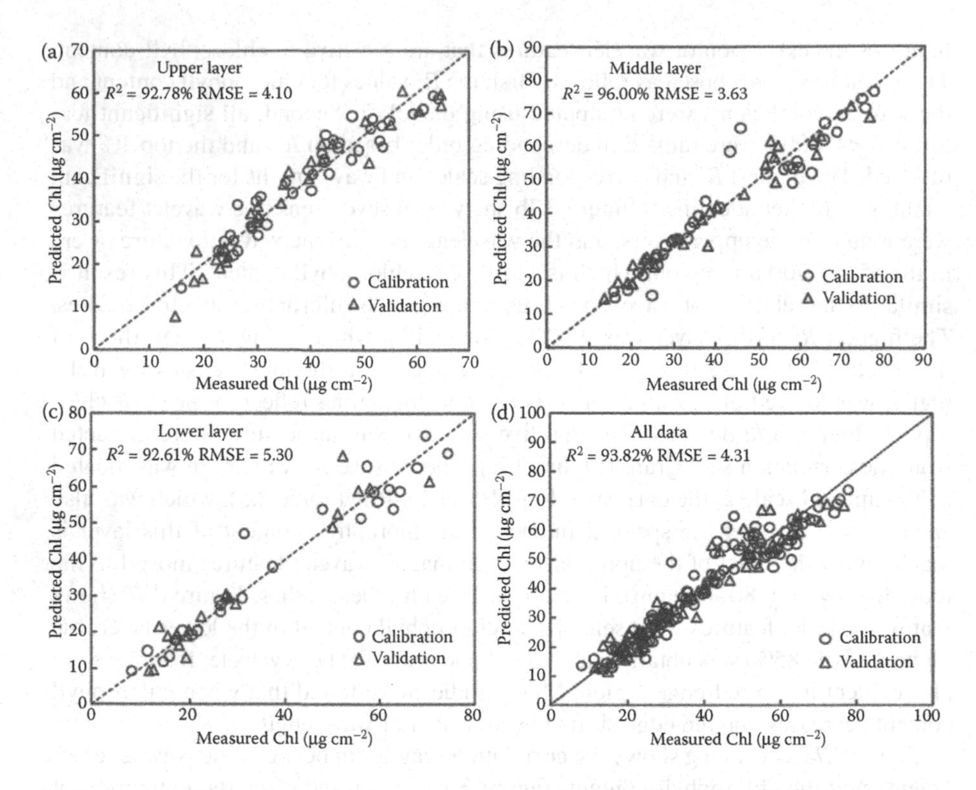

FIGURE 3.8 Plots of measured versus predicted chlorophyll content based on the models developed from the most sensitive wavelet features. (a) Upper layer, (b) middle layer, (c) lower layer, and (d) all the data. (From Liao, Q.H. et al. 2013. *Intelligent Automation & Soft Computing*, 19(3):295–304. With permission.)

The results support this conclusion and show that the position of a sensitive wavelet feature in the upper layer moved 6 nm to a shorter wavelength when compared with the middle layer. Figure 3.7h shows that in the dataset of all the layers subjected to the CWT, the highest R^2 was 90.50%, which is higher than the spectral indices of all the layers.

The position of the most sensitive wavelet feature was located at 724 nm, close to the red-edge position, indicating that this wavelength position is a good indicator to use in estimating the variation of chlorophyll content. Next, they applied the linear regression models resulting from the calibration dataset to the validation dataset. Figure 3.8a–d shows that data points dispersed close to the 1:1 line, and all the predicted R^2 values of the different layers exceeded 90%.

3.2.1.3 Evaluation of Crop Nutrition and Growth Status Based on Airborne Remote Sensing

Hyperspectral remote sensing can obtain refined spectral data on the crop canopy or leaves, from which we can determine the growth status, water and fertilizer deficiency, nutrition content, as well as grain quality and yield information on crops. Thus, precision agriculture has led to progress in hyperspectral remote sensing techniques (Moran et al., 1997). Utilizing the red-edge characteristics of operative modular imaging spectrometer (OMIS) images, statistical analysis is performed on the correlation of leaf chlorophyll, total nitrogen (TN), soluble sugar, and water content with spectral characteristics. The remote sensing model of biochemical parameters is established based on OMIS images. Maps of biochemical parameters are generated and then used for the analysis of growth status of crops (Liu, 2002).

The experiment was carried out at the National Experiment Station for Precision Agriculture in 2001 using winter wheat. OMIS images were acquired by airborne sensors on April 25, 2001. During the flight, biochemical parameters were also sampled, including chlorophyll, TN, soluble sugar, LWC, and LAI. Using the inverted Gaussian model, the red-edge parameters of the spectra were calculated, and correlation analysis was performed with the biochemical composition of the canopy. Table 3.6 lists the multiple correlation coefficients of chlorophyll, TN, sugar, and LWC with red-edge spectral parameters of OMIS images. The remote sensing model of chlorophyll, TN, sugar, and LWC was established according to the position of the red edge (Table 3.7).

Using the above model, maps of biochemical parameters such as chlorophyll, TN, soluble sugar, and LWC were plotted, as shown in Figure 3.9a–d. Figure 3.10 shows the results for color composition of the maps of three biochemical parameters, which are chlorophyll ab content (R), TN (G), and soluble sugar (B). R, G, and B in the brackets represent the three primary colors (red, green, blue) used for color composition. In the late jointing stage, the plots grown with winter wheat having high sugar, low nitrogen, and low chlorophyll are shown in blue. For the plots, the crops lack fertilizer and may enter the reproductive growth state before those with fertilizer. The plots of winter wheat having high nitrogen, high chlorophyll, and low sugar are shown in yellow. These nutritionally adequate plots are still in the vegetative growth stage. The growing status of winter wheat can be evaluated by color, facilitating decisions on agricultural management.

TABLE 3.6
Coefficients of Correlation between Biochemical Composition and Red-Edge Spectral Parameters of OMIS Images ($n = 45$)

Chemical	λ_p	λ_o	σ	IG-R^2	NDVI
Chlorophyll	0.408	0.429	0.182	0.305	0.430
Total nitrogen	0.568	0.532	0.176	0.515	0.422
Sugar	0.560	0.510	0.202	0.485	0.460
Foliar water	0.247	0.147	0.006	0.176	0.301

Source: Liu, L.Y. 2002. *Hyperspectral Remote Sensing Application in Precision Agriculture.* Postdoctoral research report of Institute of Remote Sensing Applications, Chinese Academy of Sciences. With permission.

Note: λ_o is the spectral position of the red trough corresponding to chlorophyll absorption; λ_p is the spectral position of the red edge; and σ is the variance term of the inverted Gaussian model and the difference between the spectral position of the red edge and red trough for vegetation. It corresponds to the width of the absorption trough of the red edge.

TABLE 3.7
Remote Sensing Model of Chlorophyll, Total Nitrogen, Sugar, and Leaf Water Content

Model	R^2
$N = -126.4563 + 0.3051*\lambda_p - 0.0061*\lambda_o - 84.9024*IG\text{-}R^2$	0.597
$Chlorophyll = -285.3746 + 0.1117*\lambda_p + 0.2115*\lambda_o + 61.9571*IG\text{-}R^2$	0.458
$Sugar = 93.386 - 0.3042*\lambda_p + -0.4114*\lambda_o + 421.9152*IG\text{-}R^2$	0.443
$Water = 74.7180 + 1.0473*\lambda_p + 0.6955*\lambda_o - 276.5877*IG\text{-}R^2$	0.301

Source: Liu, L.Y. 2002. *Hyperspectral Remote Sensing Application in Precision Agriculture*. Postdoctoral research report of Institute of Remote Sensing Applications, Chinese Academy of Sciences. With permission.

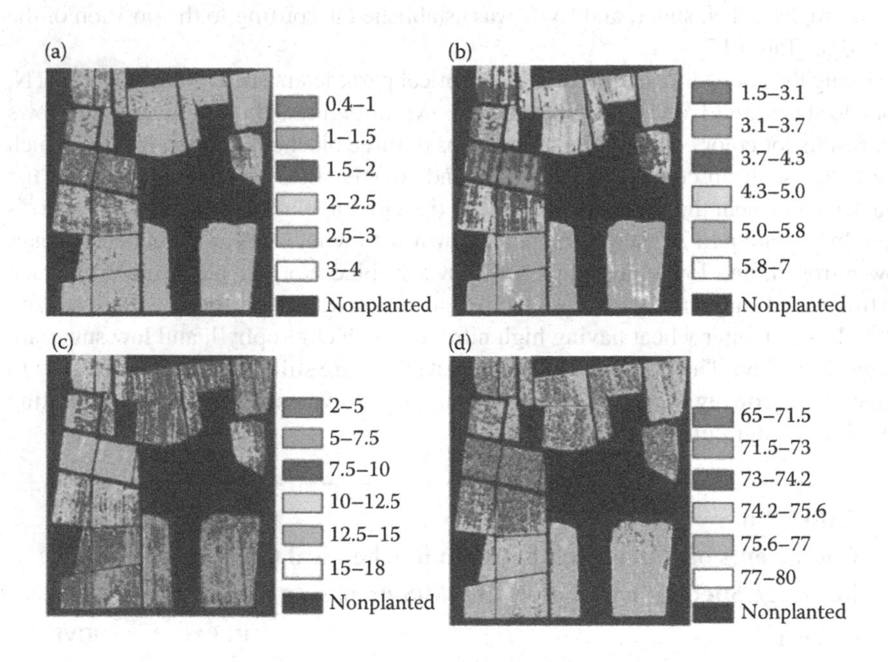

FIGURE 3.9 **(See color insert.)** Map plots of biochemical parameters, including chlorophyll, total nitrogen, soluble sugar, and leaf water content. (a) Chlorophyll concentration (mg g^{-1}), (b) nitrogen concentration (%), (c) soluble sugar concentration (%), and (d) leaf water content (%). (From Liu, L.Y. 2002. *Hyperspectral Remote Sensing Application in Precision Agriculture*. Postdoctoral research report of Institute of Remote Sensing Applications, Chinese Academy of Sciences. With permission.)

3.2.2 Crop LAI Estimation through Remote Sensing

The LAI is a major indicator for crop growth monitoring and yield estimation. Actual observed data provide statistical properties of crop LAI, whereas crop simulation models provide physical properties of LAI within the whole crop growth period. In

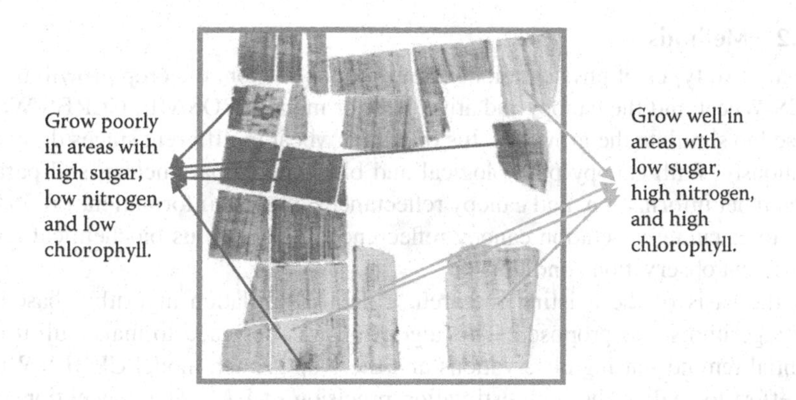
Grow poorly
in areas with
high sugar,
low nitrogen,
and low
chlorophyll.
Grow well in
areas with
low sugar,
high nitrogen,
and high
chlorophyll.

FIGURE 3.10 **(See color insert.)** Pseudocolor composition map of biochemical parameters, including chlorophyll, total nitrogen, and soluble sugar. Red lines: The crops grow poorly in areas with high sugar, low nitrogen, and low chlorophyll. Green lines: The crops grow well in the areas with low sugar, high nitrogen, and high chlorophyll. (From Liu, L.Y. 2002. *Hyperspectral Remote Sensing Application in Precision Agriculture*. Postdoctoral research report of Institute of Remote Sensing Applications, Chinese Academy of Sciences. With permission.)

the work of Dong et al. (2012), a data assimilation scheme of integrating observations and CERES-Wheat model based on the ensemble Kalman filtering (EnKF) algorithm is proposed for LAI estimation (Kalman, 1960; Evensen, 2003).

3.2.2.1 Experiment and Data Collection

The field experiments of winter wheat Jingdong 8 were conducted in 2002 at the Xiaotangshan National Experiment Station for Precision Agriculture in Changping district, Beijing. In the experiment, there were 16 testing areas, each 32.4 m × 30 m; four water treatments and four fertilizer treatments were conducted in these areas. March 25, April 2, April 10, April 18, May 6, May 17, May 24, and May 31, in 2002, were chosen as the dates in the sequence of observations covering the key growth stages of winter wheat. All of the information were obtained from the above experiments, including meteorological data, soil data, management data, time series remote sensing data of winter wheat, and so on. Meteorological data such as daily solar radiation, maximum air temperature and minimum air temperature, and precipitation were recorded by observation equipment at the DAVIS meteorological site in the Xiaotangshan National Experiment Station for Precision Agriculture in Changping district, Beijing. Soil data, such as soil moisture and soil nutrients, were recorded in the experiments. Moisture content values of soil layers (5, 20, 40, 60, 80, and 100 cm) were determined by the oven drying method. Management data were recorded during the experiments, including information on seeding, fertilizing, irrigation, and other practices. In addition, the sequential canopy spectral reflectance and LAI of winter wheat were obtained with an ASD FieldSpec Pro FR (350–2500 nm) spectrometer (ASD, USA) and the SLW method, respectively. Actual observed data provide statistical properties of crop LAI, whereas crop simulation models provide physical properties of LAI within the whole crop growth period.

3.2.2.2 Methods

There are two types of physical models used in this section, the crop growth model CERES-Wheat and the canopy radiative transfer model PROSAIL. CERES-Wheat was used to simulate the growth status of winter wheat in different stages daily and continuously. With canopy physiological and biochemical parameters, soil parameters as input information, and canopy reflectance as output information, PROSAIL served to calculate vegetation canopy reflectance under various biochemical levels and different observation conditions.

On the basis of the existing research, a data assimilation algorithm based on EnKF algorithms was proposed. This algorithm was designed to make full use of sequential remote sensing observations and the crop growth model CERES-Wheat in an effort to realize the high estimation precision of LAI. The conceptions and assimilation strategies of EnKF are described in detail as follows. EnKF is a sequential assimilation method combined with ensemble forecasting and Kalman filtering. The assimilation scheme is shown in Figure 3.11.

(1) Initialization of background field. The initial background field dataset $X^a(t_0)$ obeying Gaussian distribution and its error covariance matrix $P^a(t_0)$ of the state variable LAI is determined by remote sensed observations and the crop growth model CERES-Wheat. (2) Ensemble forecasting. At the $k+1$ moment, the forecasted dataset $X^f(t_k+1)$ and its error covariance matrix $P^f(t_k+1)$ are calculated according to Equation 3.3. (3) Analyzing. The gain factor $K(t_k+1)$ of the $k+1$ moment is calculated according to Equation 3.3. (4) Updating. At the $k+1$ moment, the analysis field

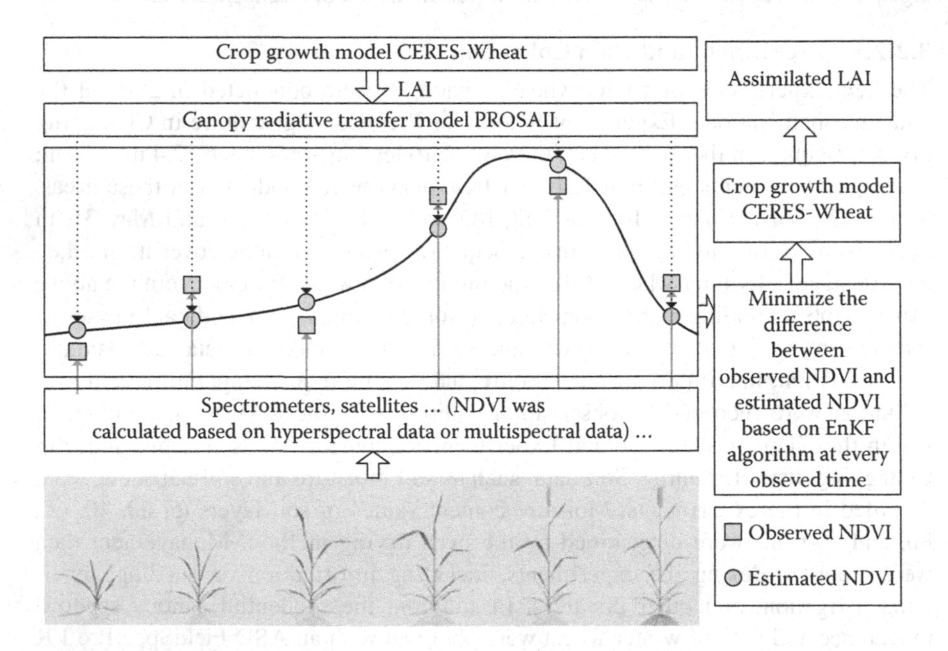

FIGURE 3.11 Data assimilation mode integrating observations with the CERES-Wheat model for LAI estimation based on the EnKF algorithm. (From Dong, Y.Y. et al. 2012. *Advances in Intelligent and Soft Computing*, 165:831–837. With permission.)

dataset $X^a(t_k+1)$ and its error covariance matrix $P^a(t_k+1)$ are calculated according to Equation 3.3. If there are still observations, the algorithm will move on to the next moment and return to ensemble forecasting. Otherwise, the assimilation process will be ended and $\overline{X^a(t_{k+1})}$ will be taken as the optimal state variable.

$$
\begin{aligned}
X^f(t_{k+1}) &= M_k X^a(t_k) + W_k;\ W_k \sim N(0,Q_k);\ P^f(t_{k+1}) \\
&= \frac{(X^f(t_{k+1}) - \overline{X^f(t_{k+1})})(X^f(t_{k+1}) - \overline{X^f(t_{k+1})})^T}{N-1}; \\
K(t_{k+1}) &= P^f(t_{k+1})H^T(HP^f(t_{k+1})H^T + R(t_k))^{-1}; \\
X^a(t_{k+1}) &= X^f(t_{k+1}) + K(t_{k+1})(Y^0(t_{k+1}) - HX^f(t_{k+1}));\ P^a(t_{k+1}) \\
&= \frac{(X^a(t_{k+1}) - \overline{X^a(t_{k+1})})(X^a(t_{k+1}) - \overline{X^a(t_{k+1})})^T}{N-1}
\end{aligned}
\tag{3.3}
$$

Regions with poor growth of wheat tend to have high sugar, low nitrogen, and low chlorophyll, whereas regions with good growth of wheat have low sugar, high nitrogen, and high chlorophyll. In Equation 3.3, N is the size of the dataset, M is the CERES-Wheat model operator, and H is the PROSAIL model operator. $X^a(t_k)$ is the analysis field dataset at moment k. $\overline{X^a(t_{k+1})}$ and $\overline{X^f(t_{k+1})}$ are the mean values of the analysis field dataset and the forecasted dataset at moment $k + 1$, respectively. $Y^0(t_k+1)$ is the observed dataset at moment $k + 1$, and Q_k is the model error. $N(0,Q_k)$ is the Gaussian white noise dataset, and $R(t_k)$ is the observation error covariance matrix. The EnKF algorithm scheme is effective in solving the nonlinear problems involved in model operation and observation by the operator.

In order to validate the feasibility and effectiveness of the data assimilation algorithm, a comparative study was made between the EnKF algorithm and the CERES-Wheat model. In the comparison experiments, the RMSE, coefficient of determination (R^2), and accuracy were selected to analyze the precision of estimates.

3.2.2.3 Results and Analysis

In the numerical experiments, sequential remote sensed observations are taken as input parameters of CERES-Wheat. The initial background field and its error covariance matrix of the state variables are calculated according to the EnKF assimilation strategy, and the model error is set as $Q_k = 3\% \times y(t_k)$ on the basis of given experience. Results of the EnKF assimilation experiments are shown in Figures 3.12 and 3.13.

In comparing the EnKF assimilations with the CERES-Wheat simulations, the three model testing indicators RMSE, R^2, and accuracy had values of 0.84, 0.87, and 0.42; 0.38, 0.92, and 1.05, respectively. In the whole EnKF assimilation process, the model simulations were effectively constrained by the observations. Confirmatory analysis showed that the EnKF-assimilated LAI not only agrees with the actual observations and the crop growth disciplines, but also reaches higher estimation precision.

For winter wheat, the NDVI became saturated when the LAI of crop canopies was greater than or equal to 3.00. The dataset of LAI was subsequently divided into

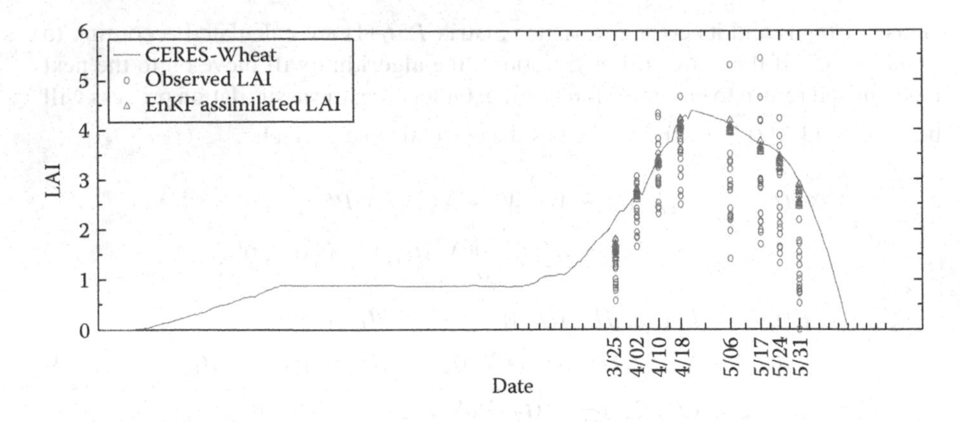

FIGURE 3.12 Estimation results of EnKF assimilation algorithms. (From Dong, Y.Y. et al. 2012. *Advances in Intelligent and Soft Computing*, 165:831–837. With permission.)

two categories: one subset included LAI values less than 3.00, and the other included LAI values greater than or equal to 3.00. Figure 3.13 shows that, for LAI ≤3.00, the EnKF-assimilated LAI is better than that of the CERES-Wheat simulations, but for LAI ≥3.00, these two methods show no improvement in LAI estimation.

In order to solve the existing problem of unsatisfactory and low-efficiency LAI assimilation, the data algorithm, the sequential remote sensed observations, and the crop growth model simulations were comprehensively utilized to enhance the precision of LAI assimilation.

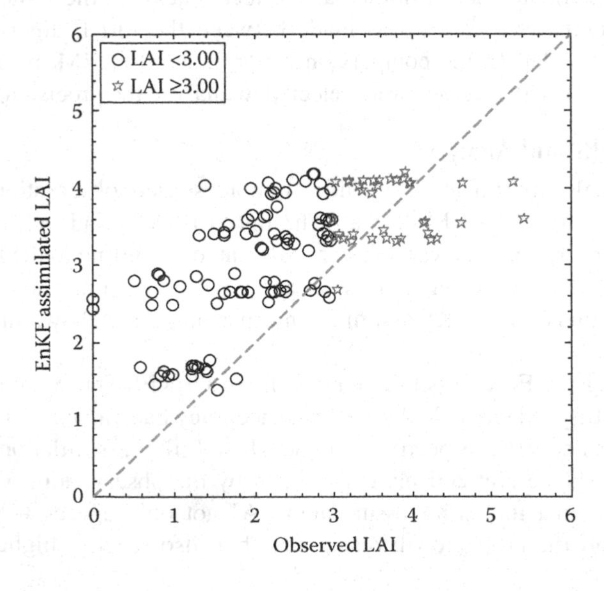

FIGURE 3.13 Scatter plots of observed LAI and EnKF-assimilated LAI. (From Dong, Y.Y. et al. 2012. *Advances in Intelligent and Soft Computing*, 165:831–837. With permission.)

In the assimilation process, the LAI dynamic changing information was provided by sequential observations, and the LAI changing tendency was constrained by the CERES-Wheat simulations; all of these results were in accordance with actual observations and crop growth principles. Theoretical analysis and numerical experiments proved that the data assimilation scheme based on the EnKF algorithm effectively improves the estimation of crop LAI values.

3.2.3 Crop Water Estimation and Inversion through Remote Sensing

In this section, regression models that are based on gray relational analysis–partial least squares (GRA–PLS), the optimal band ratio normalized difference, and three bands algorithms were developed and tested for winter wheat LWC estimation (Jin et al., 2013).

3.2.3.1 Experiment and Data Collection

The experiment site is located in Tongzhou District (39°36′–40°2′N, 116°32′–116°56′E) and Shunyi District (40°0′–40°18′ N, 116°28′–116°58′E) of the Beijing suburbs, China. Four local wheat cultivars, Nongda 195, Jingdong 13, Zhongyou 206, and Jing 9428, were planted from September 25 to 30, 2007 and from September 28 to October 2, 2008 at seeding rates of 190–225 kg ha^{-1} in 2007 and 215–265 kg ha^{-1} in 2008. Spectral measurements were performed at the following growth stages (the first date is from 2008, the second date is from 2009): jointing (15 April, 13 April), heading (29 April, 30 April), and anthesis (15 May, 18 May) of winter wheat. All canopy spectral measurements were taken using the ASD Field Spec Pro spectrometer (analytical spectral devices) mounted on a tripod boom and held in a nadir orientation 1.3 m above the canopy. Vegetation radiance measurements were taken by averaging 16 scans at an optimized integration time, with a dark current correction at every spectral measurement. A panel radiance measurement was taken before and after the vegetation measurement by two scans each time.

After the spectral positions of the biomass were collected, the aboveground biomass was sampled destructively. In each plot, average-looking plants were selected for sampling, and then 60 × 60-cm biomass sections from the scanned plants were cut at ground level. Collected plant samples were placed in a paper bag, sealed in a plastic bag, and placed in a cool, dark container to avoid as much water loss as possible. Upon returning from the field, leaves and stems were separated and weighed. All plant samples were then oven dried for 48–72 h at 85°C to constant mass, which was recorded (Woods et al., 1982; Zhang et al., 2012). LWC was calculated as Equation 3.2.

3.2.3.2 Methods and Analysis

Based on the published literature, 10 spectral parameters that better elucidate the relationship between LWC and WVIs (Table 3.8) were used. Linear and nonlinear regression analysis was conducted, with the selected spectral parameters serving as independent variables. The results indicated that relationships between LWC and all spectral parameters were significant with the exception of the water index (1300, 1450) and the normalized difference infrared index. The water index (1148, 1088), water index (1300, 1450), and vegetation dry index (VDI) were negatively correlated

TABLE 3.8
Summary of Selected Vegetation Indices, Wavebands, and References for Leaf Water Content

Spectral Parameters	Wavebands[a]	Reference
Water index (900,970)	R_{900}/R_{970}	Penuelas et al. (1993)
Water index (1148,1088)	R_{1148}/R_{1088}	Schlerf et al. (2003)
Water index (1100,1200)	R_{1100}/R_{1200}	Jin et al. (2013)
Water index (1300,1450)	R_{1300}/R_{1450}	Seeliga et al. (2008)
Water index (1300,1200)	R_{1300}/R_{1200}	This study
Normalized difference water index–hyperion	$(R_{1070} - R_{1200})/(R_{1070} + R_{1200})$	Ustin et al. (2002)
Vegetation dry index	$(R_{970} - R_{900})/(R_{970} - R_{900})$	Penuelas et al. (1993)
Normalized difference infrared index	$(R_{850} - R_{1650})/(R_{850} + R_{1650})$	Hunt and Rock (1989)
Normalized difference matter index	$(R_{1649} - R_{1722})/(R_{1649} + R_{1722})$	Wang et al. (2011)
Normalized heading index	$([R_{1100} - R_{1200}]/[R_{1100} + R_{1200}])/([R_{850} - R_{670}]/[R_{850} + R_{670}])$	Pimstein et al. (2009)

Source: Jin, X.L. et al. 2013. *Agronomy Journal*, 105:1385–1392. With permission.
[a] R_i denotes reflectance at band *i* (nm).

with LWC; the respective correlation coefficients (*r*) were −0.32, −0.14, and −0.23. The remaining parameters were positively correlated with LWC. The water index (1300, 1200) had the highest *r* value of 0.56 and an R^2 value of 0.32. The following parameters were significantly correlated with LWC.

The highly correlated area was located in the range of 1500- to 1750-nm wavelengths, occurring mainly in the water absorption wavelengths ($R^2 > 0.35$). The selected band ratio (R_{1723}/R_{1535}) performed better than did the empirical model based on reflectance spectra for the band ratio, with an R^2 value of 0.39.

A similar correlation analysis was applied to aggregated datasets with band-normalized difference indices. The highly correlated area was located in the narrow spectral region of 1600–1750 nm ($R^2 > 0.35$). The most sensitive band normalized difference was utilized to establish the LWC estimation model. Of all the water index combinations, the best normalized difference water index was $(R_{1720} - R_{1530})/(R_{1723} + R_{1530})$, with an R^2 of 0.37.

3.2.3.3 Results

The three-band algorithm was similar to the band ratio algorithm; the best three-band water index was $(R_{973} - R_{1720})/R_{1447}$, with an R^2 value of 0.60 and RMSE of 13.15% (Figure 3.14).

The final selected spectral parameters included water index (900, 970), water index (1148, 1088), water index (1100, 1200), water index (1070, 1200), water index (1300, 1200), NDWI–Hyperion, 1650, 1722, and 970 nm (Table 3.9).

These results suggested that the relationship between the nine water spectral variables and LWC were relatively stable and the influences of experimental conditions

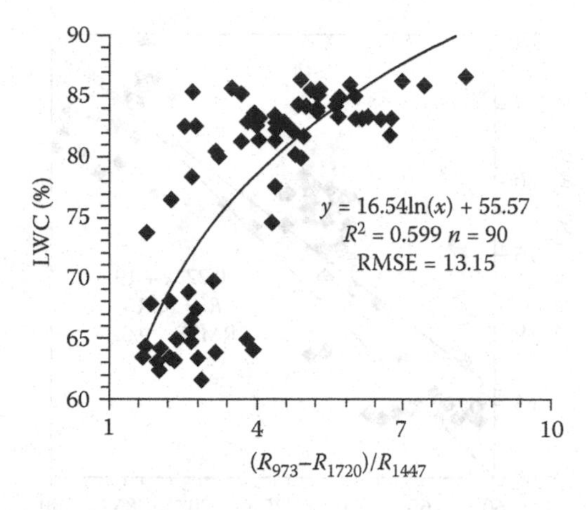

FIGURE 3.14 Nonlinear regression of leaf water content (LWC) against the optimal three-band algorithm. (From Jin, X.L. et al. 2013. *Agronomy Journal*, 105:1385–1392. With permission.)

were relatively small; so the model established in this study could be used to estimate LWC in winter wheat from 2008 data ($n = 90$). The regression equation was

$$\begin{aligned} y = {} & -10.354(R_{900}/R_{970}) + 7.089(R_{1148}/R_{1088}) + \\ & 12.524(R_{1100}/R_{1200}) + 16.624(R_{1300}/R_{1200}) \\ & -13.306[(R_{1070} - R_{1200})/(R_{1070} + R_{1200})] + 1.032(R_{1070}/R_{1200}) + \\ & 4.364R_{1650} + 5.632R_{1722} - 1.434R_{970} + 10.454 \end{aligned} \tag{3.4}$$

with an R^2 value of 0.74. To validate the model, the predicted values using the GRA–PLS model were compared with the actual values acquired during the entire growth

TABLE 3.9
Spectral Variables for the Gray Relational Analysis–Partial Least Squares Model Implementation at All Growth Stages ($n = 90$)

Evaluation Index	Gray Correlation ($\xi = 0.5$)	Orders
Normalized difference water index–Hyperion	0.9599	2
Water index (900,970)	0.9404	3
Water index (1148,1088)	0.9201	4
Water index (1100,1200)	0.9004	5
Water index (1070,1200)	0.8981	6
Water index (1300,1200)	0.9781	1
1650 nm	0.8920	7
1722 nm	0.8902	8
970 nm	0.8870	9

Source: Jin, X.L. et al. 2013. *Agronomy Journal*, 105:1385–1392. With permission.

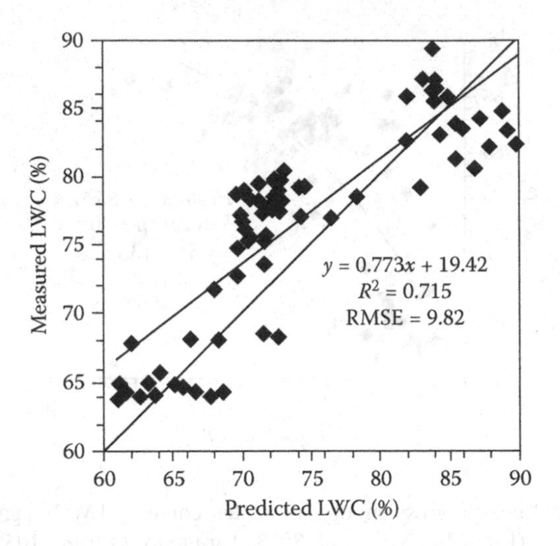

FIGURE 3.15 The relationship between the leaf water content (LWC) predicted by gray relational analysis–partial least squares and actual LWC (left) and landsat thematic mapper image data (right). (From Jin, X.L. et al. 2013. *Agronomy Journal*, 105:1385–1392. With permission.)

season of 2009 ($n = 72$). Our results showed good correlations between predicted and actual values, with an RMSE of 9.82% (Figure 3.15).

3.2.3.4 Model Application

NIR (760–900 nm) and short-wave near-infrared (SWIR) bands (1550–1750 nm) of landsat thematic mapper (TM) data were used to test the potential of the GRA–PLS models in the discrete spectral bands of contemporary spaceborne sensors. Because of the absence of water bands (e.g., 970–1300 nm) in Landsat TM data, the GRA–PLS regression equation was simplified to

$$\begin{aligned} y = {} & 30.563(\mathrm{NIR/SWIR}) + 10.245(\mathrm{NIR} - \mathrm{SWIR})/(\mathrm{NIR} + \mathrm{SWIR}) \\ & - 0.462\mathrm{NIR} + 0.843\mathrm{SWIR} + 17.454 \end{aligned} \tag{3.5}$$

with an R^2 value of 0.65. The measured value was consistent with the predicted value from the Landsat TM data, with an RMSE of 11.62%. The results indicated that GRA–PLS could be used to improve the estimation accuracy of winter wheat LWC by using Landsat TM data.

3.3 PRESCRIPTIONS FOR PRECISION MANAGEMENT

As mentioned above, the precision agriculture experiments can generate a large amount of farmland data. These data need to be managed, analyzed, and processed, and can be used to generate prescription maps for decision making. The generation of a prescription map and the decision-making process not only involves graphic processing and calculation but also includes the representation and inference of experience and knowledge. Determining how to apply information technology to support variable-rate operations is an important task in precision agriculture.

The data sources for precision agriculture include antecedent data and real-time data. Antecedent data are divided into yield data and soil nutrition data over years. Real-time data consist of airborne and spaceborne remote sensing data (for the inversion of chlorophyll and LAI) as well as data reflecting plants' growth status (by SPAD reading, LAI).

This chapter first analyzes the procedures for supporting decision making in precision agriculture management and for prescription generation. Then, based on recent studies and progresses in precision agriculture, we introduce several theoretical and methodological studies about decision making for management zone (MZ) partitioning and variable-rate operations.

3.3.1 Procedures in Decision Making and Prescription Generation

It is known that the whole process from sowing to harvest is influenced by various factors, including climate, soil, the biosphere, and cultivation. To generate a prescription for precision agriculture management, it is necessary to obtain farmland environmental information and utilize GIS, artificial intelligence technology, and simulation modeling (Figure 3.16). GIS can be used to create different graphic layers. The relationship between crop yield and the attributes from other layers is analyzed (including soil type, soil fertility, weed population, field irrigation, and drainage). Based on the analysis of yield potential of each prescription unit, the

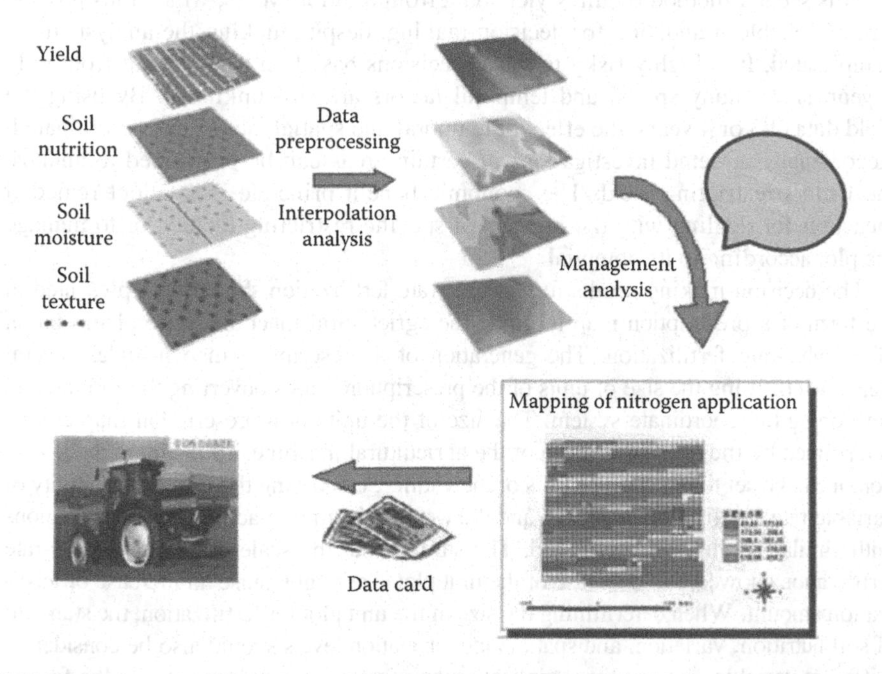

FIGURE 3.16 (**See color insert.**) Key links in decision making for precision agriculture management and prescription generation. (From Chen, L.P. et al. 2002. *Transactions of the CSAE*, 18(2):1145–1148. With permission.)

decision scheme for production management can be developed with the model. The prescription map is generated to provide guidance for decision making using intelligent precision agriculture equipment, which thus optimized the inputs of fertilizers, water, and pesticides from the economic and ecological perspectives (Chen, 2002).

The first is a scheme for variable-rate fertilization (Larscheid, 1997), which is the easiest approach to decision making. Based on the principle of nutritional balance, yield maps are used to calculate the theoretical amount of nutrients taken from the soil by crops in the current season. The input amount in the following season should equal the amount being consumed. This fertilization method is known as "supplementary fertilization." Owing to the leaching loss of nutrients, additional fertilization may be required (i.e., an additional 10%). This method assumes a constant yield restricting factor for the two growth seasons, which is based on the yield information for 1 year and a model of fertilizer supplementation. However, such a method requires less supporting information, and the prescription maps can be easily generated by software. Therefore, this method can serve as a good start for formulating the scheme for variable-rate operations.

The second approach to variable-rate fertilization decision making uses a series of yield maps over multiple years. According to the yield data from multiple years of a plot, the spatial distribution map of yield and the interannual variation map of yield can be obtained. Using these two types of maps, separate regions having stable high yield, stable low yield, and unstable low yield can be identified. Hence, a classified management map can be generated. Based on this map, decisions on the input amount of fertilization in the next season can be made (Larscheid, 1997).

This second method requires yield data from multiple years, which thus provide a more reliable foundation for decision making, despite making the analysis more complicated. It is highly risky to make decisions based on a yield map from only 1 year, since many spatial and temporal factors are still unknown. By using the yield data of 3 or 4 years, the effect of temporal and spatial factors can be mitigated. Accordingly, targeted investigations of certain areas can be performed to analyze the factors restricting yield. The economic benefit principle is to adopt remedial measures for dealing with the impacts of specific restricting factors, or to manage the plot according to its potential.

The decision-making results for variable-rate fertilization should be represented in the form of a prescription map to guide the agricultural machine for implementation of variable-rate fertilization. The generation of a prescription map includes several steps: determining the size of units of the prescription map, converting the format, and developing the coordinate system. The size of the units of a prescription map can be determined by the operation width of the agricultural machine. To facilitate the operation, it can be set to integer multiples of the width. Considering the limited sensitivity of variable-rate fertilization machines and the wear problem of machines, adjacent regions with similar results can be merged. The study about the scale effect in variable-rate fertilization showed that decrease of the unit plot size might cause an increase of fertilization amount. When determining the size of the unit plot for fertilization, the standard of soil nutrition, variation, and spatial autocorrelation levels should also be considered. Different variable-rate fertilizer applicators have different requirements for the format of prescription maps (e.g., vector or raster data). Since the results of models are almost always in raster format, a format conversion is necessary if a vector format is required by the machine. Since GPS is usually used for navigation, the coordinate system of the prescription map should be WGS-84 coordinates, which may require coordinate conversion.

3.3.1.1 Generation of the Soil Nutrition Map

3.3.1.1.1 Experiment and Data Collection

The automatic soil collection experiment was carried out on June 2007 at Xiaotangshan National Experiment Station for Precision Agriculture. The entire plot was divided into 20 × 20 m grids for sampling, with 950 sampling points arranged, using DGPS for the determination of sampling points. The ploughed layer was sampled at a depth of 0–20 cm, which is most closely related to crop growth. At each sampling point, soil drilling was performed on four points on a concentric circle with a diameter of 10 m and at the center of the circle. The soil samples were air dried within 24 h after sampling, and then sieved for nutrition determination. Indicators of TN, organic matter (OM), available phosphorus (AP), and available potassium were measured (Cui et al., 2013).

3.3.1.1.2 Methods and Analysis

3.3.1.1.2.1 Spatial Structure Analysis of Soil Nutrition Using geostatistical principles, semivariance analysis was carried out on soil nutrition in 2007. The optimal model for each nutrient was obtained; the model parameters are shown in Table 3.10. For model validation, cross validation was performed. Root-mean-square standardized (RMSS) was used as an indicator of goodness-of-fit. The closer the RMSS value to 1, the better the goodness-of-fit.

3.3.1.1.2.2 Soil Nutrition Interpolation Based on the optimal semivariance model of soil nutrition, Kriging interpolation was employed to estimate the unmeasured data points and the map was plotted. Hence, a spatial variation map of the content of each nutrient in the open field was generated (Figure 3.17). In 2007, AP was higher in the south and northeast corner, with a gradual decrease toward the middle area. The content of AP was lower in the western area. The northeastern

TABLE 3.10
Spatial Feature Values of Soil Properties

Soil Properties	Model	Sill Value (Co + C)	Nugget Value (Co)	Range (m)	Co/(Co + C) (%)	RMSS
Available phosphorus (mg kg^{-1})	Gaussian	23.6053	6.8053	830.4	28.83	1.0320
Available potassium (mg kg^{-1})	Gaussian	170.8587	77.8400	393.0	45.56	0.9392
Total nitrogen (%)	Gaussian	0.0001	0.00003	608.4	28.33	0.9880
Organic matter (%)	Gaussian	0.03815	0.010811	560.2	28.34	0.9984

Source: Cui, B. et al. 2013. *Scientia Agricultura Sinica*, 46(12):2471–2482. With permission.

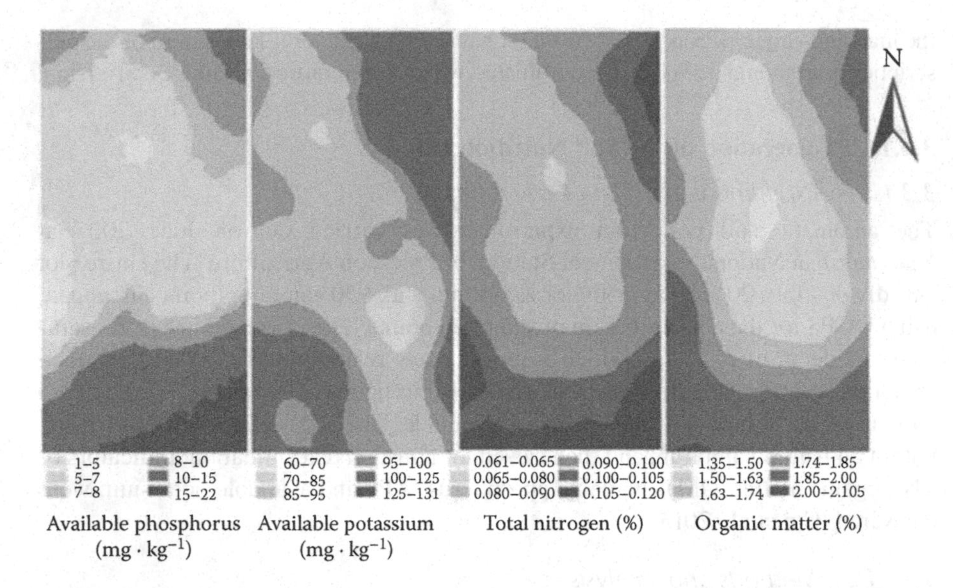

FIGURE 3.17 Kriging interpolation analysis of soil nutrition. (From Cui, B. et al. 2013. *Scientia Agricultura Sinica*, 46(12):2471–2482. With permission.)

and southwestern areas had higher content of available potassium, with a decrease toward the middle area. TN had a similar spatial distribution pattern in the whole research area to OM. The areas high in TN were mainly in the northeast corner and in the south, with a gradual decrease toward the middle area. OM content was higher in the south and in the northeast, with a gradual decrease toward the middle area.

3.3.1.2 Generation of Crop Yield Distribution with a Combine Harvester

Crop yield is the result of agricultural production, and it serves as a reference for agricultural production decision making for the next year. Obtaining crop yield information for the plot and plotting the spatial distribution map of yield are the starting points of precision agriculture, and also the basis for reasonable input and for formulating management decisions. Accuracy of the yield map is closely related to the accuracy of the decision variable.

3.3.1.2.1 Experiments and Data Collection

A CASE IH 2366 combine harvester equipped with an AFS yield monitoring system was used to acquire the yield data for wheat at Xiaotangshan National Experiment Station for Precision Agriculture in June 2001 and 2002. In addressing the errors of yield data, yield information and records of harvest processes were utilized. The statistical characteristics and spatial distribution of yield data points were combined to identify and remove abnormal velocity data and yield fluctuations.

3.3.1.2.2 Comparison of Spatial Distribution Characteristics

The semivariance of spatial distribution of yield for 2001 and 2002 was calculated. Fitting was done with a spherical surface model, and parameters of the semivariance model were obtained (Table 3.11).

TABLE 3.11
Parameters of the Semivariance Model with Yield Data for 2 Years (Spherical Surface Model)

Year	Nugget	Sill	N/S	Range	R^2
2001	365802.30	543654.2	67.29%	28.163	0.9111
2002	265950.40	551857	48.19%	23.838	0.9871

Source: Chen, L.P. 2003. *Theoretical and Experimental Studies on Variable-Rate Fertilization in Precision Farming*. Doctorate dissertation of China Agriculture University. With permission.

Kriging interpolation was performed over the filtered yield data for 2001 and 2002. The spatial distribution map (Figure 3.18) was generated with large variations in spatial distribution of yield. In the left figure (2001), the yields of plots A, B, and D were higher, and the variation was higher. The yields of plots F and G were lower and the variation was smaller. In the right figure (2002), yield showed a different spatial distribution pattern compared with 2001, especially for plots F and G.

3.3.1.3 Generation of a Yield Map Based on Remote Sensing

Crop yield data can be sampled and weighted in field. The essence of modeling yield per unit area is to treat remote sensing data as an input variable, so as to directly or indirectly represent the factors or parameters influencing the growth and yield of crops. The statistical model of the spectral index of yield is commonly used for

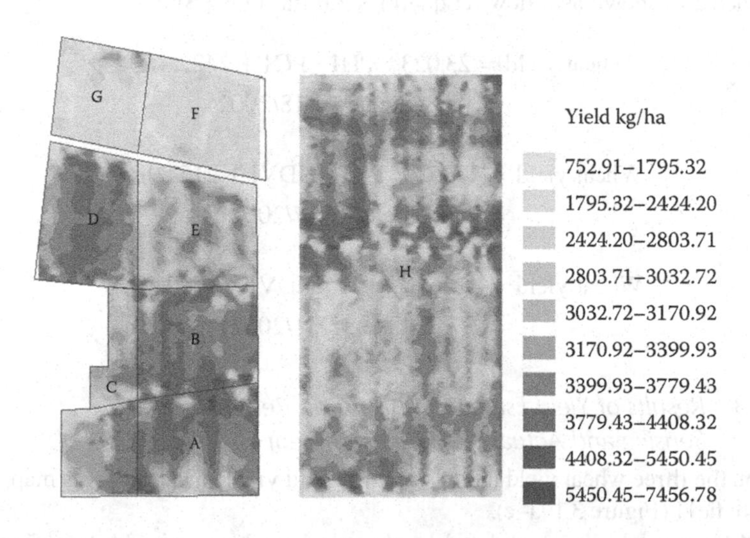

FIGURE 3.18 Spatial distribution maps of yield in each plot using filtered data. (Note: the left and right images are the yield maps of the same plot in 2001 and 2002, respectively.) (From Chen, L.P. 2003. *Theoretical and Experimental Studies on Variable-Rate Fertilization in Precision Farming*. Doctorate dissertation of China Agriculture University. With permission.)

yield estimation based on remote sensing technology. The relevant experiments are designed as follows (Song et al., 2004).

3.3.1.3.1 Experiment and Data Collection

The experiment was conducted at the China National Experimental Station for Precision Agriculture in 2002. Three winter wheat cultivars, Jingdong 8, Jing 9428, and Zhongyou 9507 were planted in 48 small plots with areas of 30 × 32.4 m^2. Wheat yield data were collected manually during the harvest season, within a sampling area of 5 × 5 m. In 2002, a push-broom hyperspectral image (PHI) sensor carried by a Yun-5 aircraft acquired data in three flights (April 18, May 17, and May 3). The height was 1000 m, and the ground resolution of subsatellite points was about 1 m.

The actual after-harvest yields were collected on a grid of 5 × 5 m in 48 subareas. Three PHI images of April 18, May 17, and May 3 were collected.

3.3.1.3.2 Construction of the Yield Estimation Model Using Remote Sensing Technology

NDVI and PRI are two widely used vegetation indices that are sensitive to crop growth conditions. Furthermore, in order to seek the widest variance parameters in the scene, principal components analysis (PCA) transformation and minimum noise fractionation (MNF) transformation are used for all spectrum bands and the first five components are kept for analysis. The transformation produces a new set of compound indices, each made of a linear combination of the original spectrum. The correlation coefficients reached 0.62 for PC1 and yield, −0.77 for yield and PC2, and 0.87 for NDVI and yield. Then, yield prediction models in three growth stages were established and shown as follows (Equations 3.6 through 3.8):

$$\begin{aligned}\text{Wheat yield} &= 23.023 * \text{PHI\_PC1} + 2422.4R^2 \\ &= 0.386\ (\text{date}: 4/18/2002)\end{aligned} \tag{3.6}$$

$$\begin{aligned}\text{Wheat yield} &= 4735.7 * \text{PHI\_NDVI} + 215.69R^2 \\ &= 0.524\ (\text{date}: 5/17/2002)\end{aligned} \tag{3.7}$$

$$\begin{aligned}\text{Wheat yield} &= 4156.8 * \text{PHI\_NDVI} + 1700.3R^2 \\ &= 0.756\ (\text{date}: 5/31/2002)\end{aligned} \tag{3.8}$$

3.3.1.3.3 Results of Yield Estimation Based on Remote Sensing and Actual Yield Measurement

Based on the three wheat yield models, within-field yield variability was mapped for the entire field (Figure 3.19a–c).

All three models were validated by the combine harvest yield data. Statistical analysis showed that R^2 values of three models are 0.112, 0.108, and 0.129, corresponding to correlation coefficients of 0.335, 0.328, and 0.360. This study demonstrated the potential of using hyperspectral airborne remote sensing in the visible and NIR regions to predict winter wheat yield.

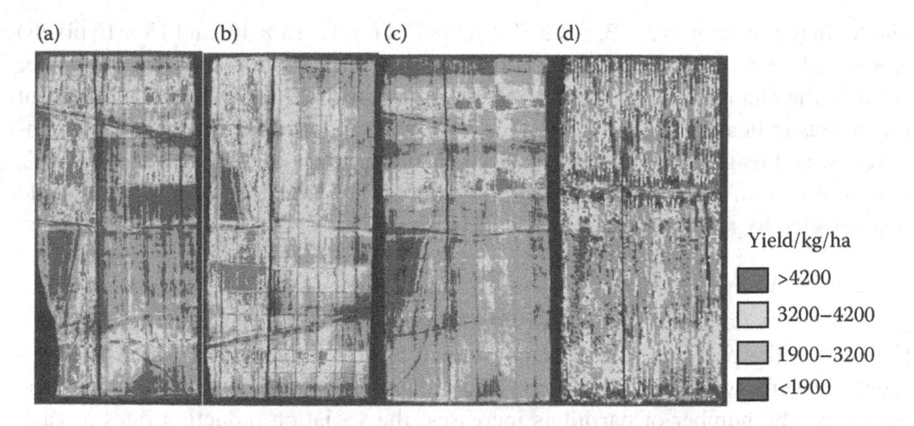

FIGURE 3.19 Wheat yield estimation maps based on PHI hyperspectral data (image dates from a to c are April 18, May 17, and May 31, 2002, respectively; image d is a vector yield data map obtained from combine harvester data). (a) The yield map estimated by PHI image collected in April 18, 2002; (b, c) the yield maps estimated by PHI images collected in May 17, 2002, and May 31, 2002, respectively; (d) the yield data map obtained from combine harvester. (From Song, X.Y. et al. 2004. *IGARSS '04. Proceedings 2004 IEEE International*, 6:4080–4083. With permission.)

3.3.2 Method for Generating Management Zones

3.3.2.1 Extraction of Precision Agriculture Management Zone Based on Multiyear Yield Data

Two yield data processing software programs Yield Editor and Yield Check were employed for error treatment of yield data from 2001 to 2004. Given the variation in wheat varieties and abnormal factors (such as natural disasters), the wheat yield for the same plot may vary across the years. To remove interannual differences, the yield data for each point were normalized by dividing the measured value by the mean value (Chen, 2003). For the sake of partitioning, ordinary Kriging was employed to interpolate the yield data points in vector form on the surface of the grid. Ordinary Kriging interpolation consists of three steps, including calculation of the semivariance of samples, establishment of the semivariance model, and spatial interpolation (Isaaks and Srivastava, 1989). For the calculation of the semivariance of samples, the minimum step length should be equal or close to the average sampling interval. This is recommended for obtaining reasonable semivariance. During the Kriging interpolation, the number of sampling points for single-point interpolation should be restricted to 10–20. Since the distance between the two sampling points in the direction of combine harvester movement is usually smaller than 4 m, the size of output pixels should be selected as 4 × 4 m during interpolation.

3.3.2.1.1 Generation of Management Zones

After a series of treatments (error treatment, normalization, and spatial interpolation), the 4-year yield data were then subjected to a grid averaging operation, so as to obtain a synthetic yield grid map. In order to remove the isolated pixels or fragments from the partition map, square filtering windows with scales of 12, 20, 28, 36, 44, 52, and 60 m (equivalent to 3 × 3, 5 × 5, 7 × 7, 9 × 9, 11 × 11, 13 × 13, and 15 × 15 pixels) were used for mode filtering on the partition image with 4-m resolution. To further analyze the changes in partitions after filtering, statistical analysis was carried out on various indicators (intraplot variance, coefficient of variation, mean, spatial consistency, and fragmentation degree) of the partition image (4, 12, 20, 28, 36, 44, 52, and 60 m). With comprehensive consideration of the indicators, a proper threshold was selected to generate the optimal partition map.

3.3.2.1.2 Analysis of Results

3.3.2.1.2.1 Changes in the Reduction Rate of Variance The smaller the variation within the partition, the more convenient it is to implement uniform management measures within the partition. The greater the variation reduction rate, the better. As the number of partitions increases, the variation reduction rates at various scales tend to be consistent, all showing a gradual increasing trend. When the number of partitions increases to four, the variance reduction rate remains basically unchanged in spite of further increases in the number of partitions. This indicates that it is useless to divide the plot into more than four partitions. The appropriate number of partitions is four. For the same number of partitions, as the scale of the filtering window increases, the variance reduction rate gradually decreases, which is unfavorable for precision agriculture management. Hence, it is necessary to adopt other methods to determine the appropriate scale of the filtering window.

3.3.2.1.2.2 Significant Changes in Lag Differences When the number of partitions is four, the variance reduction rate decreases as the scale of the filtering window increases. This means that the intraplot variance gradually increases after filtering. As the scale continues to increase, the F value decreases. When the scale is 4–44 m, there is a significant difference between the partitions. At the scale of 4–28 m, the difference between partitions is highly significant. When the scale is further increased to 52 and 60 m, the difference within the plot is no longer significant. Multicomparison between the partitions indicates that when the scale is 4–44 m, there is a highly significant difference among the four partitions at various scales. This indicates that the scale of the filtering window should be no greater than 44 m.

3.3.2.1.2.3 Changes in the Mean, Standard Deviation, and Coefficient of Variation As shown in Figure 3.20a, as the scale increases, the means of each partition remain unchanged (except that the mean of partition 1 with the lowest yield increases slightly). Accordingly, if the mean yield of each partition is used to calculate the fertilization amount, the fertilizer requirement of the partition before and after filtering does not change. However, the standard deviation and coefficient of variation increase at larger scales, as shown in Figure 3.20b and c. This indicates that it is undesirable to implement the same management scheme within the same plot. A more appropriate scale of filtering window must be selected. As the scale increases, the proportion of the area of each partition also varies (Figure 3.20d).

3.3.2.1.2.4 Changes in Fragmentation Degree of the Partition Map The landscape structure index quantitatively and intuitively reflects the differences between

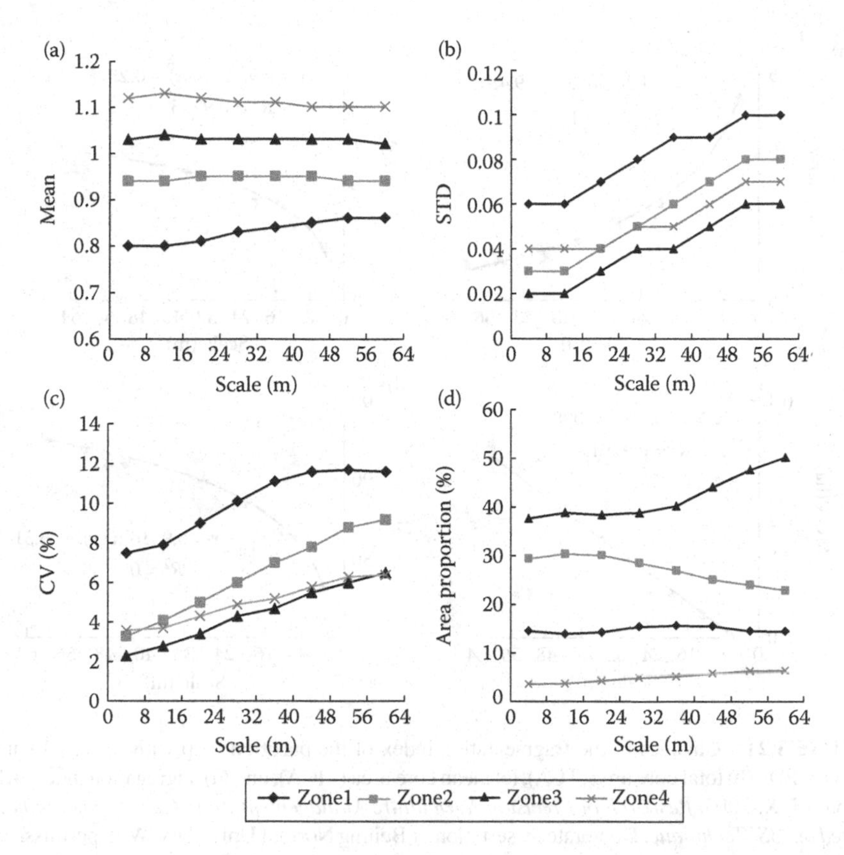

FIGURE 3.20 **(See color insert.)** Changes in the mean, standard deviation, coefficient of variation, and proportion of area with scale for each partition. (a) Illustrates mean value changes with scale increasing for each partition; (b) illustrates standard deviation value changes with scale increasing for each partition; (c) illustrates coefficient of variation value changes with scale increasing for each partition; (d) illustrates area proportion (%) changes with scale increasing for each partition. (From Li, X. 2005. *Research of Precision Agriculture Management Zone Generating Methods Based on '3S' Technique.* Doctorate dissertation of Beijing Normal University. With permission.)

partition maps at different window scales. As the scale increases, the patch density (PD) gradually decreases logarithmically. As shown in Figure 3.21a, the smaller the PD, the smaller the fragmentation degree of the partition will be, and hence the smaller the error of the decision variable with the partition as a unit. This is favorable for variable-rate operations in precision agriculture. The small patches are usually in isolated areas caused by the use of points with large deviations from adjacent points in terms of yield measurements for interpolation. Yield variation is induced by random causes, so the actual yield distribution pattern cannot be revealed. Moreover, small patch areas make it inconvenient to carry out variable-rate operations.

It is evident from Figure 3.21 that as the scale increases, the total core area increases logarithmically. As the scale of the filtering window increases, the patches with smaller areas are smoothed out, and PD decreases. As a result, the uncertain

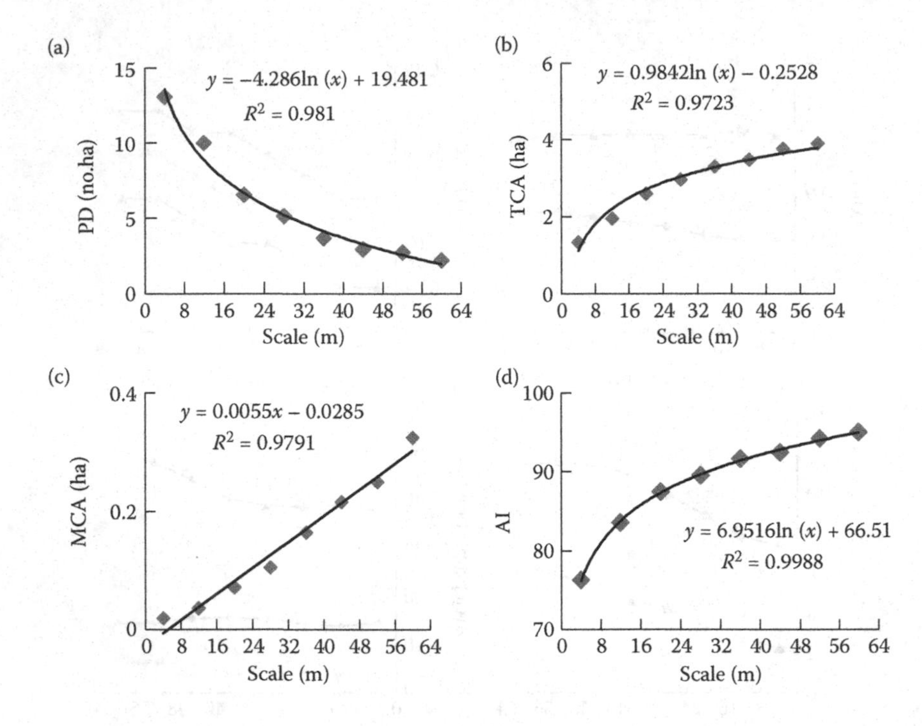

FIGURE 3.21 Changes in the fragmentation index of the partition map with scale. (a) Patch density (PD), (b) total core area (TCA), (c) mean core area (MCA), and (d) aggregation index (AI). (From Li, X. 2005. *Research of Precision Agriculture Management Zone Generating Methods Based on '3S' Technique*. Doctorate dissertation of Beijing Normal University. With permission.)

area caused by the random yield variation decreases, and the core area increases. This is especially favorable for variable-rate operations in precision agriculture. The core area refers to the central part remaining after subtraction of the buffer zone with a designated margin width. In landscape ecology, this buffer zone is the transition belt between two adjacent landscape elements, and it is called the ecotone. In precision agriculture, the core area is considered unstable because of errors in measurement and interpolation. Obviously, the larger the designated margin width, the smaller the core area will be. Here, the cutting width (6 m) during yield acquisition by the combine harvester is the margin width. The core area of a patch is the area with stable yield potential. The size of the core area is the basis to judge whether a patch can be used as a unit for the implementation of one management prescription. For example, if the area of a patch is large enough to differentiate it from the adjacent patches, but the core area cannot meet the requirements of field operations (the diameter of core area is smaller than the cutting width), then the patch is not suitable for use as a decision-making unit and it has to be merged with adjacent patches.

The mean patch area can be a criterion for choosing the optimal scale. As shown in Figure 3.21c, the mean core area increases linearly with increasing scale. When the scale is smaller than 20 m, the mean core area of the patch is smaller than 0.05 ha. A large number of small patches are caused by random yield variation across the years.

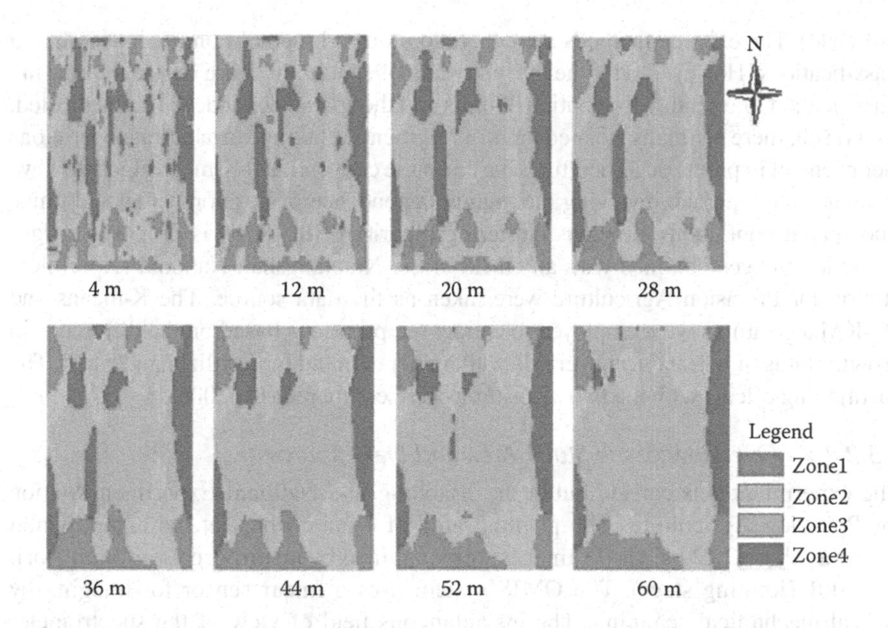

FIGURE 3.22 **(See color insert.)** Partition map after filtering with different scales of window. (From Li, X. 2005. *Research of Precision Agriculture Management Zone Generating Methods Based on '3S' Technique.* Doctorate dissertation of Beijing Normal University. With permission.)

They cannot accurately characterize the stable yield structure. However, when the spatial scale is 36 m, the mean core area of the patch increases to 0.16 ha, which is favorable for variable-rate operations.

As shown in Figure 3.21, as the scale increases, the concentration degree of the partition map increases. That is to say, there are fewer fragments or isolated pixels on the partition map. This is a favorable trend for precision agriculture management.

3.3.2.1.2.5 Changes in Spatial Consistency As shown in Figure 3.22, when the scale of the filtering window increases from 12 to 36 m, Kappa varies in the range of 0.54–0.87. The partition map after filtering shows good spatial consistency with the original partition map. After filtering, the partition map reflects the general yield distribution pattern.

When the spatial scale is increased further, Kappa becomes smaller than 0.5. The partition map after filtering has poor consistency with the original partition map (4 m). This indicates that a filtering window of the proper size can effectively remove the random yield variation (small patches), and thus better reflect the spatial distribution pattern of yield. However, if the scale is excessive, some real yield variation may be smoothed out as well, and a stable yield structure cannot be characterized.

3.3.2.2 Extraction of Management Zones Based on the Spatial Contiguous Clustering Algorithm

Many data sources have been used for partitioning of precision agriculture regions (including elevation, slope, slope aspect, electric conductivity of soil, depth of top soil, and yield). The existing methods are either monitoring-based or nonmonitoring-based classifications. However, only the attribute data of spatial units are considered during classification; the spatial distribution of units and the spatial dependence are neglected. As a result, there are many isolated units of fragments, making variable-rate operations inconvenient in precision agriculture. Based on the conventional K-mean algorithm, we introduce an approach involving the mutual dependence of positions of spatial units. The spatial contiguous K-means clustering algorithm (SC-KM) is proposed. High-resolution images obtained with an OMIS at the Xiaotangshan National Experiment Station for Precision Agriculture were taken as the data source. The K-means and SC-KM algorithms were employed to extract the partitions based on the difference in growth status of wheat during periods with higher demand for fertilizer and water. The partitioning effects with the two algorithms were compared (Li, 2005).

3.3.2.2.1 Overview of the Study Area and Data Sources

The experiment was carried out at the Xiaotangshan National Experiment Station for Precision Agriculture. The planting area of winter wheat at the experimental base was about 39.2 ha. OMIS images were obtained by an airborne sensor on April 26, 2001 (jointing stage). The OMIS system uses a linear sensor for imaging by optical mechanical scanning. The instantaneous field of view of the spectrometer was 3 mrad, and the total field of view was 70°. The visible/near-infrared, short- and medium-infrared, and thermal infrared bands (0.4–12.5 μm) were covered, including 128 spectral ranges. The visible/near-infrared region (0.46–1.1 μm) had 64 spectral ranges, with a spectral resolution of 10 nm. When the flight height was 1000 m, the resolution of subsatellite point was about 3 m. OMIS images showing the wheat plot with flat terrain and obvious variation in large field productivity were selected. This plot had an area of about 5 ha.

3.3.2.2.2 Selection of Management Zone Parameters

The NDVI is the most widely used index in remote sensing monitoring of crop growth. In this article, OMIS images after reflectance conversion were studied. Two ranges with central wavelengths of 789.2 and 675.8 nm were selected.

Because of differences in fertility and management measures, different partitions of the plot may differ in terms of the growth status of wheat. The variation (variance) over the entire plot without partitioning is the intraplot variation with the partition number being 1. With this variance as a reference (100%), the relative variances at different partition numbers were calculated. To select an appropriate partition number, it was necessary to determine a threshold for variance. When the partition number was five, the overall variance of the plot decreased to about 10% of the original. As the partition number increased further, the relative changes in variance were no longer significant. For this reason, a partition number of five was considered appropriate.

As shown in Figure 3.23, the aggregation with the SC-KM algorithm for each plot is consistently higher than with the K-M algorithm. There is less fragmentation than with the latter algorithm. This indicates that the SC-KM algorithm can greatly improve the aggregation and continuity of each plot by considering the spatial association of pixels. For these reasons, the SC-KM algorithm is suitable for open-field variable-rate operations.

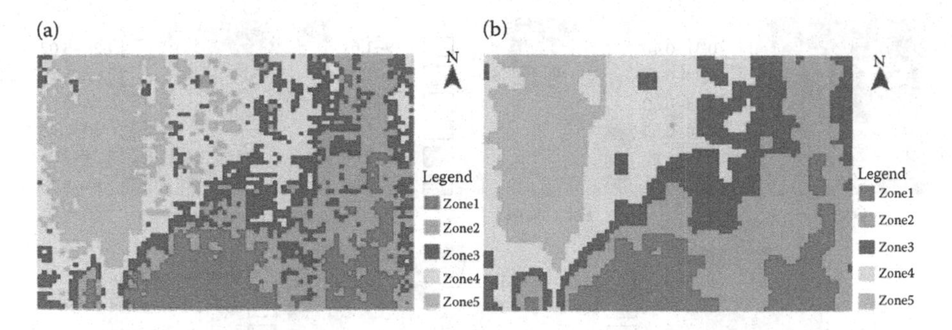

FIGURE 3.23 **(See color insert.)** (a) Partitioning results of the K-M algorithm. (b) Partitioning results of the SC-KM algorithm. (From Li, X. 2005. *Research of Precision Agriculture Management Zone Generating Methods Based on '3S' Technique*. Doctorate dissertation of Beijing Normal University. With permission.)

3.3.2.3 Delineation of Agricultural Management Zones with Remotely Sensed Data

Remote sensing images can reflect the growth status of crops in real time, whereas the spectral reflectance of crops characterizes the growth status of crops. Moreover, the growth status of crops is closely related to soil texture and soil nutrition. Hence, the monitoring of growth status and nutrition diagnosis of crops using remote sensing data and the partitioning of the plot are key components of variable-rate fertilization in modern precision agriculture.

3.3.2.3.1 Experiment and Data Collection

A field experiment at the National Experimental Station for Precision Agriculture of China was designed during the 2005–2006 winter wheat growing season (Song et al., 2009). Soil samples were taken, and levels of five crop nutrients, TN, nitrate nitrogen [NN], AP, extractable potassium [EP], and OM, were determined using standard laboratory procedures. Meanwhile, one scene of Quickbird imagery during the heading stage of wheat was acquired and processed. Spectral parameters of OSAVI were extracted from the imagery. The winter wheat was sown on September 29, 2005, with a row spacing of 15 cm. The wheat cultivar was Jingdong 8, which is one of the main winter wheat varieties in northern China. Base fertilizer was applied on September 27, 2005, and supplementary fertilizer was applied on April 22, 2006. A square plot (90 × 90 m) in the wheat field was selected as the soil sampling area.

3.3.2.3.2 Kriging of Crop Nutrients and Determining the Optimal Number of MZs

The spatial structure of soil AP and EK on April 4 and OM on June 16 was evaluated by isotropic variogram models. The model types and their parameters were calculated and ordinary Kriging was applied to the three soil properties (AP, AK, and OM) and wheat yield (Figure 3.24).

The OSAVI and Kriged values of soil AP, EK, OM, and wheat yield were analyzed with the fuzzy K-means algorithm. To determine the optimal number of classes, the

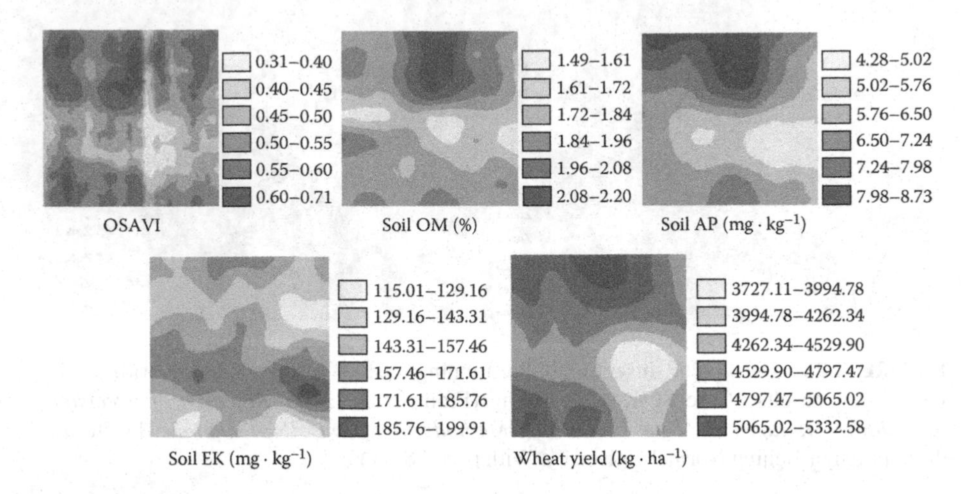

FIGURE 3.24 The spatial structure of soil AP, EK, and wheat yield was evaluated by isotropic variogram models. (From Song, X.Y. et al. 2009. *Precision Agriculture*, 10(6):471–487. With permission.)

number of classes was increased by one at a time from two to six. To determine the optimal number of classes, the fuzziness performance index (FPI) and modified partition entropy (MPE) were used. It was shown that the FPI and MPE had the same change in trend with an increase in cluster number, and the minimum FPI and MPE values were obtained with three clusters for the study area.

3.3.2.3.3 Management Zone Delineation and Evaluation

Based on the optimal number of classes, three MZ maps were generated using different data (e.g., soil, yield, and RS data). Figure 3.25 shows the resulting maps. The Kappa coefficient was then used to compare the homogeneity of the zones in the three different MZ maps. The results indicated that zones based on soil and yield (Figure 3.25a), and soil, yield, and RS data (Figure 3.25b) are the most similar; the

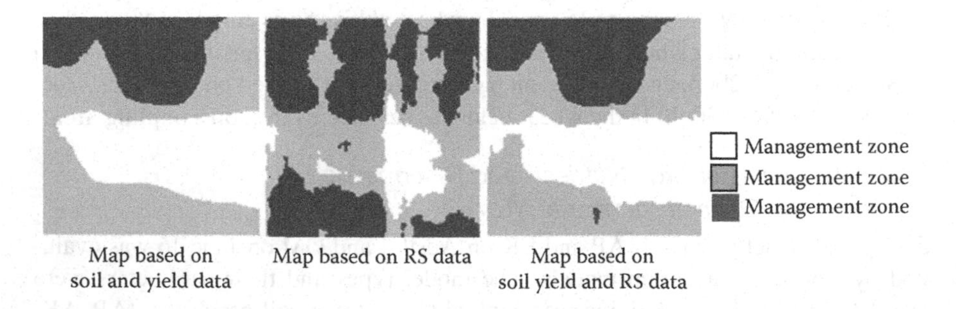

FIGURE 3.25 Maps of management zones generated by fuzzy K-means classification using different combinations of data (soil, yield, and OSAVI). (From Song, X.Y. et al. 2009. *Precision Agriculture*, 10(6):471–487. With permission.)

similarity between Figure 3.25a and c is the greatest with a Kappa coefficient of 0.91. The Kappa coefficient based on comparison of the zones in Figure 3.25a and b was 0.16.

The statistical analyses indicated significant differences between the crop nutrients and yield in each zone of the three maps. MZ 3 had the highest nutrient status and potential crop productivity, whereas MZ 1 had the lowest. The results also showed that the coefficients of variation (CVs) for wheat yield decreased in the three zones for all maps.

REFERENCES

Brantley, S.T., J.C. Zinnert, and D.R. Young. 2002. Application of hyperspectral vegetation indices to detect variations in high leaf area index temperate shrub thicket canopies. *Remote Sensing of Environment*, 115(2):514–523.

Bremnerj, M. 1965. Total nitrogen: Macro-Kjeldahl method to include nitrate. In Black, C.A. (ed.), *Methods of Soil Analysis. Part 2. Chemical and Microbiological Properties.* Agron. Monogr. 9 ASA. Madison, WI, p. 1164.

Broge, N.H. and E. Leblanc. 2001. Comparing prediction power and stability of broadband and hyperspectral vegetation indices for estimation of green leaf area index and canopy chlorophyll density. *Remote Sensing of Environment*, 76(2):156–172.

Carter, G.A. 1991. Primary and secondary effects of water content on the spectral reflectance of leaves. *American Journal of Botany*, 78(7):916–924.

Carter, G.A. 1994. Ratios of leaf reflectances in narrow wavebands as indicators of plant stress. *International Journal of Remote Sensing*, 15(3):697–703.

Chen, L.P. 2003. *Theoretical and Experimental Studies on Variable-Rate Fertilization in Precision Farming*. Doctorate dissertation of China Agriculture University.

Chen, L.P., C.J. Zhao, X.X. Liu, and X.H. Du. 2002. Design and implementation of intelligent decision support system for precision agriculture. *Transactions of the CSAE*, 18(2):1145–1148.

Cui, B., J.H. Wang, W.D. Yang, W.J. Huang, J.H. Guo, X.Y. Song, and M.C. Feng. 2013. Analysis of temporal and spatial variation of soil properties on the winter wheat–summer maize rotation field. *Scientia Agricultura Sinica*, 46(12):2471–2482.

Danson, F.M. and S.E. Plummer. 1995. Red edge response to forest leaf area index. *International Journal of Remote Sensing*, 16:183–188.

Datt, B. 1994. Visible/near infrared reflectance and chlorophyll content in Eucalyptus leaves. *International Journal of Remote Sensing*, 20(14):2741–2759.

Dente, L., G. Satalino, F. Mattia, and M. Rinaldi. 2008. Assimilation of leaf area index derived from ASAR and MERIS data into CERES-Wheat model to map wheat yield. *Remote Sensing of Environment*, 112(4):1395–1407.

Dong, Y.Y., J.H. Wang, C.J. Li, Q. Wang, and W.J. Huang. 2012. Integration of ground observations and crop simulation model for crop leaf area index estimation. *Advances in Intelligent and Soft Computing*, 165:831–837.

Dorigo, W.A. 2012. Improving the robustness of cotton status characterisation by radiative transfer model inversion of multi-angular CHRIS/PROBA data. *IEEE JSTARS*, 5(1):18–29.

Evensen, G. 2003. The ensemble Kalman filter: Theoretical formulation and practical implementation. *Ocean Dynamics*, 53(4):343–367.

Fang, H.L. and S.L. Liang. 2003. Retrieving leaf area index with a neural network method: Simulation and validation. *IEEE Transactions on Geoscience and Remote Sensing*, 41(9):2052–2062.

Filella, I. and J. Penuelas. 1994. The red edge position and shape as indicators of plant chlorophyll content, biomass and hydric status. *International Journal of Remote Sensing*, 15:1459–1470.

Fox, H., G.W. Roth, K.V. Iversen, and W.P. Piekielek. 1989. Soil and tissue nitrate tests compared for predicting soil nitrogen availability to corn. *Agronomy Journal*, 81:971–974.

Gao, B.-C. and A.F.H. Goetz. 1995. Retrieval of equivalent water thickness and information related to biochemical components of vegetation canopies from AVIRIS data. *Remote Sensing of Environment*, 52:155–162.

Hinzman, L.D., M.E. Bauer, and C.S.T. Daughtry. 1986. Effects of nitrogen fertilization on growth and reflectance characteristics of winter wheat. *Remote Sensing of Environment*, 19:47–61.

Huang, W.J., Z.J. Wang, L.S. Huang, D.W. Lamb, Z.H. Ma, J.C. Zhang, J.H. Wang, and C.J. Zhao. 2011. Estimation of vertical distribution of chlorophyll concentration by bidirectional canopy reflectance spectra in winter wheat. *Precision Agriculture*, 12(2):165–178.

Hunt Jr., R.E., P.C. Doraiswamya, J.E. McMurtrey, C.S.T. Daughtry, E.M. Perry, and B. Akhmedov. 2013. A visible band index for remote sensing leaf chlorophyll content at the canopy scale. *International Journal of Applied Earth Observation and Geoinformation*, 21:103–112.

Isaaks, E.H. and R.M. Srivastava. 1989. *Applied Geostatistics*. New York: Oxford University Press.

Jackson, R.D. and C.E. Ezra. 1985. Spectral response of cotton to suddenly induced water stress. *International Journal of Remote Sensing*, 6:177–185.

Jacquemoud, S., S.L. Ustin, J. Verdebout, G. Schmuck, G. Reoli, and B. Hosgood. 1996. Estimating leaf biochemistry using the PROSPECT leaf optical properties model. *Remote Sensing of Environment*, 56:194–202.

Jin, X.L., X.G. Xu, X.Y. Song, Z.H. Li, J.H. Wang, and W.S. Guo. 2013. Estimation of leaf water content in winter wheat using grey relational analysis–partial least squares modeling with hyperspectral data. *Agronomy Journal*, 105:1385–1392.

Kalacska, M., A. Sanchez-Azofeifa, T. Caelli, B. Rivard, and B. Boerlage. 2005. Estimating leaf area index from satellite imagery using Bayesian networks. *IEEE Transactions on Geoscience and Remote Sensing*, 43(8):1866–1873.

Kalman, R.E. 1960. A new approach to linear filtering and prediction problems. *Transactions of the ASME—Journal of Basic Engineering*, 82-D:35–45.

Kar, G., H.N. Verma, and R. Singh. 2006. Effect of winter crop and supplementary irrigation on crop yield, water us efficiency and profitability in rainfed rice based on cropping system of eastern India. *Agricultural Water Management*, 79:280–292.

Lamb, D.W., M. Steyn-Ross, P. Schaare, M.M. Hanna, W. Silvester, and A. Steyn-Ross. 2002. Estimating leaf nitrogen concentration in ryegrass (*Lolium* spp.) pasture using the chlorophyll red-edge: Theoretical modelling and experimental observations. *International Journal of Remote Sensing*, 23:3619–3648.

Larscheid, G., B.S. Blackmore, and M. Moore. 1997. Management decisions based on yield map. In Stafford, J.V. (ed.), *Precision Agriculture '97*. Silsoe: BIOS Scientific Publishers Ltd., pp. 895–903.

Li, H.L., C.J. Zhao, W.J. Huang, and G.J. Yang. 2013. Non-uniform vertical nitrogen distribution within plant canopy and its estimation by remote sensing: A review. *Field Crops Research*, 142:75–84.

Li, X. 2005. *Research of Precision Agriculture Management Zone Generating Methods Based on '3S' Technique*. Doctorate dissertation of Beijing Normal University.

Liao, Q.H., C.J. Zhao, G.J. Yang, C. Coburn, J.H. Wang, D.Y. Zhang, and Z.J. Wang. 2013. Estimation of leaf area index by using multi-angular hyperspectral imaging data based on the two-layer canopy reflectance model. *Intelligent Automation & Soft Computing*, 19(3):295–304.

Liu, L.Y. 2002. *Hyperspectral Remote Sensing Application in Precision Agriculture*. Post-doctoral research report of Institute of Remote Sensing Applications, Chinese Academy of Sciences.

Liu, L.Y., J.H. Wang, W.J. Huang, and C.J. Zha. 2004. Estimating winter wheat plant water content using red edge parameters. *International Journal of Remote Sensing*, 25(17):3331–3342.

Lu, J.L. 1994. *Plant Nutrition* (first volume). Beijing, China: Beijing Agricultural University Press (continuing China Agricultural University Press).

Manivel, L. and R.J. Weaver. 1974. Biometrics correlations between leaf area and length measurements of 'Grenache' grape leaves. *HortScience*, 9:27–28.

Marshall, J.K. 1968. Methods for leaf area measurement of large and small leaf samples. *Photosynthetica*, 2:41–47.

McKee, G.W. 1964. A coefficient for computing leaf area in hybrid corn. *Agronomy Journal*, 56:240–241.

McKinney, G. 1941. Absorption of light by chlorophyll solutions. *Journal of Biological Chemistry*, 140:315–322.

Minolta Camera Co. Ltd. 1989. *Chlorophyll Meter SPAD-502 Instructional Manual*. Minolta, Osaka, Japan.

Moran, M.S. 1997. Opportunities and limitations for image-based remote sensing in precision crop management. *Remote Sensing of Environment*, 61:319–346.

Penuelas, J., I. Filella, C. Biel, L. Serrano, and R. Save. 1993. The reflectance at the 950–970 nm region as an indicator of plant water status. *International Journal of Remote Sensing*, 14:1887–1905.

Penuelas, J., J. Pinol, and R. Ogaya. 1997. Estimation of plant water concentration by the reflectance water index WI (R900/R970). *International Journal of Remote Sensing*, 18:2869–2872.

Qin, J., S. Liang, X. Li, and J. Wang. 2008. Development of the adjoint model of a canopy radiative transfer model for sensitivity study and inversion of leaf area index. *IEEE Transactions on Geoscience and Remote Sensing*, 46(7):2028–2037.

Richter, K., C. Atzberger, F. Vuolo, and G. D'Urso. 2011. Evaluation of sentinel-2 spectral sampling for radiative transfer model based LAI estimation of wheat, sugar beet, and maize. *IEEE JSTARS*, 4(2):458–464.

Sabater, J.M., C. Rudiger, J.C. Calvet, N. Fritz, L. Jarlan, and Y. Kerr. 2008. Joint assimilation of surface soil moisture and LAI observations into a land surface model. *Agricultural and Forest Meteorology*, 148(8–9):1362–1373.

Serrano, L., I. Filella, and J. Peñuelas. 2000. Remote sensing of biomass and yield of winter wheat under different nitrogen supplies. *Crop Science*, 40:723–731.

Shadchina, T.M., V.V. Dmitrtrieva, and V.V. Morgoun. 1998. Interrelation between nitrogen status of the plants, leaf chlorophyll and grain yield in various winter wheat cultivars. *Acta Agronimica Hungarica*, 46:25–34.

Smith, J.A. 1993. Lai inversion using a back-propagation neural network trained with a multiple scattering model. *IEEE Transactions on Geoscience and Remote Sensing*, 31(5):1102–1106.

Song, X.Y., J.H. Wang, W.J. Huang, L.Y. Liu, and R.L. Pu. 2009. The delineation of agricultural management zones with high resolution remotely sensed data. *Precision Agriculture*, 10(6):471–487.

Song, X.Y., J.H. Wang, L.Y. Liu, X.Z. Xue, and C.J. Zhao. 2004. Use of airborne hyper spectral image data to assess winter wheat yield. Geoscience and Remote Sensing Symposium, 2004. *IGARSS '04. Proceedings 2004 IEEE International*, 6:4080–4083.

Thorp, K.R., D.J. Hunsaker, and A.N. French. 2010. Assimilating leaf area index estimates from remote sensing into the simulations of a cropping systems model. *Transactions of the ASABE*, 53(1):251–262.

Wang, D.W., J.D. Wang, and S.L. Liang. 2010. Retrieving crop leaf area index by assimilation of MODIS data into a crop growth model. *Science China Earth Sciences*, 53(5):721–730.

Wang, F.M., J.F. Huang, and Z.H. Lou. 2011. A comparison of three methods for estimating leaf area index of paddy rice from optimal hyperspectral bands. *Precision Agriculture*, 12(3):439–447.

Wang, Z., J. Wang, L. Liu, W. Huang, C. Zhao, and Y. Lu. 2005. Estimation of nitrogen deficiency at middle and bottom layers of winter wheat canopy by using ground measured canopy reflectance. *Communications in Soil Science and Plant Analysis*, 36:2289–2302.

Wang, Z., J. Wang, L. Liu, W. Huang, C. Zhao, and C. Wang. 2004. Prediction of grain protein content in winter wheat (*Triticum aestivum* L.) using plant pigment ratio (PPR). *Field Crops Research*, 90:311–321.

Wang, Z.J. 2004. *Prediction of Canopy Nitrogen Distribution and Grain Quality Using Remote Sensing in Winter Wheat* (*Triticum aestivum* L.). PhD thesis, Beijing, China: China Agricultural University, College of Agriculture and Biotechnology.

Woods, R.F., D.R. Betters, and E.W. Mogren. 1982. Understory herbage production as a function of Rocky Mountain aspen stand density. *Journal of Range Management*, 35:380–381.

Wu, J., D. Wang, C.J. Rosen, and M.E. Bauer. 2007. Comparison of petiole nitrate concentrations, SPAD chlorophyll readings, and QuickBird satellite imagery in detecting nitrogen status of potato canopies. *Field Crops Research*, 101:96–103.

Xiao, Z.Q., S.L. Liang, J.D. Wang, J.L. Song, and X.Y. Wu. 2009. A temporally integrated inversion method for estimating leaf area index from MODIS data. *IEEE Transactions on Geoscience and Remote Sensing*, 47(8):2536–2545.

Zhang, D., T. Li, L. Jiang, Y. Shao, W. Guo, P. Liu, and C. Li. 2012. Effect of exogenous nitric oxide on photosynthetic physiology in winter and spring wheat varieties after anthesis. *Ecology and Environmental Science*, 21:231–236.

Zhao, C., J. Wang, W. Huang, and T. Guo. 2011. Early detection of canopy nitrogen deficiency in winter wheat (*Triticum aestivum* L.) based on hyperspectral measurement of canopy chlorophyll status. *New Zealand Journal of Crop and Horticultural Science*, 39(4):251–262.

Zhao, C.J. 2009. *Precision Agriculture Research and Practice*. Beijing: Science Press.

Zhao, C.J., W.J. Huang, J.H. Wang, L.Y. Liu, X.Y. Song, Z.H. Ma, and C.J. Li. 2006. Extracting winter wheat chlorophyll concentration vertical distribution based on bidirectional canopy reflected spectrum. *Transactions of the CSAE*, 22(6):104–109.

Zhao, C.J., Z.J. Wang, J.H. Wang, and W.J. Huang. 2012. Relationships of leaf nitrogen concentration and canopy nitrogen density with spectral features parameters and narrow-band spectral indices calculated from field winter wheat (*Triticum aestivum* L.) spectra. *International Journal of Remote Sensing*, 33(11):3472–3491.

4 Control of Precision Agriculture Production

Qin Zhang

CONTENTS

4.1 INTRODUCTION

Precision agriculture (PA) is a farming management method that allows farmers to optimize their resource inputs to achieve their production potential in response to observed inter- and intrafield variability in soil properties and crop growth conditions. Motivated by the potential that PA technology can offer, farmers in the United States and other parts of the world have adopted it as a management strategy to bring data from multiple sources to assist decisions associated with crop production (Batte and Arnholt, 2003). However, owing to the biological and environmental complexity involved in crop production, farmers had difficulty incorporating their collected data into management practices to improve their production making them more profitable and sustainable. Many farmers now ask the question, "What do I do with the data?" after a few years of data collection (Carter, 2012). Such a shortcoming left a gap between the *promises* of PA technology and the *tangible results* the farmers have realized.

Precision crop production normally includes a course of action from measuring soil properties, observing crop growth conditions, selecting adequate resource inputs

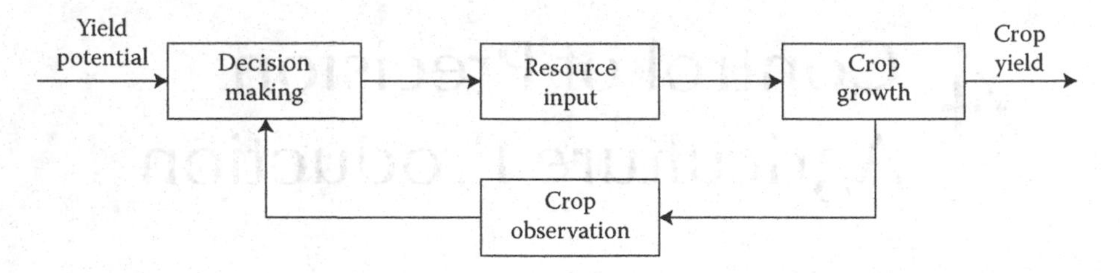

FIGURE 4.1 Representation of a typical PA system using a system block diagram.

based on field variations, and performing the required actions to deliver the right amount of selected input to the right location at the right time. The relationship between these elements in a PA system can be represented using a block diagram (Figure 4.1). This block diagram shows that decisions for PA processes are made based on observed crop variability, and applying adequate resource or cultivation actions, which could improve crop growth and therefore realize yield potential.

As represented by the block diagram, a PA process is very similar in format to the process of many industrial controls: it is concerned with understanding and controlling resource inputs in order to achieve a best possible return on inputs. Ideally, to control such a process effectively, gaining a comprehensive understanding of the responses of crop growth to resource inputs is required. However, the spatial and temporal variability of crop growth makes the task of "gaining a comprehensive understanding" very challenging. Such insufficient understanding originates from the transdisciplinary nature of implementing a precision crop production, as the understanding and controlling of such a production involves crop and soil sciences, engineering, and economics. Without finding a satisfactory solution to solve this problem, we will have to consider the control of a poorly understood system in precision crop production.

Gaining an understanding of responses of crop growth to resource inputs in a precision crop production process is similar to figuring out the dynamic behaviors of a system responding to corrections in control systems engineering. A typical control system uses sensors to monitor the performance and collect measured data of the performance of the system being controlled. Those measurements are then used to give feedback to the controller to make corrections, normally derived by mathematical modeling of the system, toward obtaining a desired performance. Similarly, precision crop production processes also use sensors to collect relevant crop growth data, and use the collected data to support making resource input decisions with respect to the goal of optimizing returns on inputs. While similar in format, a precision crop production process does present some noticeable differences from a conventional control system, and the most significant one could be its spatial variability, that is, optimizing the resource inputs based on crop yield potential at different locations within a field that forms the basis of PA (Pierce and Nowak, 1999). To cope with spatial variability in field, an automatic data gathering and storing method (Schueller and Bae, 1987) has been invented allowing PA practitioners to overlay field topography, soil fertility, and crop yield data in the form of a georeference map. Another major difference from a conventional control system is the less systematic nature of temporal crop data as it is subjected to numerous natural disturbances. The use of multiyear datasets could help to stabilize the temporal variability of PA data (Kaspar et al., 2003).

The spatial and temporal variability of field data could result from a number of factors, from the agriculturally important climatic states (such as drought, flooding, hail, and other extreme weather), soil properties (such as texture, organic and inorganic matter, moisture content, electrical conductivity, pH, and nitrogen levels), to crop growth conditions (such as water, nitrogen, or disease stress). A PA production system aims at precisely applying adequate amounts of selected resources in terms of accurate agriculturally important information.

One practical method of determining appropriate resource inputs in response to such field variability in the PA process is the control approach: measuring system parameters to determine the yield potential, controlling the resource inputs to reach the yield potential, and implementing a variable-rate application (VRA) to deliver the right amount of selected inputs precisely to target locations. This chapter will discuss the sensing, control, and implementation characteristics and solutions suitable for precision crop production practices.

4.2 SENSING FOR PRECISION AGRICULTURE

As in any control system, sensing in a PA system plays an important role in gaining an understanding of crop growth responses to soil properties and resource inputs, and providing an indication of production outcome. However, the inherent spatial and temporal variability in the PA process is attributed to the fact that much of the collectable data are not directly observable, namely, some measured data are unable to provide a direct indication on production outcome (Zhang et al., 2013). This section will discuss the technologies and methods commonly used in spatial and temporal sensing, the information spatial and temporal data could bring in, and the challenge for making such data observable in controlling a precision crop production system.

4.2.1 Spatial Sensing

The measurement of the spatial variability of crop growth affected by soil properties within a field is one of the fundamental tasks in PA. Some of the soil properties, such as soil type, topography, past usage, and organic matter content, may be unchanged or have very little change over a considerably long period of time; thus, one measurement could provide such spatial variability information for years. In comparison, the variability of some other soil properties, such as soil nitrogen and moisture content, could change rapidly and requires real-time or near-real-time measurement (Hummel et al., 1996).

Defined as a mechanism for detecting a gradient in which the property is compared at different points within and/or between fields, spatial sensing is commonly used in PA for gathering such information. In terms of the methods for obtaining the data of interest, spatial sensing can be classified into categories of remote sensing and georeferenced sensing.

Remote sensing, a technology for detecting and measuring reflected and emitted electromagnetic radiation from a distance, was quickly adopted in PA in the 1980s, and initially focused on limited wavelengths in a few visible and near-infrared (NIR) bands and then expanded to a much broader range from ultraviolet to microwave, and with different spatial resolution using satellite, aerial, or ground vehicle-mounted sensors to meet different requirements for PA management (Mulla, 2013). There are two categories of electromagnetic radiation sensors, namely, nonimaging (e.g., spectroradiometers) and imaging (e.g., spectrum cameras), available for PA applications. Spectroradiometers are devices designed to measure the spectral power distribution of a source often used in georeferenced sensing, and spectrum cameras are devices designed to acquire a spectrally resolved image of an object or scene often used in satellite-, aerial-, and ground vehicle-based remote sensing.

A remotely acquired spectrally resolved image carries two forms of resolution: the spectral resolution and the spatial resolution, with the former solely determined by the sensor design and the latter able to be changed by the distance between the sensor and the scene. The spectral resolution is specified by the number of spectral bands in which the sensor can collect reflected radiance. Both the high-resolution hyperspectral camera (up to 220 bands) and the low-resolution multispectral cameras (as low as 3 bands) covering visible to NIR spectrum of reflected radiations are commonly used in PA remote sensing. The spatial resolution is specified by the pixel size of spectral images covering the field surface. Constrained by current sensor technology, remotely sensed spectral images can only offer either a high spatial resolution associated with a low spectral resolution or vice versa.

It is noted that in acquiring spectral images for PA use, the number of bands for a sensor is not the only important aspect of spectral resolution; the position of those bands in the electromagnetic spectrum is equally, if not more, important. For example, one widely used crop sensing method in PA application is the use of a multispectral CCD (charge-coupled device) camera to measure the NIR and red bands of crop canopy reflectance to calculate a normalized difference vegetation index (NDVI). This NDVI could provide a quick determination of vegetated areas and a simple estimation of plant growth conditions (Rouse et al., 1973). However, the soil reflectance from low-density canopy areas and the insensitivity to leaf chlorophyll content change in high-density canopy areas would affect the validity of NDVI data for indicating crop conditions (Thenkabail et al., 2000). Various studies for formulating different vegetation indices (VIs) and other sensing enhancement methods using different bands have been investigated, and researchers found that the use of right combinations of different bands could improve the accuracy and robustness of the data collected for measuring crop nutrition (Noh et al., 2006), diseases or pests (Prabhakar et al., 2011), weeds (Tang et al., 2000), and water stresses (Méndez-Barroso et al., 2008).

Another common method for obtaining spatial data for precision crop production is georeferenced sensing, which could be applied to both soil and crop sensing. In georeferenced sensing, sensors are often coupled with a global positioning system (GPS) to generate field maps of measured parameters. Figure 4.2 shows an example of such a sensing system installed on a mobile sprayer for implementing VRA based on a georeferenced application map. Depending on the spacing between passes, travel speed, and measurement/sampling frequency, the number of sensed points per

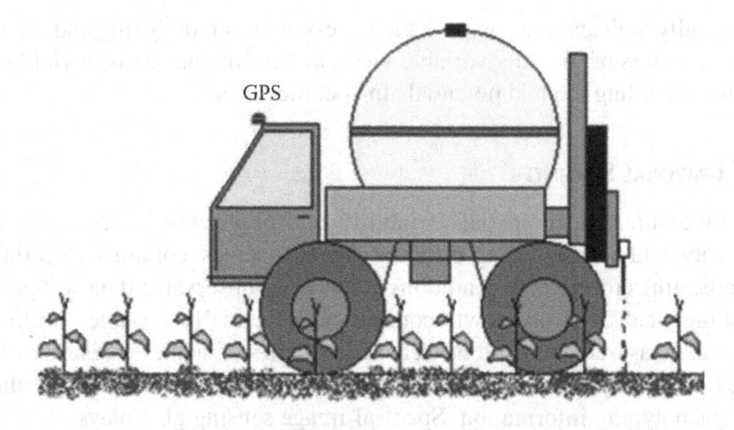

FIGURE 4.2 An example of a georeferenced sensing system installed on a mobile sprayer for implementing variable-rate application based on a georeferenced application map.

area can be varied to change the density of the sampled grid, which in turn changes the resolution of the georeferenced field map.

As most soil characteristics within a field are relatively stable over a reasonable time, soil property measurement could often be performed via georeferenced grid soil sampling and laboratory testing, or georeferenced real-time measurement using on-the-go sensors during field operations. Grid soil sampling and testing is often done by collecting a few soil samples using either a soil probe or simple hand tools within a predefined uniform grid overlaid on the field being sampled. The samples are normally sent to soil laboratories for analysis. This method requires time to complete the process, and is commonly used to support making more profitable use of fertilizers and lime (Franzen and Peck, 1995). Georeferenced real-time measurement can use some electrical and optical sensors to measure soil properties, such as soil conductivity, texture, organic matter, moisture content, and nutrient and pH levels, and generate appropriate soil property maps (Heege, 2013). A number of successful applications for collecting soil property data have been reported, such as checking soil moisture for planting depth control (Norman et al., 1992), sampling soil nutrient levels for controlling variable-rate fertilization (Bermudez and Mallarino, 2007), and variable-rate injections of herbicides based on soil texture and organic matter content (Qiu et al., 1998).

While the spatial variation data of both soil properties and crop growth conditions are collectable using different sensing technologies, such data are often not directly observable in estimating crop yield as there are many other factors, such as climate, diseases, insects, and weeds, which could affect it. One solution is to monitor the spatial variation of the yield.

Introduced in early 1990s, yield monitoring is often the first step many farmers take in practicing PA. A yield monitor coupled with a GPS receiver on a harvester can measure crop yield data combining mass, moisture, area covered, and location being harvested to generate a yield map showing the spatial variability across a field. As one of the most valuable sources of spatial data for PA, yield maps can be used in evaluating the year-to-year variation of yield distribution within a field to find areas with potentially high and low yields, which is essential for analyzing and identifying the limiting factors of spatially variable yield, and setting up realistic yield goals to vary inputs according to yield potentials in specific areas.

4.2.2 Temporal Sensing

Temporal measurement of spatial variability of crop growth information, such as color, density, and height, during different growing stages, contains essential information indicating crop growth conditions and is more observable than soil properties in estimating yield. As crop growth conditions are spatially variable within a field, the temporal measurement could be treated as a series of spatial measurements over the same field during certain periods of the growing season for obtaining either crop stress or phenotyping information. Spectral image sensing also plays an important role in temporal sensing. The resolution for temporal measurement is specified as the measurement frequency of a specific location, often in days or sometime in weeks depending on the rate of change of the variable being measured.

Temporal measurement of crop stresses during the development stage could often provide important information for predicting crop yield, thus alerting farmers to take remedial measures to mitigate yield loss. There have been many promising developments in temporal sensing technologies capable of detecting crop stresses. One well-studied technology is multispectral imagery sensing that collects light reflectance from some carefully selected spectral bands within the visible and NIR spectrums, which are sensitive to variables related to plant development and crop yield. NDVI has been widely adopted in measuring crop stresses during its development stage. Based on some extensive studies, Jackson and Huete (1991) suggested that when the NDVI value is between 0.30 and 1.00, it indicates the vegetation is most likely in healthy condition; when the value drops to between 0.10 and 0.30, it indicates the vegetation could be unhealthy or sparse; and when the value is close to zero or even negative, it often indicates there is no vegetation.

However, the capability of using NDVI to indicate healthy crop conditions based on Jackson and Huete's classification may only be applicable within a limited window of the crop growth season. Quarmby et al. (1993) found that yield estimation of wheat, cotton, rice, and maize crops using NDVI data could be stabilized at 50–100 days prior to harvest. Xiang and Tian (2011) have studied the temporal change of the NDVI readings of corn canopy within a 0.8-ha field in Central Illinois from 14 days after planting (DAP) to 122 DAP, and found that the NDVI for both fertilized and unfertilized crops began at very similar levels from below 0.2 before 25 DAP, and reached a level around 0.7 for both conditions around 31 DAP (V6 stage, defined as the corn growth stage of the collar of the 6th leaf being visible). A noticeable NDVI values disparity (greater than 0.1) was observed between fertilized and unfertilized crops after V7 stage, and then became close again around 58 DAP (VT stage, defined as the corn growth stage of the last branch of the tassel being completely visible). Figure 4.3 illustrates the NDVI variation corresponding to the DAP. This research verified an assumption that there exists an optimal window for using multispectral sensors to detect temporal changes of crop nitrogen stress by means of measuring some types of crop VIs such as NDVI.

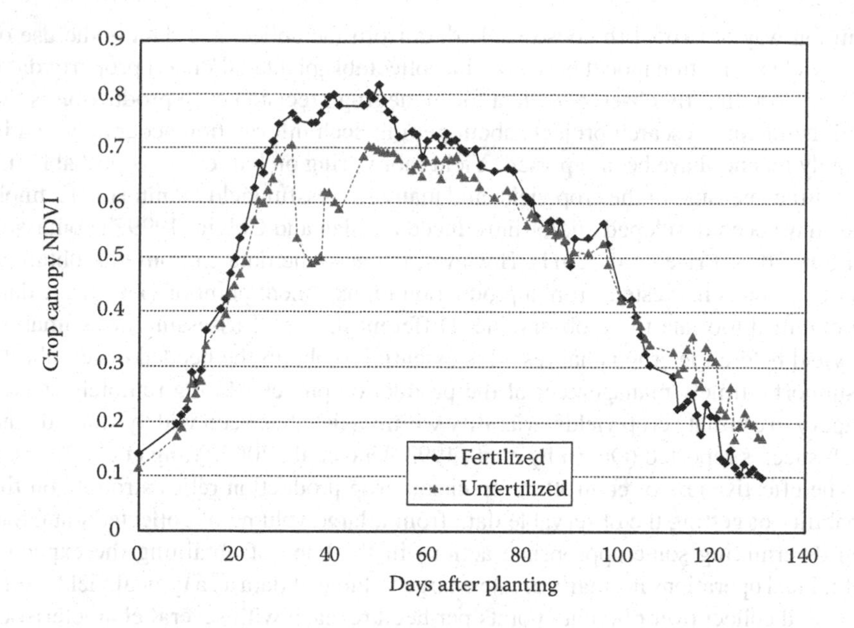

FIGURE 4.3 An example of temporal changes of canopy NDVI values for fertilized and unfertilized corn plants over a growth season. (Data courtesy of Dr. Haibo Xiang.)

Temporal measurement of soil moisture dynamics with depth plays an important role in irrigated crop production to achieve effective irrigation control. Balenzano et al. (2011) investigated the potential of using satellite remote sensing data to map temporal changes of surface soil moisture content underneath crops, and found that soil moisture content could be retrieved during the whole growing season, with accuracies ranging between 5% and 6%. Based on their studies using a wireless soil moisture sensing network to assess temporal stability patterns of soil moisture at different locations, Hedley and Yule (2009) found that the temporal stability of soil moisture could have substantial spatial variability from moderate to strong. A sensor network could provide the needed information to predict soil water status in different zones in the field to support automated variable-rate irrigation control in order to improve water use efficiency.

4.2.3 Making Data Observable

Measuring plant and soil properties continuously could gain an understanding of crop growth changes over time under different conditions. However, most of the data could not directly provide an indication of what the expected crop yield will be until being harvested (Lamb et al., 2008). To solve this problem, it is essential to make the collectable data observable. In PA management, observable data are defined as a measurement of yield-indicating crop growth conditions, which could provide an indication of expected yield, as a result of some field operations. Therefore, it is essential to obtain some observable data of a precision crop farming system for effectively controlling the process in the hope of getting the expected yield. One common way to extract the observable data from the collectable data is the use of some yield estimation model based on the collectable plant and/or soil property data.

The most directly observable data for managing precision crop production is the yield. Numerous research projects about getting such information accurately and in a timely manner have been reported. Yield monitoring on harvesters is probably the most direct measure of the crop yield, and many successful yield monitoring technologies have been developed and/or introduced (Arslan and Colvin, 1999; Thomasson and Sui, 2003; Price et al., 2011). However, because the data can only be obtained after the crop is harvested, from a production management point of view, these data are obtained too late to be observable. Different means of assessing or estimating the yield before the crop is harvested is essential to obtain the needed observability to support a timely management of the production process. Using remotely sensed imagery to estimate crop yield variability within a field has been widely studied, and much success reported (Quarmby et al., 1993; Uno et al., 2005; Yang et al., 2007).

The effectiveness of controlling precision crop production relies strongly on the capability of getting the observable data from a large volume of collected data, and then determining some appropriate actions in the hope of obtaining the expected yield. Field operations normally create a large volume of data as a typical yield monitor would collect over 600 data points per hectare, each with several characteristics (latitude, longitude, yield, moisture, etc.). The rapidly increasing and overwhelming volume of recorded production-related data, plus the need for special skills and/or tools to analyze and interpret such data, makes it very difficult for farmers to effectively make use of those collected data. This data-driven decision-making process in PA poses a number of data mining problems, and one of the fundamental ones is yield prediction. Creating some automatic means for predicting yield, based on these data, will help farmers remove a major obstacle to precision crop production, and thereby gain efficiency and economic benefits. Numerous efforts toward developing robust and trustworthy yield prediction tools, from classical regression-based approaches to machine learning-based neural networks modeling or support vector machines methods (Uno et al., 2005), have been undertaken. However, the ability to make a trustworthy yield prediction using those tools is still a big challenge due to the extreme conditional complexity induced by many factors, ranging from changes in climate patterns to individual differences among plants during growth season. A study of spatial and temporal variation yield data over a 5-year period of a commercial field located in Central Illinois showed that the spatial variation of the yield could be presented in different patterns within a field over the years (Zhang and Han, 2002). Randomly picking yield data from a few monitoring zones scattered over the field as presented in Figure 4.4, it is shown that the normalized yield in zones 5 and 8 varied marginally with different patterns around the average yield of the year over the studied period. In comparison, the normalized yield at zones 1 and 4 were noticeably lower than the average, while zone 9 was always higher over the period.

The wide range of yield variation across zones and over time shows that yield responses to the management input did not result in a consistent outcome. Such poor robustness in yield responses may be attributed to numerous factors that will require further analysis to uncover. For example, an investigation of the effect of seasonal precipitation on yield revealed that the amount of winter precipitation correlated

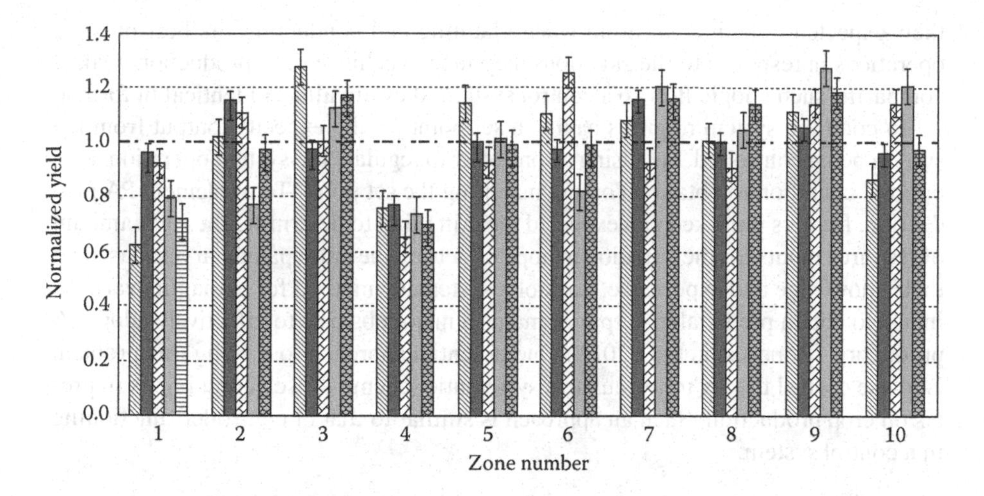

FIGURE 4.4 An example of yield variation (normalized to the average yield of the year) over a 5-year period (bars from left to right represent data of 1996–2000) from 10 randomly picked zones within a commercial field in Central Illinois.

with the yield variation most noticeably compared with other seasons. Knowing what results to expect from inputs applied is one of the greatest challenges in making the collected data observable.

In addition, as almost all field operations are performed using some type of machinery in motion, the machine drivers need actionable instructions, namely, ready-to-use commands, rather than some forms of raw or processed data to apply a controlled input in real time. For example, in performing variable-rate fertilization, what an applicator needs to know is how much a nozzle should be opened while the machine is traveling at a certain speed within a specific region in a field. Any other format of the information would induce some difficulties for the driver to effectively perform the work. Therefore, one of the basic requirements for a data-based decision-making process is the ability to extract actionable instruction from the collected data using a "transparent-to-user" method "on-the-go."

4.3 CONTROL FOR PRECISION AGRICULTURE

As a data-intensive management system, precision crop production is a site-specific management concept that observes and responds to intrafield variations to optimize returns on resource inputs. Effective implementation of PA requires having a good understanding of how to control resource inputs effectively using a transdisciplinary approach. Like many control systems in industrial processes, control of agricultural production is also applied through obtaining and processing production data, making operational decisions based on the processed data, and implementing the decision (Schueller, 2013). This section will focus on analyzing and discussing a few fundamental issues in making PA decisions.

A PA production management system shares many similar features in format with a control system. For example, in managing PA production, farmers often have their expected outcomes in mind when planning and conducting the best possible operations in response to the situations they face to achieve their production goal. If comparing such a logic flow to a control system, they are almost identical in format.

To control a system requires having a set point for the expected output from the system being controlled, and using a controller to regulate the system operation at an optimal condition to obtain an output matching the set point. To implement PA production, farmers must keep an expected yield in mind to determine the right amount of resource inputs for the situation in hope of getting the anticipated harvest based on their knowledge and experience. Lack of a systematic method for managing variable inputs to obtain predictable outputs remains a major obstacle to effectively adopt PA production (Robertson et al., 2012). One potential approach to solving this problem is to use control theory to formulate a systematic scheme for selecting inputs in precision crop production. Such an approach is similar to that of controller functioning in a control system.

4.3.1 Prescriptive Control

The theory behind control systems and the practices in managing precision crop production lays the foundation for formulating PA control schemes. The core premise of PA control is the variation of resource inputs to obtain the expected yield with minimal deviation. One of the major challenges in controlling such a process is that the actual yield is measurable only at harvest time. Therefore, the adoption of a prescriptive control is the logical choice for this application.

Typically, a prescriptive control uses a yield prediction model, either mathematics-based or knowledge-based, to estimate the appropriate resource inputs needed to get the desired yield at the specific site in terms of the yield potential of this location. Conventionally, farmers have a management strategy for their production. They manage resource inputs accordingly in the hope of achieving their production goal. Such resource input management strategy, often formed over years of experience on specific fields, constitutes the base of the prescription control. As the crop response to planned implementations will be disturbed by changes in climate conditions during the growth season, it is often difficult to manage this type of production system and to have the final yield exactly match the expected production goal. Therefore, such a prescriptive management practice could be symbolized using a basic open-loop control as the farmers conduct their field operations based only on their plans (or only on experience) as they can do very little to change the crop growth during the season. Using a block diagram presentation commonly used in control system modeling, a typical process of prescriptive crop production management could be represented as in Figure 4.5, with a transfer function expressed as follows:

$$Y_A = (GY_P - D)P \tag{4.1}$$

where Y_A is the yield from the production; G is the resource input management plan; P is the crop responses to the input; Y_P is the yield potential; and D is the climate and/or field condition disturbance.

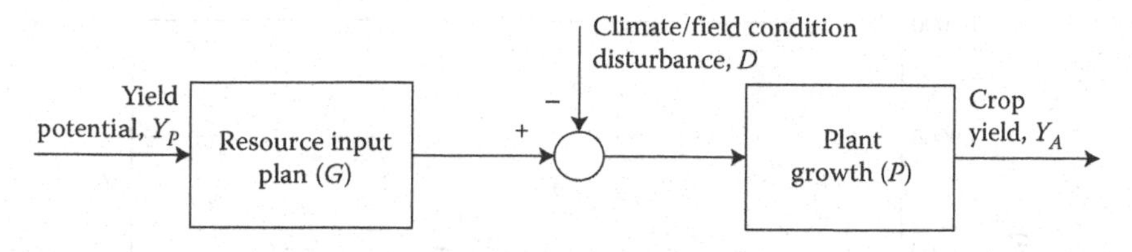

FIGURE 4.5 A system block diagram of a basic open-loop prescription control system for PA production.

This prescriptive control is very similar to gain scheduling control in control theory, which is a practical approach for controlling nonlinear systems using a family of linear controllers adequate for different operating points of the system (Levine, 1995). Because precision agricultural systems are normally nonlinear, gain scheduling control could offer an effective means for handling the nonlinear feature. However, it may require some special processes to make the control scheme be practically usable in precision crop production. One such essential process could be the creation of an input management model based on the collected input–yield relationship using statistical modeling methods. Owing to the complicity of the ecobiological process of crop growth, such input management models are normally empirical-based, which often do not intend to describe any physical or biological processes of the crop growth, but only attempt to represent the relationship between inputs and outputs using any expressible means. One of the unique issues in PA control is its extremely slow responses to the input, which are normally impossible to observe after days, or weeks, of the inputs being applied. The actual yield corresponding to a certain input could not be directly measured until after harvest, even though many of the collectable data could provide information indicating the attribution of the input(s).

The extremely slow and uncertain responses of yield to management input causes difficulty in implementing time-domain-based gain scheduling control in PA production. We could apply the format of a gain scheduling control concept, but use a non-time-domain statistical input–output relationship to formulate the control law for the prescriptive control to make it suitable for PA applications. Figure 4.6 presents an example of the yield response of hybrid corn to the amount of nitrogen (N) obtained from a set of tests conducted at the Monsanto Water Utilization Learning Center at Gothenburg, Nebraska in 2010. From this figure, it shows that the yield of this particular type of hybrid corn responded largely to the first 135 kg ha^{-1} N applied, and the yield increases diminished quickly with additional N applications (Monsanto, 2010).

This first-order-like nitrogen-to-yield model to describe the crop growth response to nitrogen input could provide the base for PA decision making, and could be defined as the transfer function of nitrogen application rate to crop yield in a PA control system. When the transfer function of a system is identified, it offers a convenient method of using its inverse transfer function to design a controller for the system. In this example of variable-rate nitrogen management for controlling hybrid corn yield, the prescriptive controller could be designed using an inverse nitrogen-to-yield model as illustrated in Figure 4.7 to determine the appropriate rate of nitrogen application for obtaining an expected yield.

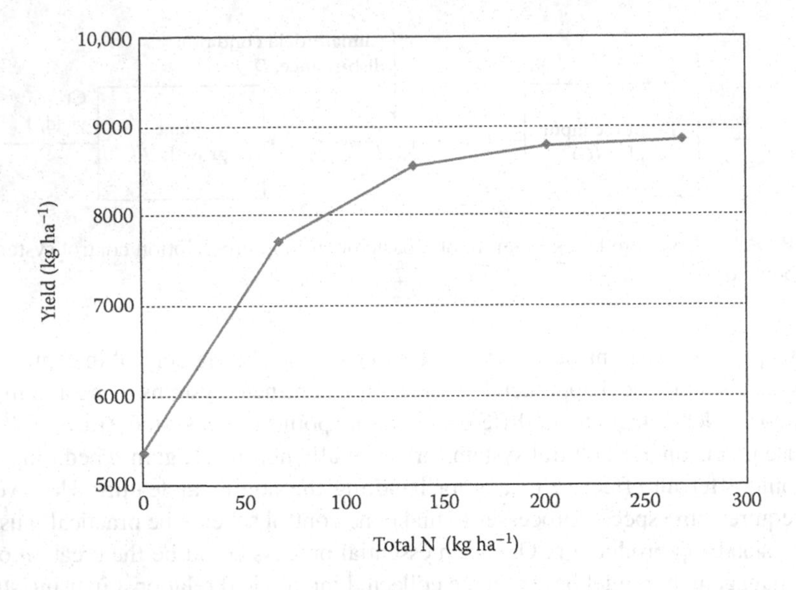

FIGURE 4.6 An example yield response model of a hybrid corn to a different nitrogen application rate based on the data published by Monsanto (2010).

This basic open-loop prescriptive control approach is suitable for site-specific resource input management for minimizing the influences of soil property variation in a field to best attain the yield potential on a site. Being inherently limited by the inability to respond to the climate or other disturbance occurring during the plant growth season, a robust enhancement method based on multiple years of yield

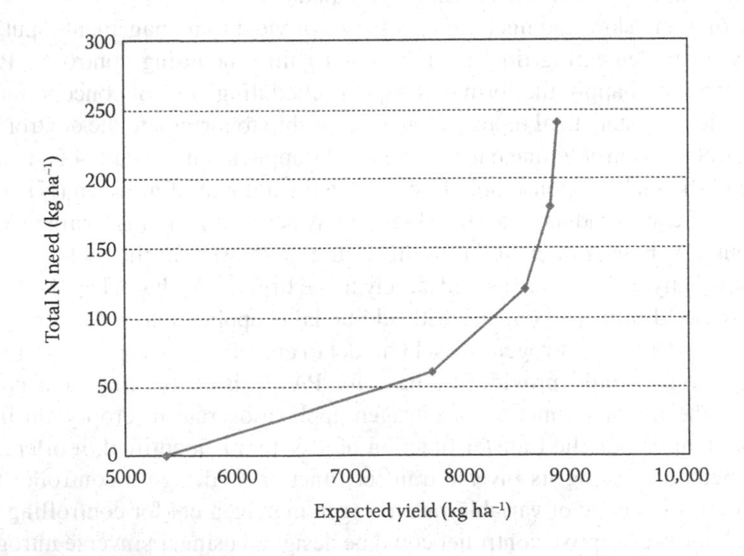

FIGURE 4.7 An example of an inverse transfer function for determining N application rate to obtain an expected yield of corn.

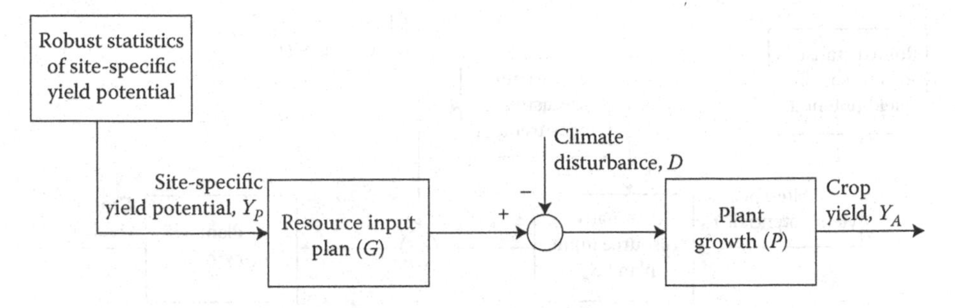

FIGURE 4.8 A system block diagram of an enhanced open-loop prescription control system for precision crop production to achieve the yield potential of a specific site.

history using some robust statistical analysis methods on the site has been applied in hope of minimizing such effect. Another improvement to the basic open-loop prescriptive control is to use the economically achievable yield potential of a zone, instead of reducing the variation between zones of high and low yield potentials, since the yield in some regions of a field could hardly reach the average yield of the field (as illustrated in zone 4 in Figure 4.4). Figure 4.8 shows the system block diagram of such an enhanced open-loop prescriptive control.

Numerous studies have been reported in support of effective decision making for profitably managing variable-rate inputs. An example reported by Havlin and Heiniger (2009) is the development of a VRA decision support tool to help make precise decisions for fertilizer inputs by determining the level of sufficiency of soil nutrient status at a specific location relative to the needed fertilizer input levels to obtain the potential maximum yield for corn, soybean, wheat, and cotton production in North Carolina.

4.3.2 Responsive Control

While the enhanced open-loop control offered a potential to have a more robust result by using on-spot yield potential determined by multiyear yield data, it is still unable to respond to any climate disturbance and other condition changes during the growth season. An enhancement capable of responding to such disturbances/changes in the decision-making process, or developing a responsive control scheme, is the logical next step. The assumption for this type of enhancement is that the difference in climate and/or field condition would require an adjustment to resource inputs for realizing the yield potential, and such climate/field condition changes could either be forecasted or measured. This type of enhancement is suitable for crops requiring most inputs to be applied at an early stage of, or even before, the growing season. Figure 4.9 illustrates a conceptual system block diagram for such a responsive control scheme. Its transfer function could then be expressed as follows:

$$Y_A = (G_A Y_P - D)P \tag{4.2}$$

where G_A is an optimal input management plan corresponding to the identified situation, and Y_P is the yield potential of the specific zone.

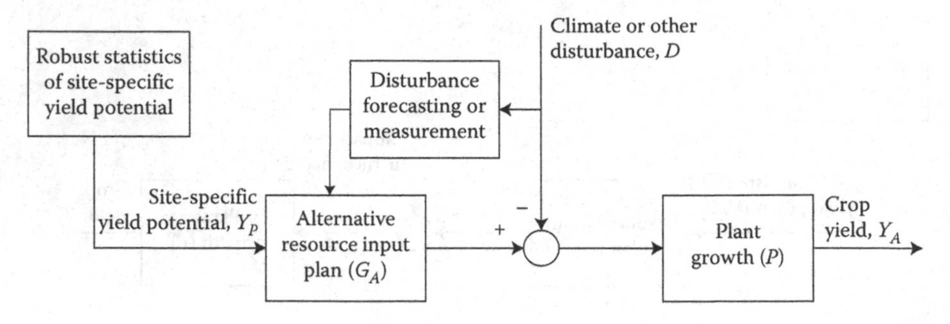

FIGURE 4.9 A system block diagram of an enhanced climate/field condition responsive control system for PA production.

The core of this enhancement would allow farmers to include either climate forecasts or the latest measurable site-specific field condition data into the resource management decision-making process by best utilizing the historic yield data in similar conditions to adjust the input(s) responsively to the situation. Mid- to late-vegetative growth-stage variable-rate nitrogen side-dress application is a good example of responsive control. Either based on the data obtained from in-season canopy reflectance sensing or from late spring soil nitrate tests, N-deficient crop plants will respond to additional nitrogen fertilizer being side-dress applied. It could potentially achieve higher yield efficiency with a smaller amount of total nitrogen fertilizer being applied if the amount of side-dressed fertilizer could be properly determined.

Differential spraying in weed control is another example of applying a responsive control. As weeds are normally irregularly distributed in cereal crop fields, to achieve an operation goal of minimizing the use of herbicides, precision weed control management requires making responsive application decisions to differentiate the location and amount of herbicides to be sprayed corresponding to the detected weed quantity and distribution in the fields.

Another application of responsive control for crop production is in precision irrigation. For example, Goumopoulos et al. (2014) have developed a proactive closed-loop irrigation control system using an adaptive decision-making layer, which employed a machine learning approach to determine significant thresholds of plant-based parameters to optimize irrigation control in response to detected plant growth conditions.

4.3.3 Feedback Control

With advances in PA technology, different sensing technologies can provide farmers with effective means of detecting crop growth indices which then can be used to estimate final yields and make necessary adjustments to inputs in response to the estimated yield (Shanahan et al., 2008). As this management practice could improve crop growth by modifying the resource application plan in the hope of getting the production output closer to the expected level, it can be considered a closed-loop feedback control. As previously pointed out, the control goal of precision crop production is to achieve the yield potential of a site. Since the actual yield cannot be measured until it is harvested, it should be noted that in this closed-loop control system, the feedback information to the controller is not the actual system output, but an estimated output based on collected data that indicates the state of crop growth. Such feedback contains inherent uncertainty in representing the actual output. Figure 4.10 shows a block diagram of such a closed-loop feedback control scheme for sensor-based PA production with its transfer function expressed as follows:

$$Y_A = [G(Y_P - Y_E) - D]P \tag{4.3}$$

where Y_E is the estimated yield based on the sensed crop growth condition.

In control theory, system response is used to describe how a system is responding to the changes in inputs and/or disturbances to the system and making an estimation of system output based on some measurable parameters. A set of time behaviors, such as delay time, rise time, peak time, settling time, overshoot, and steady-state error, is commonly used to determine a response. Figure 4.11 shows typical responses of system output to a change in set point for both the first- and second-order (with a high damping ratio) systems. The two curves in the figure demonstrate how systems respond to a change in the control set point and approach its final value within a finite period of time. A first-order system normally responds to the input slowly to gradually reach a stabilized output. In comparison, a second-order system often responds to the set point change at a much faster rate but will overshoot before the output is stabilized. However, owing to the biological nature of crop production, the time behavior in PA control is quite different from many industrial controls. From knowledge of agronomy, we know that the yield corresponding to the amount of fertilizer input exhibits a similar behavior to a first-order system with regard to the input amount (not time) domain. Such behavior implies that the yield control could be achieved through analyzing the system behavior similar to a first-order system and creating a system response constant analogous to the time constant in a time-domain system for predicting yield responses to resource inputs. Meanwhile, for

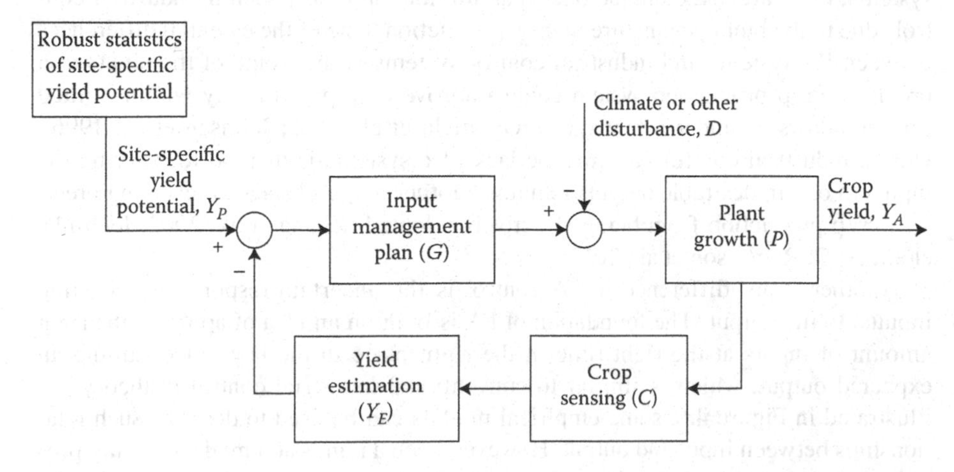

FIGURE 4.10 A system block diagram of a closed-loop feedback control system for precision crop production.

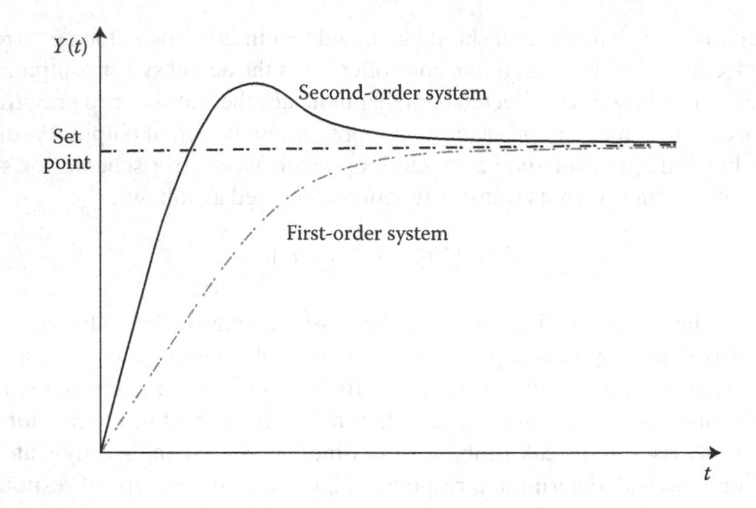

FIGURE 4.11 An example of a dynamic response of system output to a set point. The solid line is the response from a second-order system and the double dashed line represents a first-order system.

many field operation controls, such as canopy reflectance sensing-based side-dress nitrogen application to correct crop N stress, the crop responses to resource inputs are often presented similar to those of a second-order system with a large damping ratio in the time domain just as shown in Figure 4.11.

As many unexpected disturbances originated from a natural production environment may draw the output away from the desired set point, a study on how to effectively reduce the influence of those disturbances is a particularly important research goal for agricultural system controls.

While a PA control shares many features similar in format to industrial control systems, there are some unique ones that are not commonly seen in industrial controls due to the biological nature of crop production. One of the essential differences between PA systems and industrial control systems is that some of the inputs to a precision crop production system could improve crop growth only within limited time windows or under a few states (Ogunlela et al., 1982; Johnson et al., 1996). Unlike industrial control systems, the lack of a systematic method for varying the inputs to obtain desirable outputs remains another major obstacle, preventing precision crop production from being effectively adopted (Zhang et al., 2002; Jochinke et al., 2007; Robertson et al., 2012).

Another major difference in PA control is the uncertain response of adjusting input(s) to the output. The foundation of PA is built on an idea of applying the right amount of inputs at the right time in the right place in the hope of obtaining an expected output, which is similar to conventional industrial control in theory. As illustrated in Figure 4.6, some empirical models can be used to describe such relationships between input and output. However, such a regression model can only provide some information supporting an uncertain estimate of the yield as it is normally formulated based on collected data from previous seasons, and the predicted yield response is not a tie behavior. These are some of the major obstacles for farmers in effectively managing their precision crop production since the collected data during the crop growth season is not directly useful as feedback to support making control decisions. One essential requirement for making feedback control of precision crop production manageable is making the collected data observable.

In adopting control theory in PA management, researchers have started to apply state-space modeling to understand and explain the spatial correlation of crop and soil in the hope of gaining a more reliable prediction of yield (Wendroth et al., 1992; Timm et al., 2000). In state-space analysis, *observability* is a measure of the ability to tell what is going on inside the system and whether the desired output could be obtained from the system through observing system behaviors. Formally, a system is said to be observable if its current state can be determined in finite time in terms of only its outputs. If a system is not observable, the current values of some state variables cannot be determined by the sensed output data. This implies that the controller cannot adjust those parameters to an appropriate level to obtain the desired output. Referring to control of crop production systems, this observability is reflected in the relationship between collectable crop growth data and the yield. To attain such observability is a practical challenge in effectively controlling crop production since the yield is not measurable until harvested (Lamb et al., 2008). The good news is that it is technically possible to detect the crop growth condition during its growing season using certain types of crop sensors. Figure 4.12 illustrates an example of visually detectable difference in corn plant growth conditions randomly

FIGURE 4.12 An example of visually detectable growth condition differences among corn plants randomly found in a commercial corn field in the Midwest Corn Belt of the United States. The noticeably smaller plants could lead to a distinguishable lower yield.

captured in a commercial corn field in the Midwest corn belt of the United States. While there is still a lot of uncertainty, the noticeably smaller plants often lead to a measurably lower yield.

One method to improve observability is the use of data fusion, a process of integrating multiple data sources to obtain a more accurate and robust estimation of crop yield. A study based on remote sensing data from 37 fields in Texas showed that it could improve the average yield estimation from about 30% underestimation using only the raw satellite imaging data to a 2% overestimate after using three state variables (stage of crop development, green leaf area index, and aboveground dry mass) (Maas, 1988). In a study on the spatial dependence between crop yield, effective soil N and N_2 fixation, Wendroth et al. (1992) proved that it was possible to use a state-space approach to determine spatial variability of yields from local field observations.

Another major challenge in practicing PA is the uncertain responses of adjusting inputs(s) to the output. While PA requires employing a responsive management strategy based on detailed, site-specific information, it also requires determining how much of the observed yield variability was caused by natural variation in yields, how much by variations in management practices, and thereby to determine what management practices are most appropriate for what conditions, both edaphic and climatic, in a management zone. To solve this problem, a study on the controllability of precision crop production systems is essential. As the collectable data from the production process is often indirect measurement of factors that could affect the final yield, we can use two definitions of controllability: state controllability and output controllability. The former describes the ability of an external input to change the internal state of a system from any initial state to any other final state in a finite time interval; and the latter defines the ability to change the output. As a precision crop production system is often not directly observable with regard to the yield, but observable in terms of the crop growth condition (plant growing status) or crop yield potential (soil fertility status), the controllability in precision crop production typically means the state controllability if not specifically notified. Many variable-rate resource input decisions are made based on crop growth condition or yield potential, which is a state variable rather than the production output.

One essential assumption for variable-rate-based site-specific crop production is that it could attain a higher crop yield by varying the resource(s) input to bring crop everywhere in a field to its yield potential (McKinion et al., 2001). However, such an assumption does not always lead to the best solution in practice. Based on an extensive study, Peng et al. (2010) found that there was no correlation between grain yield and total N input in rice production in China, and that such poor correlation could be attributed to many factors, including location, season, variety, pest damage, and other crop management practices. Raun et al. (2011) investigated the relationship between grain yield and its response to N in long-term wheat and corn experiments, and also found no clear relationship between response to N and grain yield. Upon further study, they found that both yield and response to N were consistently independent of one another. As both affect the demand for fertilizer N, estimates of both should be combined to calculate realistic in-season N rates. As N management is a common practice in realizing state controllability, such findings present a challenge to understand whether state controllability is sufficient for PA management or a study on output controllability is essential to advancing PA. Control system theory has proven that a state controllable system is not necessarily output controllable, and vice versa. Numerous reports have been published on predicting grain yield based on crop growth conditions (Diker and Bausch, 2003; Wendroth et al., 2003; Tremblay et al., 2010). Such research offers a method for utilizing the controllable crop growing state to achieve the controllability on final yield.

A common precision crop production practice that many farmers are following today is first collecting yield maps of their fields during harvest, and then utilizing these maps combined with other relevant information, such as weather data, to make field management decisions for implementing variable-rate preplant fertilizer applications, precision planting, and/or postplanting fertilizer applications (McKinion et al., 2010). However, the use of yield maps in making accurate decisions on production management for the next season is always difficult due to many mitigating factors (Kaspar et al., 2003). Research revealed that the standard deviation for crop yield at different parts of a field under the same management practice could surpass 20%, with the possibility of the yield from one zone being less than 60% of the yield from a neighboring zone (Zhang and Han, 2002). Such inconsistency in final yield adds another layer of complexity to PA, the lack of robustness in agricultural system control under the same management actions.

Crop fields often vary between and within themselves in landscape position, terrain attributes, erosion class, and soil properties (Stone et al., 1985). Such frequent and random variations, along with uncertain weather changes, play a major role in affecting the lack of robustness in controlling precision crop production. A successful precision agricultural control system should be observable, controllable, and robust. However, this is difficult due to the uncertainty inherent to crop production, often entering an agricultural control system through uncertain states caused by spatial or temporal variation and actions constrained by technical or economic difficulties (Adams et al., 2000). This presents challenges to creating robust agricultural control systems. For example, some transient spatial factors, such as insect or disease pathogen spreading, planter or applicator malfunctions, and measurement error in yield monitoring, can substantially affect the yield or yield observation in specific areas in 1 year but not every year (Colvin et al., 1997; Lark et al., 1997). How to robustly manage an uncertain crop production system with unknown dynamics subject to unknown disturbances is still the key problem a precision agricultural control system needs to solve.

4.4 PRECISION AGRICULTURE IMPLEMENTATION

Another essential element in controlling a PA process is the reliable implementation of planned field operations as any management plans will never result in any effectiveness unless such plans are accurately implemented. A typical field crop production process normally includes some or all the following operations: planning, planting, resource input management (often implemented in the forms of either VRAs or targeted applications), and harvest. To provide farmers with reliable tools to effectively implement those operations, many automated technologies have been developed in the past few decades.

4.4.1 Planning and Planting

Effective PA management begins with planning. An appropriate site-specific production plan based on soil property and yield potential plays a vital role in practicing precision crop production. Farmers have doubts about investing in precision agricultural technologies because of their inability to directly apply the collected data as feedback for improved management practices. While this remains one of the biggest obstacles for farmers to gain the promised benefits, a transformation of a long-term yield-map dataset into profit maps based on economic thresholds for profitability at different zones could help create a profitable site-specific management plan (Massey et al., 2008). This method uses actual input costs, crop prices, published custom rates for field operations, and region-specific land rental prices to transform yield maps into profitability maps, which could map yield into profitability metrics for different management options to support farmers in planning a profitable precision crop production.

After a production plan is made, it needs to be implemented precisely following the implementation plan, and sometimes also requires having the capability of adapting to scenario changes by modifying or changing some specific actions during field operation. In the preplanting to planting process, there are a few typical field operations that could be controlled precisely in implementation. Normally, the first field operation in crop production is field preparation. A zone tillage, a form of modified deep tillage in which only narrow strips are tilled, requires positioning the plows precisely to target strips to agitate the soil to reduce soil compaction and improve soil internal drainage. To achieve precise tillage depth control, both an auto-steering system and a tillage depth control system would be required. An auto-steering system could accurately guide a tractor following the target strips to achieve improved operation efficiency, accuracy, and speed, therefore gaining financial benefits from practicing precision zone tillage. Using an automated tillage depth control system, the plows could automatically follow a predetermined tillage control plan to adjust the tillage depth. Xie et al. (2013) have developed a depth control system capable of adjusting tillage depth from −100 to −200 mm within a 3.5 s response time and with ±8 mm depth control accuracy. Wells et al. (2005) verified that deep tillage in general could result in a yield improvement for corn, soybean, and wheat compared to those receiving no deep tillage. The use of an RTK-DGPS-guided auto-steering system could accurately navigate a tractor to perform controlled-traffic farming by traveling only on a few fixed traffic lanes in a field to create nontrafficked cropping zones with optimum soil structure. In addition, the auto-steering function has dramatically improved operator comfort by those who have adopted this technology.

Planting or seeding is another critical operation in precision crop production. It requires putting the exact number of seeds or seedlings precisely at the right place, and is implemented using machines in modern mechanized agricultural productions. Precision planting or seeding often requires having an accurate control of the number, as well as the location and depth of seeds or seedlings being planted. Numerous commercial products of planters and seeders capable of attaining the required seeding/planting numbers and depth accuracy are available in today's market. One addition to improve the position accuracy is the increasing adoption of GPS-based auto-steer technology for those planters/seeders. As it could help to reduce overlap and eliminate skips, and therefore could result in a reduction of input costs for labor, seeds, and fuel, the auto-steer planting or seeding technology has been praised by farmers as the most effective PA technology. Based on Deere & Co., a high-accuracy auto-steering system could reduce overlap by up to 90% through accurate and repeatable guidance in both curved and straight tracks in crop fields (Deere & Co., 2013).

Yuan et al. (2011) have developed a variable seeding–fertilizing planter suitable for no-tillage cultivation practices to realize yield potential. This precise seeding and fertilizing system could adjust the seeding distance and control the amount of fertilizer being applied in terms of a predeveloped seeding and fertilizing plan (or prescription), supported by a GPS. In adjusting the seeding space, a seed releasing mechanism was automatically controlled according to the planned seeding rate and the detected planter traveling speed at the location. Meanwhile, a predetermined amount of fertilizer would be applied according to a prescription using an automatic variable-rate applicator. Yuan's prototype could control the seeding space from 10 to 20 cm with a maximum error of 4.5%. Variable-rate fertilization accuracy was within ±3.3%.

4.4.2 Variable-Rate Applications

Another critical PA operation is VRA of fertilizers, herbicides, or pesticides. Much success has been reported in developing applicators capable of implementing VRAs based on predetermined plans. While the specific designs could vary from one machine to another, the core element of variable-rate technology (VRT) normally involves an integrated sensor and rate controller system. The sensor system often includes a GPS receiver and is typically used to provide georeferenced information for setting the site-specific set points for rate controllers to deliver different rates of agrochemicals to the location. Most existing commercially available applicators use either pulse width modulated (PWM) actuated fixed orifice nozzles (PWM applicator) or fast close (FC) valve controlled variable orifice nozzles (FC applicator) to implement the VRT applications. A study of the response time of those designs showed that the PWM applicator resulted in a slightly faster response time than the FC applicator, but the latter could maintain a more stable flow rate and pressure with less application error either under sensor-based or map-based controls (Bennur and Taylor, 2010).

Applying the right amount of fertilizer in response to detected crop nutrient stress on-the-go is one of the primary implementation mechanisms in precision crop production. An on-the-go crop nutrient stress sensing system capable of self-calibrating to environment changes could provide the necessary machine intelligence to support more trustworthy intelligent variable-rate fertilization. Figure 4.13 shows an example of sensor-based intelligent applicator. In this intelligent machine, a model-based yield potential estimator based on sensed crop nitrogen stress using an applicator-mounted multispectral imaging sensor was used to determine the application rate (Noh et al., 2006). A core element in this intelligent machine is the self-calibration system capable of automatically calibrating the detected crop canopy reflectance according to the current light conditions to remove all soil background and over/underexposure canopy surfaces for a more consistent measure of canopy reflectance (Noh et al., 2005).

FIGURE 4.13 An example of an intelligent variable-rate applicator. The machine-mount crop sensing system is capable of self-calibrating to natural lighting conditions.

One of the fundamental requirements for VRA is the dose accuracy. However, there are a few factors that could make it difficult to achieve accurate dosage in field operations. For example, an uneven dose could result from some undesired sprayer boom vibrations of the extra-wide boom (Figure 4.14) caused by the sprayer traveling on uneven ground surface at different speeds or under different wind effects. Based on a study reported by Langenakens et al. (1995), the spray deposit could vary between 0% and 1000% induced by vertical boom vibrations and between 20% and 600% from the horizontal ones, which would lead to greatly reduced spray efficiency, and therefore reduced yield. The control of undesired sprayer boom vibrations is therefore essential for achieving precise VRA. Tahmasebi et al. (2012) have designed an iterative learning active force control for an active suspension system of spray boom, and reported having the potential to improve undesired vibration under given parameters and conditions.

Yet another technical challenge originates from the uneven amount of chemicals at individual nozzles across an individual applicator. An unconfirmed applicator operator's observations indicated that the difference in injected anhydrous ammonia could be over 50% of the norm between the nozzles on the same applicator, which could be caused by an imperfectly designed distribution manifold. To have sufficient

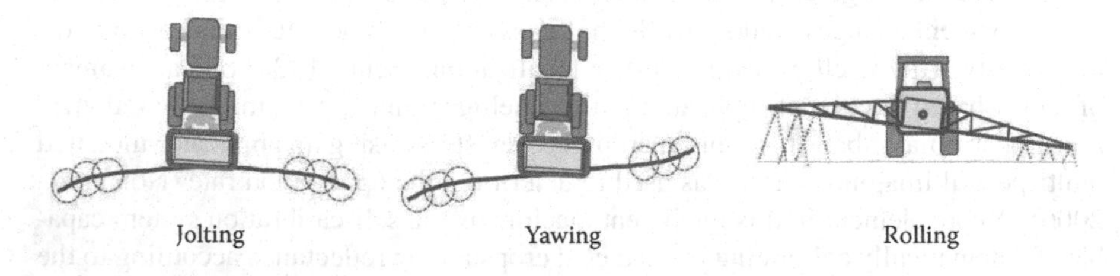

FIGURE 4.14 A few scenarios of unwanted sprayer boom vibrations or waving. (Data courtesy of Professor J. De Baerdemaeker.)

nitrates in the field, it is not uncommon for farmers to apply up to 20%–30% more fertilizer than necessary to compensate for this uneven application. An improved controllability of the ammonia application rate could effectively reduce the consumption of nitrogen fertilizer and consequently reduce nitrate leaching.

A precision irrigation system is commercially available now. The system allows spatially variable delivery of water and fertilizers to different zones in a field. One example of such a system was a microsprinkler system with individually addressable nodes developed by Coates et al. (2006) for tree fruit orchard use. One microsprinkler node, assembled with a standard microsprinkler emitter, a latching solenoid valve, and a control circuit, was installed at each tree in the orchard, and a drip line controller was used to store the irrigation schedule and issued commands to individual nodes. A master computer allowed remote access to the drip line controller using a wireless modem to update the schedule and monitor the implementation. The delivery of prescribed variable-rate water and fertilizer was implemented by operating the emitters for different durations at individual nodes.

4.4.3 Pest and Weed Control

Accurate target pest control using target sprayers is a promising pest management method for precision horticulture production, especially for control of some specific pests in tree fruit/grape production. For example, control over cutworms, a primary pest in vineyards, could be accomplished using traditional broadcasting application methods, which could use a conventional canopy sprayer, or by a targeted barrier application, which requires using a robotic self-targeting sprayer. Kang et al. (2011) have developed a robotic target sprayer for vineyard pest control (Figure 4.15). This sprayer integrated an efficient target recognition system and a rapid and precision sprayer control system to ensure an adequate coverage of pesticide on grape trunks for effectively repelling climbing cutworms to attain the crop protection goal. Field efficiency tests revealed that a targeted application in a robotic precision operation

FIGURE 4.15 A robotic target sprayer for vineyard pest control. It can use less than 10% of pesticides to achieve similar efficacy with certain pest control compared to conventional broadcast applications.

could reduce pesticide usage 90% or more compared with a traditional broadcasting application and achieve a similar efficacy.

Precise weed management using target sprayers is another promising method for reducing labor dependency, decreasing chemical inputs, improving food safety, and lowering production costs for many crops, especially for vegetable crops. Supported by auto-guidance technologies, an auto-tracked tractor could run a cultivator closer to the crop row within a centimeter precision to achieve a high speed and high efficacy in interrow (between crop rows) weed control (Han et al., 2002). As precise mechanical weed management also requires effective intrarow (within the crop row) weed control, and intrarow weeds are in general much more difficult to eliminate mechanically than the interrow weed due to their proximity to the crop rows, great effort has been directed to developing practical automated or robotic solutions for precise inter- and intrarow weed control. Such an effort requires bringing in expertise in plant sciences, engineering, and economics together to address the challenge.

A few core technical barriers to effective target application exist: a robust real-time sensing technology capable of detecting and mapping weeds and differentiating them from crops; a high-speed and high-accuracy weeding mechanism capable of removing both inter- and intrarow weeds. Numerous research projects focused on developing aforementioned core technologies have been initiated in the past few decades. Tang et al. (2000), among a few early researchers in this field, had studied the use of machine vision-based weed detection technology for applications in outdoor environments, and successfully developed a supervised color image segmentation method usable for field weed detection under natural lighting conditions of both sunny and cloudy days. Today, a few weeding robot products are becoming commercially available. For example, a research team from the University of Southern Denmark has been working with their manufacturer partners to convert their research outcomes to a commercial product for a field weeding robot by integrating control systems, tractive mechanism, and weeding tools in one mobile platform, and had made it available to farmers performing more environmentally friendly crop production (Jensen, 2013).

4.4.4 Harvest Automation

Harvest automation, from site-specific yield monitoring, operation management, and machine control to selectively harvest, is an essential operation in control of agricultural production. Since it was introduced in the early 1990s, yield monitoring has become a standard automation function for modern agriculture, and a large selection of yield monitors can be purchased either from the combine manufacturer or an independent yield monitor manufacturing company. A yield monitor, consisting of at least a grain flow sensor, a grain moisture sensor, and a GPS receiver, is simply an electronic data collection system for harvesters collecting yield data at a specific location. Collected georeferenced yield data are used either to build field yield maps for a given year or yield frequency maps over multiple years. As such spatial-temporal variation maps could be predictive of yield potential, yield monitoring and mapping is considered the starting point for implementing precision crop production management.

Efficient harvesting needs to have a harvester operating at its optimal condition. One of the most straightforward ways is to make the harvester very easy to operate so that every driver can operate the machine optimally, and machine automation is the key to providing such capability. Harvest efficiency is strongly influenced by the biological variability of the crop, which could be changed during harvest due to variations in weather, soil type, and environment. To keep a harvester operating at its optimal condition under varying field conditions, a standard practice is to automatically regulate the forward speed, which directly changes the feeding rate based on measurable variables such as engine load and grain mass flow.

The harvest efficiency could also be increased by improving the overall operation efficiency. An innovative harvest system automation technology, based on auto-guidance technology to synchronize grain carts in automated grain unloading from a working combine, has been extensively studied by both the academy and industry for filling the gap. Originated from a master–slave navigation, often using a manned combine harvester (the master) to control an unmanned grain cart (the slave) following the master at a designated angle and distance, the core of multimachine synchronization is a model of communication, which creates an in-field, high-speed wireless machine control network to facilitate synchronized speed and location control between neighboring machines. Similarly, machine synchronizing control systems have also been successfully marketed by major agricultural equipment manufacturers. This multimachine coordinating and synchronizing harvesting technology could help farmers to increase their efficiency, reduce operation costs and improve safety.

Ideally, crops should be harvested at an optimal stage for the best quality and yield. However, owing to the biological and environmental complexity involved in agricultural production, it is almost impossible for all crops to mature uniformly. Mapping and monitoring the spatial variation of crop maturity during the harvest season will provide time-critical information for farmers to selectively harvest crops at their optimal maturity. Compared to the primary area of interest being the VRA of resource inputs in large-scale grain production, selective harvest is more attractive in fruit and vegetable production as the prime quality produce will often bring in a better economic gain. Selective harvesting of fruits or vegetables is often based on a maturity or quality sensing evaluation.

4.5 SUMMARY AND DISCUSSION

The PA process can be viewed as a control process of crop production with some unique features. It offers farmers the possibility for making the best use of resource inputs for reaching the yield potential from a specific site. However, in a recent international PA forum, the world research leaders collectively identified one of the major obstacles preventing farmers from gaining promised benefits of PA as the lack of a systematic and automated method for supporting them in making optimal and trustworthy operation decisions based on the collected data.

This chapter intends to use control system theory to lay a foundation for creating some systematic methods for making optimal and trustworthy decisions for more effective precision crop production. Technology development in PA over the past 20 years has made the data collection and processing technology able to robustly obtain necessary crop and soil data for estimating yield potentials and/or monitoring actual yield on specific locations in a field. Many agricultural equipment manufacturers have made machinery capable of implementing different site-specific precision operations. This chapter introduces a new concept of formulating a systematic method for making appropriate operation decisions based on yield potential or estimated yield in a format similar to a control system. It could provide an opportunity to integrate collected information of precision crop production in supporting more consistent precision management.

It is also worth pointing out that there are fundamental differences between a typical PA management system and the one described by traditional control theory. The most important difference is the system response to the input: in PA management, the system response such as the yield of a specific site to the amount of fertilizer being applied often describes how the final output would respond to the amount of input applied to the system, and present an input–output relationship; while in conventional control theory, the system response is used to describe how fast a system is responding to the changes in inputs and/or disturbances to the system and presents a time-domain reaction. Such a fundamental difference presents the first challenge in applying control theory to create systematic methods for making trustworthy decisions for supporting profitable precision crop productions.

The second major challenge is making the collected data observable as discussed in the text, and the development of some reliable and robust yield prediction tools, which could offer a solution to this problem. The ability to predict corn yields based on collectable production-related system parameters data, such as soil property, plant morphology, and weather data, would provide a useful tool to utilize control theory in making reliable precision crop production decisions. The lack of such models is one of the obstacles preventing PA production from being effectively adopted. Owing to the attribution of collected data to various fields of sciences and technologies, a transdisciplinary study could be pivotal in developing such models.

The last, but not the least, major challenge is the uncertain controllability of the system as discussed in the text. Such uncertainty could be mainly attributed to the numerous factors that can influence the response of crop growth to resource inputs, and to the timing of applying such inputs. The application of data fusion, a process of integrating multiple data sources for obtaining more accurate and robust information to gain a more confident estimation of responses to certain inputs within a definite time window, is a possible approach to solving this problem. However, finding the solution would require transdisciplinary research.

REFERENCES

Adams, M.L., S. Cook, and R. Corner. 2000. Managing uncertainty in site-specific management: What is the best model? *Precision Agriculture*, 2:39–54.

Arslan, S. and T.S. Colvin. 1999. Laboratory performance of a yield monitor. *Applied Engineering in Agriculture*, 15:189–195.

Balenzano, A., F. Mattia, G. Satalino, and M.W.J. Davidson. 2011. Dense temporal series of C- and L-band SAR data for soil moisture retrieval over agricultural crops. *IEEE Journal of Selected Topics in Applied Earth Observations and Remote Sensing*, 4:439–450.

Batte, M.T. and M.W. Arnholt. 2003. Precision farming adoption and use in Ohio: Case studies of six leading-edge adopters. *Computers and Electronics in Agriculture*, 38:125–139.

Bennur, P.J. and R.K. Taylor. 2010. Evaluating the response time of a rate controller used with a sensor-based, variable rate application system. *Applied Engineering in Agriculture*, 26:1069–1075.

Bermudez, M. and A.P. Mallarino. 2007. Impacts of variable-rate phosphorus fertilization based on dense grid soil sampling on soil-test phosphorus and grain yield of corn and soybean. *Journal of Production Agriculture*, 8:568–574.

Carter, P.G. 2012. Agriculture Technology Adoption in Inland Pacific Northwest Dryland Crop Production. Oral Presentation at International Forum on Precision Agriculture, March 16, 2012, Richland, WA, USA.

Coates, R.W., M.J. Delwiche, and P.H. Brown. 2006. Design of a system for individual micro-sprinkler control. *Transactions of the ASABE*, 49:1963–1970.

Colvin, T.S., D.B. Jaynes, D.L. Karlen, D.A. Laird, and J.R. Ambuel. 1997. Yield variability within a central Iowa field. *Transactions of the ASAE*, 40:883–889.

Deere & Co. 2013. *Guidance Systems*. Available at http://www.deere.com/en_CAF/products/equipment/agricultural_management_solutions/guidance_systems/guidance_systems.page, Accessed on September 30, 2014.

Diker, K. and W.C. Bausch. 2003. Potential use of nitrogen reflectance index to estimate plant parameters and yield of maize. *Biosystems Engineering*, 85:437–447.

Franzen, D.W. and T.R. Peck. 1995. Field soil sampling density for variable rate fertilization. *Agronomy Journal*, 99:822–832.

Goumopoulos, C., B. O'Flynn, and A. Kameas. 2014. Automated zone-specific irrigation with wireless sensor/actuator network and adaptable decision support. *Computers and Electronics in Agriculture*, 105:20–33.

Han, S., Q. Zhang, and H. Noh. 2002. Kalman filtering of DGPS positions for parallel tracking application. *Transactions of the ASAE*, 45:553–559.

Havlin, J.L. and R.W. Heiniger. 2009. A variable-rate decision support tool. *Precision Agriculture*, 10:356–369.

Hedley, C.B. and I.J. Yule. 2009. A method for spatial prediction of daily soil water status for precise irrigation scheduling. *Agricultural Water Management*, 96:1737–1745.

Heege, H.J. 2013. Sensing of natural soil properties. In Heege, H.J. (ed.), *Precision in Crop Farming*. Berlin, Germany: Springer, pp. 51–102.

Hummel, J.W., L.D. Gaultney, and K.A. Sudduth. 1996. Soil property sensing for site-specific crop management. *Computers and Electronics in Agriculture*, 14:121–136.

Jackson, R.D. and A.R. Huete. 1991. Interpreting vegetation indices. *Preventive Veterinary Medicine*, 11:185–200.

Jensen, K. 2013. *Weeding Robot Ready for Agriculture*. Available at http://www.sdu.dk/en/om_sdu/fakulteterne/teknik/nyt_fra_det_tekniske_fakultet/radrenserrobottillandbruget, Accessed on September 28, 2014.

Jochinke, D.C., B.J. Noonon, N.G. Wachsmann, and R.M. Norton. 2007. The adoption of precision agriculture in an Australian broadacre cropping system—Challenges and opportunities. *Field Crops Research*, 104:68–76.

Johnson, P.A., F. Seeney, and D. Williams. 1996. The effect of physiological age and planting date on the response of potatoes to applied nitrogen and on levels of residual nitrogen post-harvest. *Potato Research*, 39:561–569.

Kang, F., F.J. Pierce, D.B. Walsh, Q. Zhang, and S. Wang. 2011. An automated trailer sprayer system for target control of cutworm in vineyards. *Transactions of the ASABE*, 54:1–9.

Kaspar, T.C., T.S. Colvin, B. Jaynes, D.L. Karlen, D.E. James, and D.W. Meek. 2003. Relationship between six years of corn yields and terrain attributes. *Precision Agriculture*, 4:87–101.

Lamb, D.W., P. Frazier, and P. Adams. 2008. Improving pathways to adoption: Putting the right P's in precision agriculture. *Computers and Electronics in Agriculture*, 61:4–9.

Langenakens, J.J., H. Ramon, and J. De Baerdemaeker. 1995. A model for measuring the effect of tire pressure and driving speed on the horizontal sprayer boom movements and spray patterns. *Transactions of the ASAE*, 38:65–72.

Lark, R.M., J.V. Stafford, and H.C. Bolam. 1997. Limitations on the spatial resolution of yield mapping for combinable crops. *Journal of Agricultural Engineering Research*, 66:183–193.

Levine, W.S. 1995. *The Control Handbook*. Boca Raton, FL, USA: CRC Press.

Maas, S.J. 1988. Using satellite data to improve model estimates of crop yield. *Agronomy Journal*, 80:655–662.

Massey, R.E., D.B. Myers, N.R. Kitchen, and K.A. Sudduth. 2008. Profitability maps as an input for site-specific management decision making. *Agronomy Journal*, 100(1):52–59.

McKinion, J.M., J.N. Jenkins, D. Akins, S.B. Turner, J.L. Willers, E. Jallas, and F.D. Whisler. 2001. Analysis of a precision agriculture approach to cotton production. *Computers and Electronics in Agriculture*, 32:213–228.

McKinion, J.M., J.L. Willers, and J.N. Jenkins. 2010. Spatial analyses to evaluate multi-crop yield stability for a field. *Computers and Electronics in Agriculture*, 70:187–198.

Méndez-Barroso, L.A., J. Garatuza-Payán, and E.R. Vivoni. 2008. Quantifying water stress on wheat using remote sensing in the Yaqui Valley, Sonora, Mexico. *Agricultural Water Management*, 95:725–736.

Monsanto. 2010. *Demonstration Report: Corn Hybrid Response to Different Nitrogen Rates*. Available at http://www.monsanto.com/products/Documents/learning-center-research/2010/Summary%20GL C%202010%20-%20Corn%20Hybrid%20Response%20to%20Different%20N%20Rates.pdf, Accessed on September 1, 2013.

Mulla, D.J. 2013. Twenty five years of remote sensing in precision agriculture: Key advances and remaining knowledge gaps. *Biosystems Engineering*, 14:358–371.

Noh, H., Q. Zhang, B. Shin, S. Han, and L. Feng. 2006. A neural network model of maize crop nitrogen stress assessment for a multispectral imaging sensor. *Biosystems Engineering*, 94:477–485.

Noh, H., Q. Zhang, B. Shin, S. Han, and D. Reum. 2005. Dynamic calibration and image segmentation methods for multispectral imaging crop nitrogen deficiency sensors. *Transactions of the ASAE*, 48:393–401.

Norman, M., I. Schirmer, and N.H. Hancock. 1992. Development of electronic moisture tracking for automatic control of planting depth. In *National Conference Publication—Institution of Engineers*, Australia, 92:115–119.

Ogunlela, V.B., G.L. Lombin, and S.M. Abed. 1982. Growth response, yield and yield components of upland cotton (*Gossypium hirsutum* L) as affected by rates and time of nitrogen application in the Nigerian savannah. *Fertilizer Research*, 3:399–409.

Peng, S., Q. Tang, R. Hu, Y. Liu, A. Dobermann, F. Zhang, K. Cui et al. 2010. Improving nitrogen fertilization in rice by site-specific N management. A review. *Agronomy for Sustainable Development*, 30:649–656.

Pierce, F.J. and P. Nowak. 1999. Aspects of precision agriculture. *Advances in Agronomy*, 67:1–85.

Prabhakar, M., Y.G. Prasad, M. Thirupathi, G. Sreedevi, B. Dharajothi, and B. Venkateswarlu. 2011. Use of ground based hyperspectral remote sensing for detection of stress in cotton caused by leafhopper (Hemiptera: Cicadellidae). *Computers and Electronics in Agriculture*, 79:189–198.

Price, R.R., R.M. Johnson, R.P. Viator, J. Larsen, and A. Peters. 2011. Fiber optic yield monitor for a sugarcane harvester. *Transactions of the ASABE*, 54:31–39.

Qiu, W., G.A. Watkins, C.J. Sobolik, and S.A. Shearer. 1998. A feasibility study of direct injection from variable-rate herbicide application. *Transactions of the ASABE*, 41:291–299.

Quarmby, N.A., M. Milnes, T.L. Hindle, and N. Silleos. 1993. Use of multi-temporal NDVI measurements from AVHRR data for crop yield estimation and prediction. *International Journal of Remote Sensing*, 14:199–210.

Raun, W.R., J.B. Solie, and M.L. Stone. 2011. Independence of yield potential and crop nitrogen response. *Precision Agriculture*, 12:508–518.

Robertson, M.J., R.S. Llewellyn, R. Mandel, R. Lawes, R.G.V. Bramley, L. Swift, N. Metz, and C. O'Callaghan. 2012. Adoption of variable rate fertiliser application in the Australian grains industry: Status, issues and prospects. *Precision Agriculture*, 13:181–199.

Rouse, J.W., R.G. Hass, J.A. Schell, and D.W. Deering. 1973. Monitoring vegetation systems in the great plains with ERTS. In *Proceedings of the 3rd Earth Resources Technology Satellite (ERTS) Symposium*, December 10–14, Washington DC, USA, Vol. 1, pp. 309–317.

Schueller, J.K. 2013. Agricultural automation: An introduction. In Zhang, Q. and F.J. Pierce (eds.), *Agricultural Automation: Fundamentals and Practices*, Boca Raton, FL, USA: CRC Press.

Schueller, J.K. and Y.H. Bae. 1987. Spatially attributed automatic combine data acquisition. *Computers and Electronics in Agriculture*, 2:119–127.

Shanahan, J.F., N.R. Kitchen, W.R. Raun, and J.S. Schepers. 2008. Responsive in-season nitrogen management for cereals. *Computers and Electronics in Agriculture*, 61:51–62.

Stone, J.R., J.W. Gilliam, D.K. Cassel, R.B. Daniels, L.A. Nelson, and H.J. Kleiss. 1985. Effect of erosion and landscape position on the productivity of Piedmont soils. *Soil Science Society of America Journal*, 49:987–991.

Tahmasebi, M., R.A. Rahman, M. Mailah, and M. Gohari. 2012. Sprayer boom active suspension using intelligent active force control. *World Academy of Science, Engineering and Technology*, 6:1051–1055.

Tang, L., L. Tian, and B.L. Steward. 2000. Color image segmentation with genetic algorithm for in-field weed sensing. *Transactions of the ASABE*, 43:1019–1027.

Thenkabail, P.S., R.B. Smith, and E. De Pauw. 2000. Hyperspectral vegetation indices and their relationships with agricultural crop characteristics. *Remote Sensing of Environment*, 71:158–182.

Thomasson, J.A. and R. Sui (2003). Mississippi cotton yield monitor: Three years of field-test results. *Applied Engineering in Agriculture*, 19:631–636.

Timm, L.C., L. Fante Jr., E.P. Barbosa, K. Eichardt, and O.O.S. Bacchi. 2000. A study of the interaction soil-plant using state-space approach. *Scientia Agricola*, 57:751–760.

Tremblay, N., C. Belec, and Z. Wang. 2010. Performance of dualex in spring wheat for crop nitrogen status assessment, yield prediction and estimation of soil nitrate content. *Journal of Plant Nutrition*, 33:57–70.

Uno, Y., S.O. Prasher, R. Lacroix, P.K. Goel, Y. Karimi, A. Viau, and R.M. Patel. 2005. Artificial neural networks to predict corn yield from Compact Airborne Spectrographic Imager data. *Computers and Electronics in Agriculture*, 47:149–161.

Wells, L.G., T.S. Stombaugh, and S.A. Shearer. 2005. Crop yield response to precision deep tillage. *Transactions of the ASAE*, 48:895–901.

Wendroth, O., K.C. Kersebaum, and H.I. Reuter. 2003. Predicting yield of barley across a landscape: A state-space modeling approach. *Journal of Hydrology*, 272:250–263.

Wendroth, O., D.R. Nielsen, A.M. Al-Omran, C. Kirda, and K. Reichardt. 1992. State-space approach to spatial variability of crop yield. *Soil Science Society of America Journal*, 56:801–807.

Xiang, H. and L. Tian. 2011. An automated stand-alone in-field remote sensing system (SIRSS) for in-season crop monitoring. *Computers and Electronics in Agriculture*, 78:1–8.

Xie, B., H. Li, Z. Zhu, and E. Mao. 2013. Measuring tillage depth for tractor implement automatic using inclinometer. *Transactions of the CSAE*, 29:15–21.

Yang, C., J.H. Everitt, and J.M. Bradford. 2007. Airborne hyperspectral imagery and linear spectral unmixing for mapping variation in crop yield. *Precision Agriculture*, 8:279–296.

Yuan, Y., X. Zhang, C. Wu, J. Zhang, and L. Zhou. 2011. Precision control system of no-tillage corn planter. *Transactions of the CSAE*, 27:222–226.

Zhang, N., M. Wang, and N. Wang. 2002. Precision agriculture—A worldwide overview. *Computers and Electronics in Agriculture*, 36:113–132.

Zhang, Q. and S. Han. 2002. An information table for yield data analysis and management. *Biosystems Engineering*, 83:299–306.

Zhang, Q., Y. Shao, and F.J. Pierce. 2013. Agricultural infotrnoic systems. In Zhang, Q. and F.J. Pierce (eds.), *Agricultural Automation, Fundamentals and Practices*. Boca Raton, FL: CRC Press, pp. 41–62.

5 Intelligent Agricultural Machinery and Field Robots

Shufeng Han, Brian L. Steward, and Lie Tang

CONTENTS

5.1 INTELLIGENT MACHINE DESIGN FRAMEWORK

An "intelligent machine" can be defined in a variety of ways. One line of thought is that an intelligent machine is one that exhibits the same type of behavior as a human in the same context. Such a definition requires attributes such as reasoning, perception, learning, control, and supervision to be present for machines classified as intelligent (Jain et al., 2007). Another line of thinking results in a more functional definition: an intelligent machine is one that achieves a particular goal in the context of uncertainty and variability (Rzevski, 2003; Jarvis and Grant, 2014). This definition has a lower threshold for a machine to be considered intelligent. Probably almost any automated machine in agriculture would fit this definition because of the high uncertainty and variability associated with agriculture. The current technology level of agricultural machines is somewhere in between machines being automated, since they can repeat specific tasks with a decreasing requirement of human intervention, and being intelligent with higher-level behavior than just doing specific tasks repeatedly.

Automated agricultural machines have a long history of development. Much progress was made during the 1970s when electronics for monitoring and control was introduced to agricultural machines. However, the most significant advance toward machine autonomy started in the 1990s when precision agriculture (PA) became the key driver for developing more intelligent machines. PA requires intensive management of spatial and temporal variability of fields. Therefore, automated or autonomous operation of machines becomes necessary. As an example, variable-rate application of inputs, one of the major PA practices, needs the application rate to be changed on-the-go and sometimes within every square meter of a field. Manual operation of the machine and its control is infeasible. Thus, automatic steering and map-based rate control have to be implemented on the machine. Use of small and smart machines (robots) is desired for many PA practices, such as soil sampling, crop scouting, site-specific weed control, and selective harvesting. Robotic applications are not only desirable but are also more economically feasible than conventional systems for some agriculture applications (Pedersen et al., 2006).

In building on the current state of technology toward machines that are more intelligent, conceptual frameworks have been developed to categorize the required technologies for intelligent agricultural machines. Some authors have thought of these categories as a set of building blocks for agricultural machines (Reid, 2004). However, an alternative framework with different technology layers naturally dependent upon one another may be helpful. In this chapter, a framework, consisting of four layers that tend to build on each other, will be used (Figure 5.1). These layers

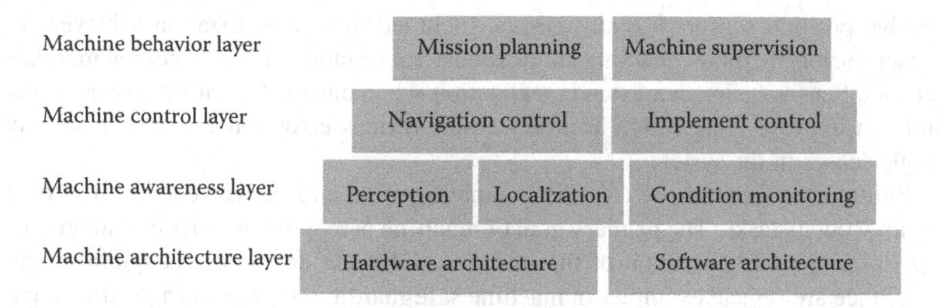

FIGURE 5.1 A multilayer design framework for intelligent agricultural machines.

are, starting from the bottom, (1) machine architecture, (2) machine awareness, (3) machine control, and (4) machine behavior.

At the lowest level, the machine's system architecture, consisting of both hardware and software, must be in place to build the higher-level layers of an intelligent machine. Since the machine must interact with the physical world, physical hardware architecture must be in place. For an intelligent agricultural field machine, or field robot, the hardware must enable mobility within the crop field, as well as provide the capability to perform field operations in an automated or intelligent manner. The hardware architecture must be mechatronic to support intelligent operations; that is, be an integration of mechanical, electrical and electronic, fluid power, and computational systems. The necessary interconnections between systems must be included to communicate both data and power. Other hardware that must be present are the sensors that transduce physical or biological signals into electrical signals, and actuators that provide force and motion to interact with the crop or the environment.

Complementary to the hardware architecture, software and communications architectures must also be in place so that the development of higher-level layer technology can be built on preestablished software components enabling communication and reusing lower-level computational solutions. The ISOBUS standard (ISO 11783), for example, has enabled major technological advances in agricultural equipment. This standard was released over the period 2007 to present, and has had a major impact on the agricultural machinery industry enabling electronic control units (ECU) from different manufacturers to communicate with each other and generic virtual terminals to serve as user interfaces. The impact of the ISOBUS standard on the current state of agricultural automation cannot be overstated. Other examples of software architectures include the robot operating system (ROS) and the joint architecture for unmanned systems (JAUS), among others, which will be described further in Section 5.7.

The next layer, machine awareness, is built on the machine architecture layer. This layer mainly consists of localization and perception technologies. In field robotics, localization is often accomplished through the global navigation satellite system (GNSS) with inertial sensors. However, many agricultural applications require the machine to follow existing crop rows. In this case, machine localization using relative position sensors has advantages. Included in this localization sublayer are sensor fusion methods enabling more robust localization through complementary sensors. Sensor fusion can extend localization when one of the sensor signals is lost and can improve localization accuracy when various error sources exist from any single sensor in the system.

Before a machine can be classified as intelligent, it must perceive its environment to carry out its tasks. The primary goal of machine perception is machine safeguarding to ensure safe operation of the machine. Obstacle detection, recognition, and avoidance are typical examples in machine safeguarding. Perception algorithms and strategies are built on top of the perception sensors in the hardware architecture to achieve safeguarding functions.

Agricultural machines are designed to accomplish field operations such as tillage, seeding, fertilizer and chemical applications, cultivation, and harvest. During these operations, the machine must interact with crop, soil, field topography, and weather conditions. Thus, for machine intelligence, perception systems must support machine awareness of these factors. Because agricultural machines are operated in unstructured environments and interact with highly variable bioproducts, perception system development is challenging. Section 5.3 describes different perception sensors and sensor selection.

Another aspect of machine intelligence needing consideration is the condition of the machine itself, which to increase machine autonomy, must be monitored. In a human-operated machine, the operator is not only controlling the operation of the machine, but also monitoring the machine through visual, audio, or vibration cues to ensure that the machine is functioning correctly. Thus, machine health awareness is necessary, along with machine supervision. Section 5.4 outlines possible approaches to machine health awareness.

Once the machine is aware of its location, environment, and health, the machine control layer must be in place to navigate the vehicle through the field and to control the implements to accomplish the field operations. Navigation control of agricultural machines is highly developed and has progressed through several generations of automatic guidance technologies as applied to conventional agricultural vehicles and implements. However, for smaller, next-generation field robots, research questions exist since several vehicle platforms provide additional degrees of mobility freedom, through independent four-wheel steering (4WS) and four-wheel drive (4WD), that can be utilized for novel navigation control strategies (see Section 5.6).

Implement control has also been implemented commercially for various machine operations. For example, in the case of liquid chemical application, chemical application rate control was first developed and commercialized in the late 1970s, upon which variable-rate application systems were developed in the 1990s. Since that time, more and more aspects of machine operations are controlled such as individual spray nozzles, boom sections, boom height, planter row unit, plant population, harvester feed rate, and harvester header height. More details are provided in Section 5.6.

The highest layer of the design framework, machine behavior, includes mission planning and machine supervision. Mission planning includes the optimization of vehicle or implement path based on criteria such as the shortest time to accomplish a given field operation. Some mission planners will also optimize machine functions associated with the vehicle path, such as vehicle speed, implement position, and power-take-off (PTO) speed. Mission planning can be accomplished off-line prior to mission execution, or it can be done on-line, leading to adaptations of the mission based on new field and crop condition information being perceived during mission execution. Examples of mission planning include path planning, vehicle routing, and machine coordination, are further described in Section 5.5.

Machine supervision is similar to mission planning but focuses more on machine conditions and behaviors required in reaction to unforeseen events. In response to machine conditions in which either machine or environmental states are outside expected conditions, machine supervision will control the machine to a fail-safe condition. For example, if an autonomous vehicle detects a moving obstacle in front of its planned trajectory, the machine supervision algorithm will decide if the vehicle should stop, wait, or take a detour. As a first step of machine supervision, human operators will monitor the automated machines, but with continued development, increasing amounts of machine supervision will be done via higher-level intelligent supervisory control.

In addition to the technology layers mentioned above, this chapter also discusses classification of intelligent machines (Section 5.2), examples of autonomous vehicles and field robots (Section 5.8), and summary and discussion of future directions (Section 5.9).

5.2 INTELLIGENT MACHINE CLASSIFICATION

Machine intelligence and automation technology found in agriculture are varied. Thus, to engage in a focused discussion about an intelligent machine, one must find a way to classify various machine systems. At the highest level, intelligent machines in agriculture can be classified according to the agricultural production systems in which the machines are used. Agricultural production systems include irrigation systems, animal facilities, fruit production systems, greenhouses, and field machinery (Figure 5.2).

In this classification scheme, the automation of irrigation systems is used to improve water use efficiency. Generally, automation technologies will site-specifically vary the rate at which water is applied to the crop based on current crop and soil status. This irrigation automation strategy is called variable-rate irrigation and is accomplished by either varying the speed that the irrigation system is passing over the crop or by controlling the flow rate of nozzles (LaRue, 2014).

Automation technology is also used in animal facilities. One application is maintaining indoor animal environmental variables such as temperature, humidity, and gas concentrations at levels that maximize feed conversion efficiency and maintain animal health and welfare (Purswell and Gates, 2013). Feed distribution systems can also be automated to control and monitor feed for individuals or groups of animals (Aerts et al., 2003; Frost et al., 2003; Tu et al., 2011). Robotic or automatic milking systems (AMS) also fit in this automation class. AMS automatically harvest the milk from dairy cows without the need for human labor traditionally associated with milking. These systems are being adopted rapidly in North America and Europe (de Koning and Rodenburg, 2004; Lely, 2014) and are changing dairy production

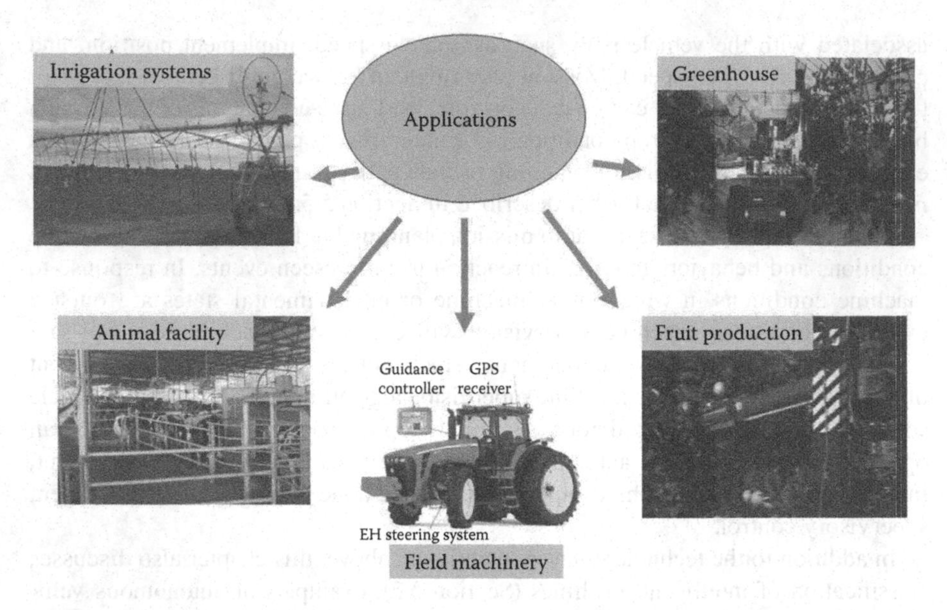

FIGURE 5.2 Intelligent machine classification based on the production systems. (From Edan, Y., S. Han, and N. Kondo. 2009. Automation in agriculture. *Springer Handbook of Automation*, 1095–1128. With kind permission from Springer Science+Business Media.)

in many ways—including the role of the farmer, dairy management systems, and farmer and dairy cow relationships (Butler et al., 2012).

Greenhouse plant production systems, another class of agricultural automation systems, have numerous environmental variables that can be controlled to optimize plant growth and health. These variables include air temperature, relative humidity, light intensity, and CO_2 concentration. Many control strategies exist that may use artificial intelligence and physiological plant growth models (Ferentinos, 2006). In this class, we could also include robots designed for greenhouse use.

In developed countries, because of limited labor availability and high labor costs, automation and mechanization technologies are being rapidly developed and adopted for fruit production in orchard crops. Particular cultural practices developed include automated pruning and hedging systems, fruit thinning, precision chemical application, and harvesting. Because robotic fruit harvesting is particularly challenging yet has potential for substantial impact to fruit production, research and development efforts have been undertaken in this area. Technologies have also been developed to monitor tree crops (Burks et al., 2013).

The last major class in this high-level classification is the automation for field crop production, including row crops such as corn, soybeans, and cereal grains. This class has seen much development as it was one of the early foci of PA research, particularly in North America and Europe. Characteristic of this type of agriculture are large machines and large field sizes.

Within this high-level classification scheme, we can further classify automation technology according to the machine or operation being automated. Many automation technologies that have been commercialized and made available to agricultural producers are in this category. These technologies are found in tractors and implements, combine harvesters, chemical applicators including sprayers, and planters.

The main focus of this chapter is on machine intelligence as applied to field crops and orchard crops. Technologies that are currently applied to agricultural machines, which generally tend to be larger equipment, will be surveyed as well as field robots, which are generally at the research stage of development. A main consideration for the development of machine intelligence for field and orchard crops is the semi-structured or semicontrolled nature of the operating environment, as well as the crop, which presents many challenges to the development of this type of technology. Because of these challenges, the automation of machinery in agricultural fields has been, until recently, a slow evolution, starting first with the automation of machine function on conventional field machinery.

5.3 PERCEPTION SENSING TECHNOLOGIES

Human perception is the organization, identification, and interpretation of a sensation in order to form a mental representation (Schacter et al., 2011). In this definition, two aspects of human perception are involved: sensing of the environment (sensation) and interpretation of the sensory information (mental representation). A conventional machine requires the human operator to perform both of these perception tasks to ensure its safe operation in the field. An intelligent machine, however, is equipped with sensors and processors to achieve some level of perception to reduce human intervention or to even completely eliminate the human operator.

In general, a perception system for an intelligent mobile machine requires one or more of the following capabilities: localization (where the machine is relative to the world), object recognition (what is around the machine), navigation and collision avoidance (how the machine can safely interact with the environment), and learning and inference (how the perception system can solve new problems). Abundant literature relates to these topics. For intelligent agricultural vehicles, navigation and safeguarding (obstacle detection) are two of the most important tasks in field operations. This section discusses perception sensors and their selection for agricultural applications primarily in vehicle navigation and vehicle safeguarding.

5.3.1 PERCEPTION SENSORS

Human beings receive stimuli detected by our five senses: sight, hearing, taste, smell, and touch. Accordingly, perception sensors have been developed in each of these sensing categories, for example, vision sensors as sight, acoustic sensors as hearing, and tactile sensors as touch. However, modern perception sensors can respond to environment stimuli in the electromagnetic spectrum at a much wider range than a human being can (Figure 5.3). Each type of perception sensor in the electromagnetic spectrum will be briefly discussed below.

5.3.1.1 Monocular Vision

Although cameras were invented and used in photography centuries ago, their industrial application as perception sensors did not start until the 1960s and 1970s when

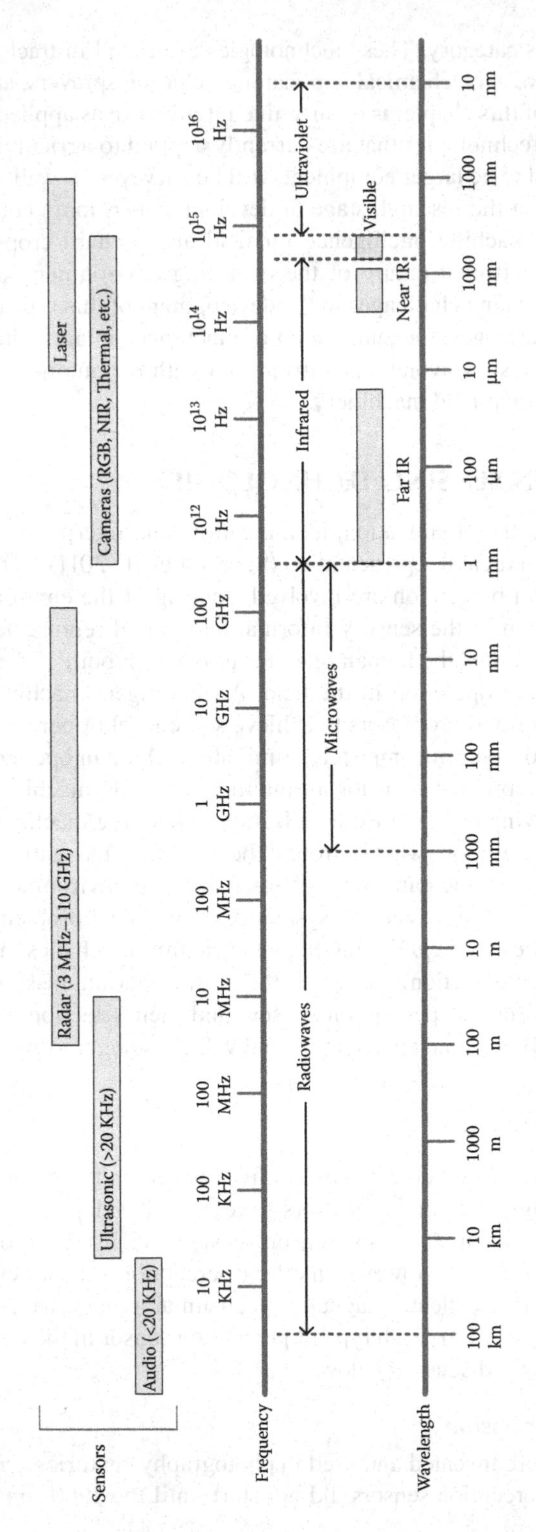

FIGURE 5.3 Electromagnetic spectrum and potential perception sensors.

Larry Roberts and David Marr undertook breakthrough research at MIT's Artificial Intelligence Laboratory. Computer vision, or machine vision, the science of teaching a computer how to identify a physical object in its surroundings, was born during that time. In the 1980s, computer vision took off and saw great expansion with its mass adoption by semiconductor manufacturers. Presently, two technologies can be used for the image sensor in a camera: charge-coupled device (CCD) and complementary metal–oxide semiconductor (CMOS). To produce a color image, a filter in front of the image sensor allows the sensor to assign color tones to each pixel. Traditionally, a CMOS camera is less expensive and consumes less power, but a CCD camera produces better image quality. In recent years, however, these differences have disappeared.

The success of a perception application using image sensors is heavily dependent on image processing algorithms. Grayscale machine vision algorithms have been widely investigated (e.g., optical flow, motion detection, and pattern recognition), but at best, the results have been mixed. The main difficulty is that computer vision algorithms are almost all brittle; an algorithm may work in some cases but not in others (Huang, 1996).

Agricultural applications of computer vision were first studied in the late 1980s. Typical applications include guiding a tractor for row crop cultivation, or guiding a combine for harvest operation. In such applications, finding guidance information from row crop structure is the key to achieving accurate control of a vehicle. A number of image processing techniques have been investigated to find the guidance line (directrix) from row crop images. As examples, Reid et al. (1985) developed a binary thresholding strategy using Bayes classification to effectively and accurately segment crop canopy and soil background for cotton at different growth stages. Gerrish et al. (1985) concluded in that the combination of noise filtering, edge detection, thresholding, and rescaling was the most promising technique. Image analysis using the Hough transform to find crop rows was reported in several studies (Marchant and Brivot, 1995; Marchant, 1996). Billingsley and Schoenfisch (1997) reported on a vision guidance system relatively insensitive to additional visual "noise" from weeds, while tolerating the fading out of one or more rows in a barren patch of the field. They showed their system is capable of maintaining an accuracy of 2 cm. In terms of vehicle safeguarding using a monocular vision system, the most successful application is perhaps the lane departure warning system in the automobile industry (e.g., Mobileye, 2015; TRW, 2015). No literature has been reported for agricultural vehicle safeguarding.

A monocular vision system can provide rich information, including color and shape of objects. The cost is low. It can be easily integrated onto a vehicle due to the small footprint. However, it is not robust to illumination variance and cluttered background.

5.3.1.2 Stereo Vision

Stereo vision may be passive or active. In a passive stereo vision system, two or more cameras are used to acquire different images of the same object from slightly different viewpoints in space. The depth information of the object can be calculated by the differences in these monocular views of the scene and by the geometry of the imaging system. In an active stereo vision system, one of the cameras is replaced by a projector, which projects structured light, such as parallel lines and grids, onto the object surface. The structured light is distorted by the object geometry, and a new distorted pattern is formed. An object's three-dimensional (3D) shape can be recovered by analyzing images containing the distorted light pattern. Active stereo vision has been successfully used for applications in constrained indoor environments such as industrial inspection. However, the short detection range and the vulnerability to strong ambient light make it hardly useful in outdoor environment.

The most challenging task in using stereo vision to determine the 3D depth information of objects is stereo correspondence matching—finding pairs of matched points corresponding to a single point on the 3D object. Stereo matching is one of the most active research areas in computer vision. Two general methods for stereo matching are intensity-based and feature-based. The intensity-based approach attempts to establish correspondence by matching pixel intensities of the image pair. With the feature-based approach, features such as edges, corners, lines, and curves are first extracted from the images, and the matching process is applied to these features. The selection of the best matching algorithm depends on the applications but unfortunately most algorithms are not robust for outdoor applications.

Several studies of stereo vision for agricultural vehicle navigation have been reported. For example, Kise et al. (2005) developed a stereo vision-based crop row detection system to automatically navigate a tractor in a soybean field with a lateral deviation of less than 0.05 m at speeds up to 3.0 m/s. They used stereo images to create an elevation map (i.e., a map of crop height). Since the search of the guidance parameter was based on the elevation map, not on color or intensity, their algorithm was robust under weedy field conditions. Wang and Zhang (2007) developed a stereo vision-based trajectory tracking method for automated navigation of an agricultural vehicle in an unstructured environment based on 3D feature tracking and motion estimation. Recently, Lin et al. (2014) reported an object tracking and collision avoidance system utilizing a stereo vision system. For vehicle safeguarding applications, Wei et al. (2005) tested the safeguarding capability of a stereo vision system using a person standing in front of a vehicle as the potential obstacle. Obstacle detection in short-ranges (less than 12 m) was repeatable. In other research, Rovira-Más et al. (2007) showed that, in real-time applications, ranges up to 15 m can be sensed with acceptable accuracy using compact off-the-shelf binocular stereo cameras.

A stereo vision system has an advantage over a monocular vision system. It more effectively represents distance, size, and spatial relationships between different objects in the camera's field of view. It is less sensitive to changing external environments because it relies on size, shape, and distance, which are invariant under lighting changes. However, the lack of robust stereo matching algorithms and the high computation cost requirement have historically made stereo vision costly and impractical. Recently, low-cost stereo processors are more available on the market. Stereo vision systems are expected to replace monocular vision systems in the future.

5.3.1.3 Laser, Ladar, and Lidar

A laser is a device that emits light through a process of optical amplification based on the stimulated emission of electromagnetic radiation (Gould, 1959). A laser system used for perception purposes is often based on the time-of-flight (ToF) principle.

The ToF method measures the time for the electromagnetic wave to travel to a target and back. The distance (range) is calculated as half of this time multiplied by the velocity of the wave. When a laser is used as the probe, the device is called Ladar or Lidar, which stands for laser detection and ranging or light detection and ranging, respectively. The light is short-pulsed for time measurement.

Ladar or Lidar are also called laser range finder. A laser range finder can only measure the distance to a single point on the object. For 3D object recognition and object modeling, the laser beam is often rotated, either by mechanical or by optical methods, to achieve two-dimensional (2D) or 3D range measurements. Those systems are called laser scanners or scanning lasers. One of the commonly used scanning laser brands in agricultural applications has been SICK (SICK AG, Waldkirch, Germany), primarily due to their performance and affordable cost. For example, a SICK outdoor scanning laser range finder, LMS151, is a long-distance measurement type series (at 75% reflectance) with a 50 m maximum measuring length. On the high end, Google has been using a roof-mounted Lidar (HDL-64E, Velodyne Acoustics, Inc., Morgan Hill, California), which spins a unit containing 64 fixed-mounted lasers to capture a full 360° horizontal field of view. However, agricultural use of this sensor has not been reported due to its high cost.

Lidar has become well recognized in terrain model building since the late 1990s. It has advantages in measuring surfaces with accuracy and density (Ma, 2005). In agricultural applications, Ryo et al. (2004) used a laser scanner to automatically guide a robotic gator in an orchard. They concluded that control by the laser scanner was more accurate and stable than control by a global positioning system (GPS) and an inertial measurement unit (IMU). Lateral and heading error were 0.1 m and 0.7°, respectively.

The major strengths of Lidar include accurate 3D shape information, accurate 3D position, and performance independent of varying illumination. However, Lidar does not produce color information, is dependent on weather conditions, and is expensive.

5.3.1.4 Radar

Radar stands for radio detection and ranging and uses radio waves in the range of 3 MHz to 110 GHz (Figure 5.3) reflected from the surface of an object to determine the range, direction, and speed of the object. Radar signals are reflected especially well by materials of considerable electrical conductivity—especially by most metals and by wet ground. Radar can provide accurate distance information but no shape information. As such, it can be primarily used for object detection but not object identification. It is well suited for safeguarding applications.

Recent advancement in silicon germanium (SiGe) technology has made the high-frequency millimeter-wave applications practical. Automotive radar (77 GHz) is now readily available at a low cost. The 77 GHz sensor may soon be integrated into agricultural vehicles for safeguarding applications.

5.3.1.5 Ultrasonic Sensor

Ultrasonic sensors work on the ToF principle similar to Lidar, but they use sound waves at a frequency just above the range of human hearing (>20 kHz, Figure 5.3). Most ultrasonic sensors operate at frequencies between 40 and 250 kHz. Because the speed of sound travels much slower than the speed of light, an ultrasonic range sensor has a much shorter object detection range than Lidar. The detection range is typically less than 10 m.

Ultrasonic sensors were studied to detect a moving object in the vicinity of agricultural machinery in real time under an outdoor environment (Guo et al., 2002). They were also used to measure the relative position between tree canopies and a vehicle for tractor navigation in an orchard (Iida and Burks, 2002). Because the speed of sound changes due to variations in air temperature and humidity, ultrasonic sensor measurement errors tend to be large. Thus, ultrasonic sensors alone may not be a good choice for vehicle navigation and safeguarding. However, they can be used as complementary sensors due to their low cost. One potential application is using an ultrasonic sensor array to safeguard a slow-moving automated vehicle.

5.3.1.6 Active 3D Range Camera

Active 3D range cameras share several traits with both scanning lasers and cameras. Like lasers, they measure distances with modulated light based on the ToF principle. Similar to cameras, distance measurements are obtained with a 2D array of pixels without any moving parts. The photonic mixer device (PMD) is one of the promising technologies for active 3D range cameras (Schwarte et al., 1998; Xu et al., 1998; Ringbeck and Hagebeuker, 2007). Recently, PMD-based cameras have been developed that are compact, affordable, and capable of capturing reliable-depth images directly in real time. The technology has been successfully demonstrated for tracking people in surveillance applications (e.g., Gokturk and Tomasi, 2004; Grest and Koch, 2007).

Active 3D range cameras are a competing technology with stereo vision-based surface reconstruction. Under optimal conditions, the PMD system outperformed the stereo vision system in terms of achievable accuracy for distance measurements (Beder et al., 2007). Agricultural applications of PMD cameras include classification of plants (Klose et al., 2009) and mapping of apple trees for automatic pruning (Adhikari and Karkee, 2011). However, no research was found investigating PMD cameras for vehicle navigation and safeguarding in agriculture.

5.3.2 Selection of Perception Sensors

All the perception sensors mentioned above have been studied for agricultural applications. However, no single type of sensor has been shown to have clear advantages or disadvantages over the others. Compared with indoors manufacturing applications or even outdoors automobile applications, agricultural field operations are exposed to a much more challenging environment for perception sensors. Many factors need to be considered in the selection of perception sensors in agriculture. The following are some of the major performance and cost requirements for perception sensors.

5.3.2.1 Range

The requirement for range, or detection distance, depends on the type of application. In the case of perception-based vehicle navigation, only a small look-ahead distance is needed for straight crop rows, but a larger look-ahead is desired for curved rows at higher speed. All the sensors discussed above should meet the range requirement for navigation applications. For vehicle safeguarding, a safe distance—the distance from the vehicle to the detected obstacle—is a function of vehicle speed, system response time, and vehicle stopping distance. Gray (2002) stated that the obstacle detection sensor used on the tractor must have a maximum detectable range of at least 15 m. Most of the sensors discussed above, except the ultrasonic sensors, can meet this range requirement.

5.3.2.2 Robust to Lighting

Unlike manufacturing operations under controlled environments, agricultural operations are normally performed under all kinds of ambient lighting conditions, such as dim light in the early morning and evening, bright sunlight during the day, overcast sky, partial clouds, or even complete darkness during night operations. Radar and ultrasonic sensors are not affected by lighting conditions since they use radio and ultrasonic waves, respectively. On the other hand, vision sensors (mono and stereo) are highly sensitive to changing lighting conditions since they operate in the visible light spectrum, which is directly affected by the ambient lighting.

5.3.2.3 Robust to Dust

Agricultural machinery is often exposed to very dusty conditions in field operations. Almost all the perception sensors based on the ToF principle will return some false "echoes" from dust particles. If the sensor is able to penetrate those dust particles, it is more robust to dust. The dust penetration capability of a sensor depends on the type of wave, power, etc. In general, vision sensors are the most robust sensors to dust, if an appropriate image processing algorithm is implemented. Lasers are most sensitive to dust.

5.3.2.4 Spatial Resolution

Spatial resolution refers to the size of the smallest possible object or feature in space that can be detected and identified. For image sensors, the spatial resolution is related to the pixel size of the imager. For Lidar sensors, the spatial resolution is dependent on the point density, which is a function of the scanning frequency. The spatial resolution of the system is also dependent on the instantaneous field of view (IFOV) of the sensor. Thus, an image sensor mounted at a lower height will give a better spatial resolution of the ground surface than the same sensor mounted higher. Monocular vision provides the best spatial resolution, which is very helpful in applying feature-based algorithms for object identification. Stereo vision and Lidar are often used in parallel with monocular vision through point cloud rendering to provide better object identification capability. Both radar and ultrasonic sensors have the lowest spatial resolution. They are commonly used only for object detection, not for object identification.

5.3.2.5 Maintenance

Reliability is essential for perception-based applications. Sensor performance will quickly degrade when exposed to harsh agricultural environments, and frequent sensor maintenance is required. The ease of sensor maintenance is another consideration in selecting a perception sensor. A sensor with an inherently stable design will require less maintenance. Monocular camera, radar, and ultrasonic sensors all require low maintenance due to their small package size and simple design with no moving parts. Vibration may cause changes in the relative orientation between two stereo pairs leading to frequent calibration of stereo vision sensors. Lidar sensors have internal moving parts, which are vulnerable to sensor failure—making it hard to maintain long-term operation in agricultural fields.

5.3.2.6 Cost

The range of costs for perception sensors is wide. Cost ranges from over $100,000 for a high-end Lidar to under $100 for a monocular vision sensor. Cost is often a limiting factor for agricultural perception applications. The cost target depends on the application, and is often defined by the ratio of the sensor cost to the machine cost. A high-end machine such as a large tractor can tolerate a higher cost for the add-on perception system.

Table 5.1 summarizes the ratings for each of the six types of perception sensors. These ratings can be used for the selection of particular sensors for specific applications. The conclusion is that no single type of perception sensor can meet all the requirements of agricultural operations. Compromise is required to find a sensor that can meet most of the requirements. Another approach is to use multiple low-cost sensors, either redundant or complementary to each other, in a sensor-fusion system.

5.3.3 Challenges and New Development

Since the late 1970s, industrial robots have been widely used in manufacturing operations for fixed automation. Such robots can perform repetitive tasks in a carefully controlled environment, and the perception needs for these robots can be kept to a minimum. Machine automation in agriculture is significantly different from automation in the manufacturing industry. An automated agricultural machine needs the ability to sense its world and change its behavior on the basis of what it perceives. The workspace of agricultural machinery is typically a large open field, unstructured,

TABLE 5.1
Perception Sensor Ratings

Capability / Sensor	Monocular Vision	Stereo Vision	Lidar	Radar	Ultrasonic	3D Range Camera
Range	Average +	Average –	Good	Good	Poor	Average –
Robust to lighting	Average –	Average +	Good	Good	Good	Good
Light dust penetration	Good	Good	Average	Good	Average +	Average +
Heavy dust penetration	Poor	Poor	Poor	Good	Poor	Poor
Spatial resolution	Good	Average	Average +	Poor	Poor	Average –
Maintenance	Good	Poor	Poor	Good	Good	Average
Cost	Good	Average	Poor	Average	Good	Poor

and with large topographical change. Numerous types of obstacles in the workspace are often unknown *a priori* and hard to classify. In addition, agricultural machinery is exposed to harsh environmental conditions, such as extreme weather, extreme temperature, dust, rain, and vibration. As such, operating agricultural machinery without human operators becomes extremely difficult. Agriculture has been one of the last industries to use robotics and intelligent machines mainly because of the lack of machine perception capabilities.

Machine vision is perhaps the most promising technology for agricultural perception applications because it can provide a vast amount of data at a relatively low cost. However, many challenges exist in the interpretation of these data. To derive shape information from images, the commonly accepted bottom-up framework developed by Marr (1982) is being challenged, as it has limitations in speed, accuracy, and resolution. A new approach, called purposive vision (Aloimonos, 1992), has been suggested. The purposive vision paradigm does not attempt to generate a complete, detailed, symbolic 3D model of the environment. Rather, it is task-oriented and focuses only on the parts of the environment relevant to its task. Purposive vision complements general machine vision techniques with domain-specific information. Collision avoidance for autonomous vehicle navigation is an appropriate application of the purposive vision approach because precise obstacle shape description is unnecessary.

Future growth in machine vision is likely with smart camera technology. A smart camera is a stand-alone vision system with a built-in image sensor and processor. It is capable of extracting application-specific information from captured images, along with generating event descriptions or making decisions that are used in an intelligent and automated system (Belbachir, 2010). Stereo vision capabilities can also be built in the smart camera due to its processing power. Smart cameras have just recently become small and affordable enough to justify their use for agricultural machine automation. Several commercial applications of smart cameras in agricultural machinery have already been developed.

Autonomous machines cannot be commercialized without proper safeguarding. Safe operation of agricultural machines is the single most critical requirement. Since no single type of perception sensor can meet all the requirements of agricultural operations, development of sensor fusion systems for machine perception will continue to be a major effort. Sensor fusion developments include not only the selection of multiple low-cost sensors, but also appropriate sensor fusion algorithms.

5.4 MACHINE HEALTH AWARENESS

With any machine, with or without intelligence, failures or breakdowns will occur. Thus, monitoring machine conditions may lead to preventative maintenance before a catastrophic failure occurs or corrective action after a malfunction has occurred. Because of increasingly higher-level complexity with automated machinery, it has become more difficult for human operators to detect faults. In addition, accompanying machine automation may be lower-skill operators without the background knowledge to manually diagnose machine problems. Automation technology can also create conditions where the operator is more isolated from the working of the machine, either by physical isolation (e.g., a more comfortable cab with climate control and more sound isolation) or by greater distraction as the operator's attention has moved to other farm management tasks such as marketing produce or making decisions about input purchases.

As agricultural machinery becomes more automated, machine condition monitoring, along with fault detection and diagnosis, also needs to be more automated, although it can still have some reliance on human intervention when a human operator is present. For driverless, autonomous agricultural machines, machine intelligence to monitor machine health must be in place to produce machine health awareness with no human assistance—a considerable requirement for the development of these machines. Little attention has been given to the automation of condition monitoring and fault detection system in agricultural machinery (Craessaerts et al., 2010; Khodabakhshian, 2013).

Machine health awareness requires a high degree of intelligence, perhaps higher than all other requirements for an intelligent machine. At the heart of machine health awareness are technologies often referred to as condition monitoring systems or fault detection and diagnosis systems. Condition monitoring is typically part of an overall maintenance strategy for a process, machine or machine system, which will involve a human manager. Condition monitoring uses signals from a machine acquired with sensors to provide some indication of the condition of machine components. Based on these signals and their changes over time, with some signal processing and pattern recognition analysis, managers can make decisions about what maintenance interventions should be taken and when they should be scheduled.

As shown in Figure 5.4, implementing a machine health awareness system for an autonomous machine requires several layers of technology, which can be structured in a format similar to the intelligence machine framework presented in Section 5.1. For machine health awareness, there must first be a hardware layer consisting of sensors that are measuring physical signals known to be related to machine component condition. Several sensing modes have been used for condition monitoring and will be described in greater detail below.

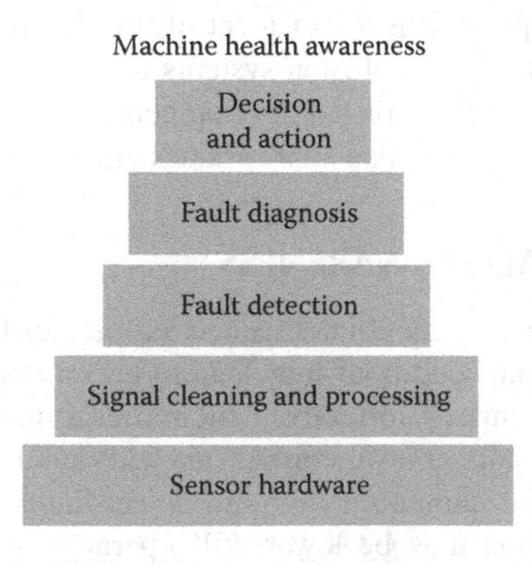

FIGURE 5.4 Machine health awareness technology will require at least these five layers of technology as a part of any autonomous, driverless system.

Second, the signals from the sensors must be cleaned to remove outliers and then processed to extract the features that are correlated to machine component conditions. Next, fault detection applies automatic pattern recognition processes to determine if a fault has occurred in the system. Generally, this step involves finding a deviation from patterns associated with normal operation. Once a fault has been detected, it must be diagnosed to identify what the fault is and what might have caused it.

The last layer might be the most important for an autonomous machine, that is, to decide what action should be taken next and then execute it. Several possible actions can be taken when a fault occurs, including (1) initiate a graceful shutdown and remain at current position, (2) stop operations and move to a designated location for maintenance, (3) stop operations, alert remote human supervisor for further instructions, or (4) continue operations, and send a warning message to human supervisor. Blackmore et al. (2002) identified six safety modes similar to those listed above.

Khodabakhshian (2013) surveyed the sensing modes found in the literature for machine condition monitoring and discussed those used for agricultural machines. Temperature measurement can be used to detect increased friction in bearings that are moving into a failure mode, but it is not typically applied to agricultural machines. Dynamic monitoring includes analysis of vibration signals or acoustic signals associated with rotating machines and relating vibration signatures to wear and machine life. While substantial work has been done in this area for rotating machinery in general, with ISO standards developed for it (ISO 13373-1:2001; ISO 17359:2003), not many efforts specifically for agricultural machinery have been reported. Exceptions include Heidarbeigi et al. (2009, 2010) who sought to diagnose faults in a Massey–Ferguson gearbox with the power spectral density of the vibration signals.

Another approach to condition monitoring is monitoring internal wear debris or particle contamination of oil. While light blockage particle sensors are available, they are typically not applied directly to off-road machines because of cost and robustness limitations. Typically, this approach involves sampling oil from the machine being monitored and oil analysis done in a laboratory setting. This approach does not lend itself to automated machine health awareness. However, recent investigations into dielectric spectroscopic sensing technology for oil contaminants have produced an on-line sensor to be used continuously during machine operation (Kshetri et al., 2014). This type of robust sensor technology could be applied to intelligent agricultural machines.

Another promising approach to sensing the condition of a machine is to monitor machine performance variables, such as power consumption or hydraulic pressure, searching for anomalies in those dynamic variables. Craessaerts et al. (2010) took this approach applying self-organizing maps and neural networks to detect failures in a New Holland combine harvester.

The application of these technologies to intelligent agricultural machines has limitations. They are more easily applied to rotating machinery such as planters, engines, and grain harvesters but not suitable to machines involving lateral motion or limited rotational motion. They are also better suited for more controlled machine operating environments found in factories rather than in fields. While some agricultural machine monitoring technologies exist (e.g., planter monitors and combine yield monitors), development has focused typically on machine performance in affecting agronomic factors such as plant population or harvester crop loss. Little has been done to more broadly monitor agricultural machine health.

For autonomous operations, the correct operation of the implement must also be monitored. If anomalies occur, the machine must self-correct or transition to an appropriate fail-safe mode, and the remote operator must be notified to take further corrective action. A good example of this type of technology is planter monitors in which, if a seed tube is plugged, the planting operation must be stopped immediately and corrected.

New technologies may be on the horizon to more broadly monitor the health of agricultural machine implements. For example, the German auto supplier Continental AG (Hanover, Germany) recently introduced a surround view system, ASL360, to create a fully 360°, 3D bird's-eye view of the vehicle and its surrounding local environment. The system consists of four fisheye cameras mounted on the front, rear, and sides of a vehicle and a special processor that stitches four images into a single 3D image. Although the system was originally intended for automotive applications such as assisted parking and safe maneuver, it can be easily adapted for agricultural applications (Continental AG, 2013).

5.5 MACHINE BEHAVIOR

5.5.1 Robotic Behavior

For a robot to be considered intelligent, it must exhibit behaviors similar to those observed of humans. Some of these behaviors include planning by determining the best plan of action to achieve a particular goal and supervision by monitoring the work environment and making modifications to the planned actions based on new information.

Blackmore et al. (2007a,b) promoted a structure for defining the behaviors field robots need to perform agricultural operations autonomously. At the highest level, a field operation is the action that a robot will carry out to meet the needs of a crops' cultural practices. Within an operation, certain tasks must be carried out—either deterministic or reactive. Deterministic tasks can be planned before the operation starts, are goal-oriented to achieve the objective of the operation, and can be optimized to best draw on the resources available. Reactive tasks are foreseen responses to uncertain situations that may occur during the operation. They are captured in terms of behaviors that the robot should do in response to new situations. For example, when an unknown obstacle is perceived in the current path of the robot, the robot should behave according to the type of obstacle. If a tree is perceived in the path, the robot could alter its path to go around it. If an animal is detected in the path, the robot might wait until it moves away, or produce stimuli to scare the animal away, or stop and seek guidance from a human supervisor. An example deterministic task is field coverage where the robot covers a field by navigating through a predetermined coverage path. Several examples of intelligent machine behavior, including coverage path optimization research, are presented below.

5.5.2 Optimal Path Planning

With trends toward larger field sizes, lower-skilled operators of agricultural machinery, and rapid automatic guidance systems adoption, automatic path planning will be an important farm management tool for optimizing field efficiencies and minimizing soil erosion. Field coverage is a deterministic task of an intelligent agricultural machine; path planning can be applied to both conventional agricultural machines with automatic guidance or to autonomous field robots (Oksanen and Visala, 2007, 2009). The time and travel over field surfaces associated with field operations should be minimized within constraints associated with machine characteristics, field topography, and field operation-specifics characteristics. To achieve these goals, optimized coverage path planning algorithms are needed for both planar surfaces and fields with 3D terrain features.

5.5.2.1 Optimized Coverage Path Planning on 2D Planar Surface

Research has been done on coverage path planning of planar surfaces, but results have some limitations in being applied to agricultural fields. Fabret et al. (2001) framed the coverage path planning problem as a traveling salesman problem (TSP), and first chose a "steering edge" that provided the direction to guide successive swaths. In the field headland, characteristic points were then collected. Those points were connected by lines in the steering direction via an associated graph constructed by a TSP solver. It was not clear how the steering edge was chosen. Neural networks have also been applied to this problem (Yang and Luo, 2004). Their approach planned collision-free complete coverage robot paths. The collision-free requirement is of low importance, however, for agricultural field coverage planning. Turning cost at field edges were not investigated in approach, so it may have limited application in agriculture.

Field decomposition has potential to further improve the efficiency of field operations before determining the best path directions in fields, particularly those with irregular field boundaries. Field decomposition must take place simultaneously with the path direction search for cases where the field can be decomposed into several subregions that can reduce the whole field coverage time. The trapezoidal decomposition method has been investigated as an approach to field decomposition (Berg et al., 2000). First, a direction was chosen, and lines parallel to this direction were drawn through all the field boundary vertices. The field was then divided into trapezoids according to these lines. Choset and Pignon (1997) explored trapezoidal decomposition for coverage path planning. However, they were not clear about how the direction of the trapezoidal decomposition lines was determined and if these parallel lines led to the field decomposition that minimized coverage costs.

Determining the best path direction is the main goal of coverage path planning. Whole fields can usually be covered by boustrophedon paths (straight parallel paths with alternating directions) parallel to the optimal coverage path direction for each given field. Several approaches to optimal path direction discovery have been investigated. Following the longest edge of the field is a simple strategy (Fabret et al., 2001), but it is only suitable for fields with simple convex shapes such as a rectangles.

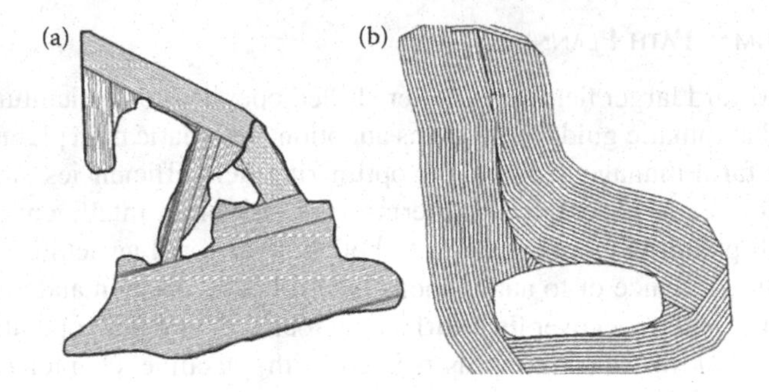

FIGURE 5.5 A coverage path planning in planar field. (a) A top-down trapezoidal decomposition algorithm. (b) A bottom-up approach using prediction and brute-force method. (From Oksanen, T. and A. Visala. 2009. *Journal of Field Robotics*, 26:651–668. With permission.)

Field boundary irregularities must be considered for general coverage path planning solutions. Oksanen and Visala (2009) explored greedy search algorithms to find coverage paths of planar (2D) field surfaces. Their search algorithm iteratively found the optimal trapezoidal field decomposition and path direction using a split and merge strategy (Figure 5.5). Optimal decomposition was not guaranteed, but they demonstrated the important of simultaneous decomposition and direction search needed to minimize headland turning cost.

Jin and Tang (2010) developed an algorithm that optimally decomposed planar fields and planned optimized operational patterns (Figure 5.6). Their algorithm used a geometric model that represented the coverage path planning problem. The objective function accounted for operational costs, including turning costs and resulted from analysis of different headland turns. To reduce the total turning cost, the number of turns is minimized and turns with high operational costs are avoided. Their path planner was applied to planar fields with complexity ranging from simple convex shapes to irregular polygons with multiple obstacles. Their algorithm produced better solutions than farmers' solutions and showed good potential to improve field equipment efficiency on planar fields.

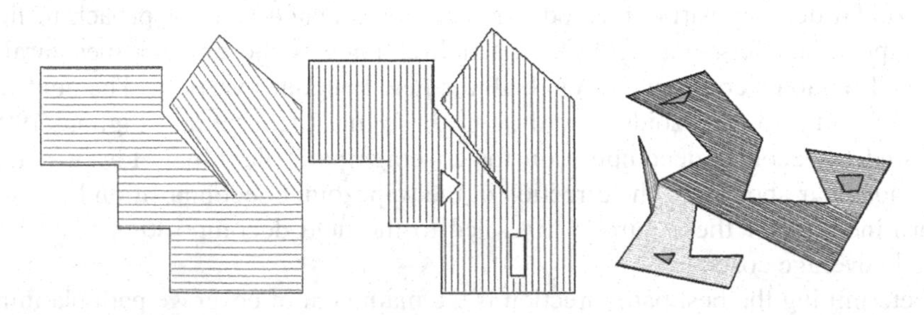

FIGURE 5.6 Examples of an optimized coverage path planning algorithm based on a headland turning cost function and a divide-and-conquer strategy for 2D terrains, where the inner polygons indicate nontraversable obstacles. (From Jin, J. and L. Tang. 2006. *Optimal Path Planning for Arable Farming, ASABE Paper Number 061158*, ASABE, St. Joseph, MI, USA; Jin, J. and L. Tang. 2010. *Transactions of the ASABE*, 53:283–295. With permission.)

5.5.2.2 Optimized Coverage Path Planning on 3D Terrain

More factors must be considered when optimizing the coverage path over terrain with 3D topographic features. The main factors are headland turning, soil erosion, and skipped area. Jin and Tang (2011) approached the problem by first developing an analytical 3D terrain model with B-splines surface fits to facilitate the computation of various path costs. They then analyzed different coverage costs on 3D terrains and developed methods to quantify soil erosion and curving path costs of particular coverage path solutions. Similar to the planar field approaches, they developed a terrain decomposition and classification algorithm to divide a field into subregions with similar field attributes and comparatively smooth boundaries. The most appropriate path direction of each region minimized coverage cost.

A "seed curve" search algorithm was successfully developed and applied to several practical farm fields with various topographic features (Figure 5.7). The 3D path planning algorithm performed better on 3D terrain fields compared to the 2D planning algorithm. In one field, the 3D planning algorithm generated a result with 69.5% reduction in estimated soil loss as compared with that of the 2D algorithm. Typically, the skipped area was also much smaller.

5.5.3 Optimized Vehicle Routing

After optimal field decomposition and coverage path planning, the vehicle route, which is the sequence of an agricultural vehicle following individual paths, can

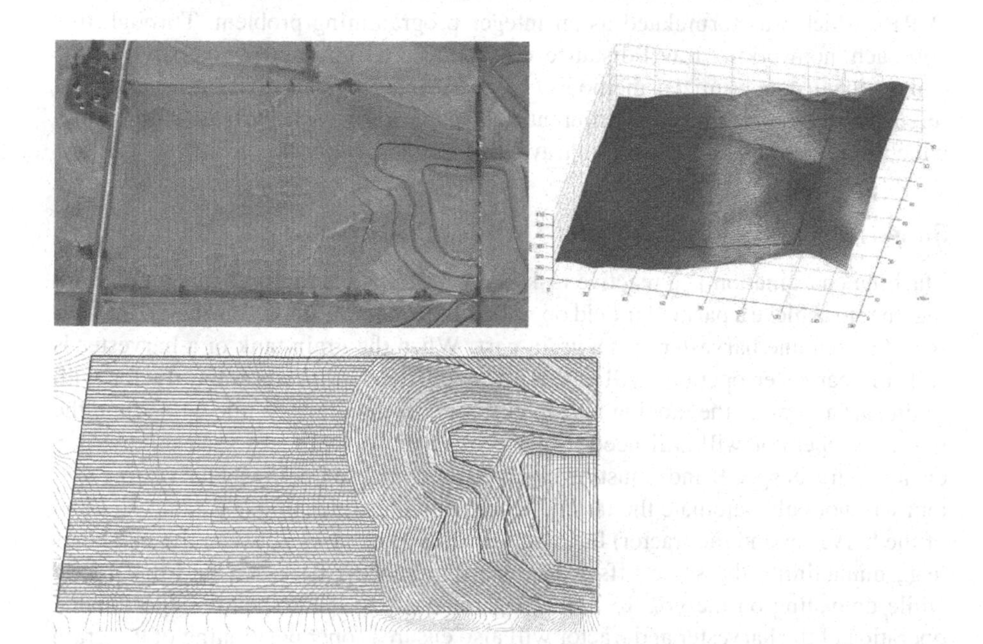

FIGURE 5.7 **(See color insert.)** Examples of an optimized 3D coverage path planning algorithm for a 3D terrain where terraces and valleys exist. (From Jin, J. and L. Tang. 2011. *Journal of Field Robotics*, 28:424–440. With permission.)

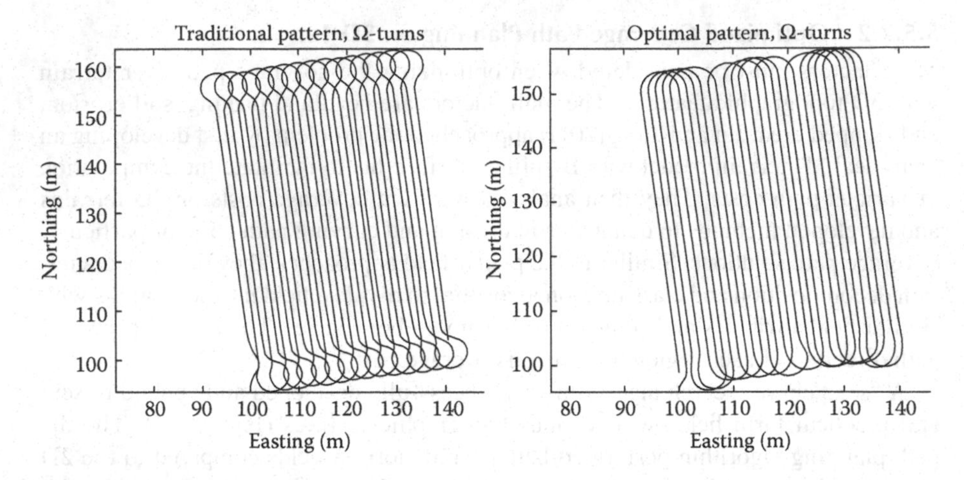

FIGURE 5.8 Difference between traditional and optimized routes of a mowing operation. (Adapted from Bochtis, D.D., S.G. Vougioukas, and H.W. Griepentrog. 2009. *Transactions of the ASABE*, 52:1429–1440.)

be further optimized to minimize the distance traveled in headland turning and improve field efficiency when performing an agricultural operation. Bochtis et al. (2009) developed a mission planner based on an algorithmic approach where field coverage planning was transformed and formulated, as a vehicle routing problem (VRP), which was formulated as an integer programming problem. Through this approach, nonworking travel distance was reduced by up to 50% compared to the conventional nonoptimized method. They also incorporated different operational requirements and produced a different field pattern for each particular operation, which were optimal in nonworking travel distance (Figure 5.8).

5.5.4 Machine Coordination

Machine coordination is a reactive task behavior in which multiple machines work together to achieve a particular field operational goal, for example, on-the-go unloading of a combine harvester into a grain cart. When the grain tank of a harvester is full, the harvester operator will call the tractor driver to position the tractor with grain cart alongside the moving harvester to unload the grain. While unloading, the harvester operator will still need to perform other normal tasks such as steering, changing travel speed, and adjusting machine settings. An intelligent harvesting system will not only automate the tasks of each individual machine (e.g., auto-steering of the harvester and the tractor) but also coordinate the tasks between the machines (e.g., maintaining the same offset distance between the harvester and the tractor while unloading on-the-go). To reduce operator stress and errors, the coordinated operation of the harvester and tractor will also ensure proper positioning of the grain cart without reducing harvesting speed.

In recent years, several equipment manufacturers have commercialized guidance systems that allow a tractor-and-grain-cart unit to be driven without operator input

FIGURE 5.9 John Deere's Machine Sync that can synchronize the operation of combines and tractors with grain carts for "on-the-go" harvesting. (Photo courtesy of Deere & Company.)

while unloading combines. One example is the Machine Sync product from John Deere (Figure 5.9; Deere, 2015). Machine Sync creates an in-field wireless network that can include 10 machines (combines and tractors with grain carts). When a combine's grain tank is full, a "ready-to-unload" signal is sent to the network, and one of the tractors in the network will be driven to the location alongside the combine based on its proximity to the combine to minimize the wait-time. The combine operator then automatically controls the tractor's speed and location while unloading.

5.6 NAVIGATION AND MACHINE CONTROL TECHNOLOGIES

An important layer of any intelligent agricultural machine is the control layer, which uses knowledge from the machine perception layer, and controls physical machine actuators and power systems to achieve the machine behaviors required to meet field operation goals. In many respects, agricultural machine control is the most developed technology of all technology layers required for machine intelligence. Closed-loop control systems have been a part of agricultural machines for many years, from sprayer rate controllers being introduced in the late 1970s, to automatic guidance being commercialized in the early 2000s. Recent years have seen an explosion of newly commercialized controller technology on agricultural machines. There is no reason to expect this trend to diminish in the near future. In this section, we will thus provide examples of control technology in agricultural machines.

5.6.1 Controls Background

Control systems automatically regulate machine output variables such as a motor shaft speed or actuator position in the presence of uncertainty and disturbances.

Control systems can be open-loop, which means with knowledge of the system being controlled, called the plant, a control signal to a plant input will result in desired plant output without measurement of the output. However, if the actual plant changes from how it is represented in the plant model, the output will deviate from the desired output value. In addition, the output will be sensitive to disturbances to the system. Deviation from the desired output is called controller error.

Closed-loop control measures the plant output and compares it with the desired output. The difference between the two, the error signal, is then fed as input to the controller, which modifies and amplifies this input and provides a control input to the plant. The closed-loop technique overcomes plant model deficiencies and makes the overall system less sensitive to disturbances. Closed-loop control is necessary for most intelligent agricultural machine applications, because of limited fidelity in the plant models and because of the uncertain environment in which the system must operate with many disturbances.

However, often for agricultural applications, the loop is closed at the point where the plant output can continually be measured with available sensing technology. The relationship between the measured value and the final output must be calibrated and operated according to a calibration relationship. For example, droplet size controllers on sprayers measure nozzle pressure and provide a controller input to the system to affect the size of droplets. Nevertheless, droplet size is not measured directly because of the high cost of droplet size measurement equipment. Similarly, the application rate of dry fertilizer is not measured directly. Rather the speed of the fertilizer metering system is measured, fed back to the controller, and compared with the desired speed. Based on the calibration curve relating the meter speed to the application rate, the meter speed is controlled to provide the desired application rate. So this system is closed loop to the meter speed, but the application rate is actually running open loop.

There are many intelligent machine control examples in agriculture. van Straten and van Willigenburg (2006) provide an overview of control systems, including a classification of different control methods. The subsections below will provide examples as applied in the areas of vehicle navigation, boom section control, and implement control.

5.6.2 Navigation Control

The main goal of navigation controls is to automatically guide or steer a vehicle along a path and to minimize the error between the actual trajectory that the vehicle takes and the desired path. Automatic guidance of mobile agricultural field equipment improves the productivity of many field operations by improving field efficiency and reducing operator fatigue. The idea of automatically guiding vehicles is by no means new, and relevant literature can be found from several decades back (Grovum and Zoerb, 1970; Parish and Goering, 1970; Smith et al., 1985; Tillett, 1991; Stombaugh et al., 1999; Wilson, 2000; Hagras et al., 2002; Zhang and Qiu, 2004; Zhou et al., 2008; Gomez-Gil et al., 2011; Tu, 2013). The advent of GPS in the early 1990s led to a flurry of research investigating the use of GPS as a positioning system for automatic guidance (Larsen et al., 1994; Elkaim et al., 1997; Griepentrog et al., 2006;

Burks et al., 2013). Commercialization of GPS-based automatic guidance occurred in the first decade of this century, and was adopted very quickly to become one of the most highly adopted PA automation technologies.

A typical navigation controller design process involves modeling, simulation, implementation, and field test and tuning steps. Different navigation control algorithms have been developed based on different vehicle models. The selection of a proper kinematic or dynamic model is the necessary first step in controller design, as it will greatly influence the computation in system identification, the order and complexity of the system, and the dynamic and steady-state performance of the derived navigation controller. Kinematic model-based controllers are more suitable for lower-speed vehicles—for example, under 4.5 m/s—since they cannot represent dynamic effects such as side slip (Karkee and Steward, 2010). Since the majority of farm field tractors are front-wheel steered, bicycle models were typically used to develop various closed-loop feedback control laws, among which proportional, integral, and differential (PID) controllers are the most common. Zhang and Qiu (2004) developed a dynamic path search algorithm for tractor navigation based on a bicycle model (Figure 5.10). They achieved a lateral offset error of less than 0.1 m on straight paths, but experienced a noticeable degradation of tracking accuracy on curved paths. Other vehicle models have also been used in tractor controller development. Stombaugh et al. (1999) created a double-integrator transfer function to relate vehicle lateral deviation to steering angle. They developed a proportional controller to auto-steer a tractor at speeds up to 6.8 m/s with less than 16 cm of lateral path

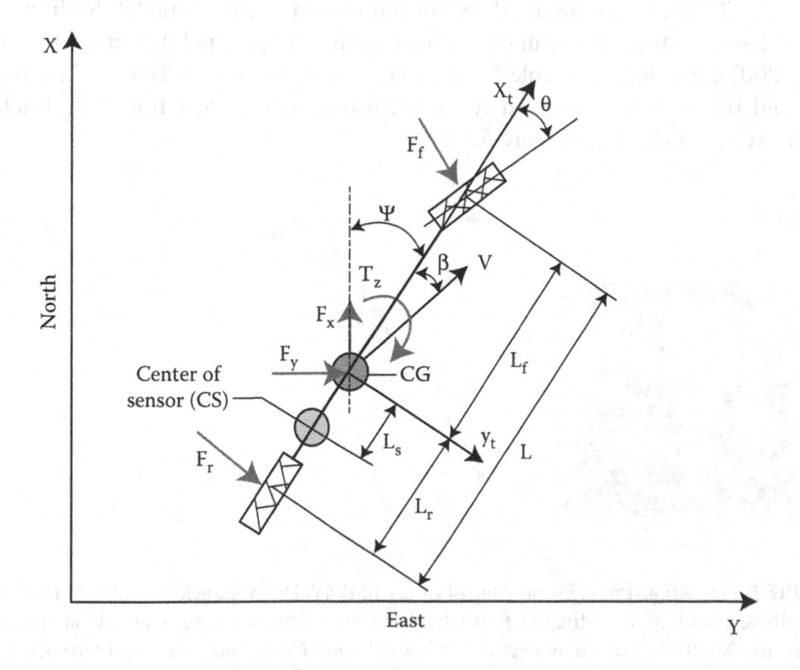

FIGURE 5.10 Bicycle model of a tractor in field and tractor–body coordinates. (From Zhang, Q. and H. Qiu. 2004. *Transactions of the ASAE*, 47:639–646. With permission.)

tracking error. Other methods for designing navigation controller for agricultural vehicles include linearization followed by linear quadratic regulator (LQR) optimization (O'Connor et al., 1996; Thuilot et al., 2002), fuzzy logic, and neural network-based approaches (Cho and Ki, 1999; Hagras et al., 1999; Ashraf et al., 2003; Zhou et al., 2008).

Like other nonholonomic nonlinear systems, agricultural robotic vehicles have system uncertainties and time-varying parameters, especially when working in off-road environments.

External factors such as soil conditions also affect vehicle dynamic characteristics. Both unpredictable internal perturbations and external disturbances create a great challenge. In their early work to develop a self-tuning navigation controller for farm tractors, Noh and Erbach (1993) used a variable forgetting factor in an adaptive steering controller based upon a minimum variance control strategy to cope with nonlinear time-varying dynamics. More recently, Gomez-Gil et al. (2011) developed two control laws: one for tracking straight lines and the other for tracking circular arcs. These control laws were shown to have global asymptotic stability with no singularity points.

4WS and 4WD designs provide maneuverability and traction control advantages to a field robot. Tu (2013) reported on the development of a 4WD/4WS vehicle, AgRover, and developed a sliding mode control-based robust navigation controller. A backstepping method was used to decompose the complex nonlinear system to lower-dimension subsystems that were controlled through pseudocontrol variables. When compared with the feedback linearization method (Kim and Oh, 1999; Wang and Yang, 2005), backstepping does not require an accurate model. Sliding mode control has robustness to parameter perturbations and external disturbances (Cheng et al., 2007), making it suitable for off-road environments. In Tu's work, errors of 0.08 and 0.13 m were observed for straight-line and curved trajectory tracking, respectively, in field tests (Figure 5.11).

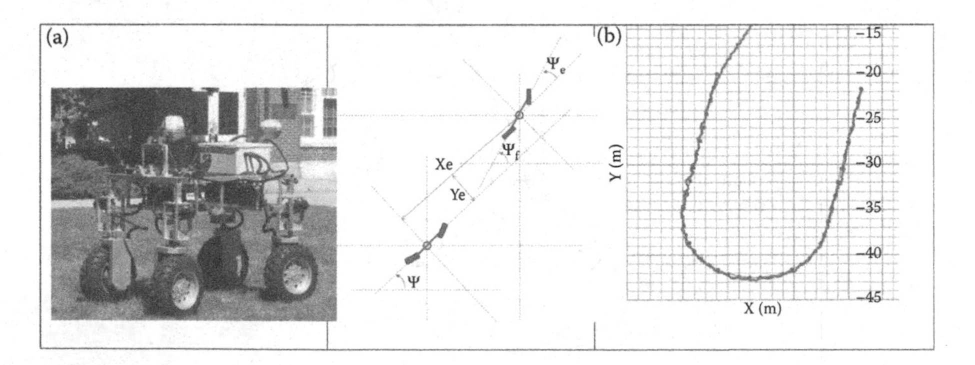

FIGURE 5.11 Robust navigation control of a small 4WD/4WS agricultural robotic vehicle. (a) AgRover; middle: coordinated four-wheel steering. (b) Tracking over a U-shaped path. (From Tu, X. 2013. *Robust Navigation Control and Headland Turning Optimization of Agricultural Vehicles*. Graduate thesis and dissertation. Iowa State University, Ames, Iowa, USA. With permission.)

5.6.3 Sprayer Boom Section Control

A widely adopted control system application in PA is automatic section control, which ranks second only to automatic guidance technology in terms of its commercial success. To implement section control, the width of a field sprayer boom is divided into multiple sections, with individual sections controlled in an on/off fashion. Currently, the smallest unit of section control is the individual nozzle, but in most cases, several adjacent nozzles will be grouped into the minimum controlled section resolution. Boom section control can enable more efficient spraying by reducing pass-to-pass overlap as well as preventing application to off-target areas. Each section is controlled independently of the rest of the system based on the section's location within the field or canopy. These systems rely on a task computer to control the desired state of each section based on the section's location within the field, taken from accompanying sensors and/or on-board maps.

Map-based boom section control systems are commonly used in row crops. GIS maps are the central components in this system, which contain no-spray zones determined by the operator prior to spraying, as well as a dynamically updated "as-applied" history where product has been applied to the field. The task computer uses vehicle location and orientation from a GPS receiver to determine the location of each of the boom's sections and decides if each boom section should be on or off. If the area under the boom has not yet been sprayed and is an acceptable spray location (within the defined boundaries), the section is turned on. If the area has already been sprayed or is outside the field boundaries, the section is turned off.

A main benefit of using the boom section control technology for a sprayer is the reduction in the chemical application overlap (Figure 5.12). As the sprayer enters the headland (Figure 5.12a), the sections (nozzles) are sequentially turned off from (a) to (b); when the sprayer exits the headland (Figure 5.12b), they are sequentially turned on from (a) to (b). The savings realized through the adoption of section control technology is mainly based on the number of control sections and the field shape (Luck et al., 2011a,b).

Luck et al. (2010) found that automated section control of row crop spraying with a seven section resolution reduced overapplication from a 12.4% overlap with a manually controlled five section system to only 6.4%. Savings typically increase with increases in the number of control sections. The greatest potential occurs with irregularly shaped fields and with fields containing inclusions such as grassed waterways. Average savings by using boom section control were estimated at 5% for a typical size machine and field shape. Section control can also be used for planters in a similar fashion. As the size of modern agricultural equipment increases, section control becomes a more important part of PA.

Nevertheless, boom section control can lead to substantial dynamic variation in nozzle pressures and flow rates as sections are turned on and off. Sharda et al. (2010) showed over a 10% increase in nozzle flow rate during boom and nozzle section control without rate control compensation. The nozzle pressure variations also occurred when exiting and reentering point rows, leading to overapplication when exiting point rows and underapplication during reentry (Sharda et al., 2011). Even with the integration of rate control with section control, the variation in application

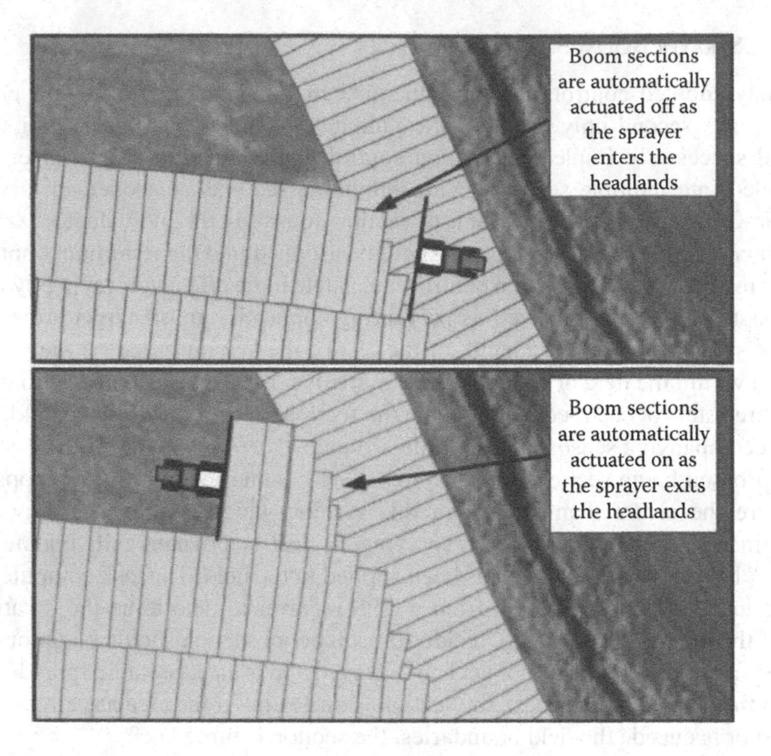

FIGURE 5.12 Section control technology for a spray boom. (Courtesy of Luck, J.D., T.S. Stombaugh, and S.A. Shearer. 2011a. *Basics of Automatic Section Control for Agri cultural Sprayers*. Available at http://www2.ca.uky.edu/agc/pubs/aen/aen102/aen102.pdf. Accessed on January 31, 2015. With permission.)

rate remains an issue due to the slower response of the rate control system (on the order of seconds) as compared to that of a section control system (on the order of milliseconds). New controller technologies, such as a feed-forward control system using boom pressure and flow-rate measurements, boom section states, and a boom model could reduce these application rate errors.

5.6.4 Implement Control

Implement control is also available so that the burden on the operator to control implement settings can be moved to automatic control. This reduces stress and fatigue on the operator and gives the operator freedom to take on more of a supervisory role of the overall machinery system. In addition, implement control often leads to the reduction of errors in the field operation such as turning on or off the seeding at the wrong location and overlap of adjacent swaths or skips in chemical application. There are many implement control examples for field crop machines.

One good example is iTEC Pro, or intelligent Total Equipment Control, from John Deere (Figure 5.13). iTEC Pro is used for handsfree turns and implement control at headlands with a focus on optimizing implement field efficiency. The iTEC Pro integrates AutoTrac automatic steering and implement management systems (IMS) on

FIGURE 5.13 John Deere's iTEC Pro for hands-free turns and implement control at headlands. (Photo courtesy of Deere & Company.)

certain tractors to control tractor speed, power-take-off (PTO) engagement, hydraulic valve position, front and rear mounted implement height, and differential lock engagement during headland turns.

5.7 SYSTEM AND SOFTWARE ARCHITECTURE

In intelligent systems engineering, an architecture is a means for managing complexity. Intelligent agricultural machines and field robots of necessity are complex systems comprised of various components and subsystems; many of which are complex systems themselves. Since individual humans and teams are limited in their time and resources, as well as their ability to keep track of details, they need a way to manage system complexity during development. The development of automatic guidance systems would be very slow, for example, if for each instance of development, the team would need to develop a new GPS receiver for that specific application. Rather, the complexity of a GPS receiver is physically encapsulated in the GPS package, the interface to the GPS is well defined through a common electrical connector, and the electrical signals conform to standardized communications protocols. The receiver accuracy can also be documented through measurements from standard test procedures and well-defined performance metrics. This principle of abstracting complexity through encapsulation of components and clearly defined interfaces to the components is generally what is meant by the phrase "robotic system (or software) architecture."

Intelligent agricultural machines require, to varying degrees, architecture for both hardware and software. Just thinking about the components required for a particular robotic application and how those components are connected to one another and are interacting with one another is an example of "architecture." The potential is excellent for leveraging work across research teams through system architectures that can be shared, standardized, and distributed so that different researchers can build on the efforts of others. These architectures can be proprietary so that development teams within a company can work together more effectively and can internally manage complexity. Architectures can also be open and public to facilitate more rapid development across development teams, as well as to facilitate the interconnectivity of components and subsystems available on the market.

Many examples of robot system architectures exist. Kramer and Scheutz (2007) surveyed nine open-source robotic development environments, or system architectures, for mobile robots, and evaluated their usability and impact on robotics development. Jensen et al. (2014) surveyed available robotic system architectures, including CARMEN, CLARAty, Microsoft Robotics Developer Studio, Orca, Orocos, Player, and ROS. They also found examples of lesser-known architectures, which may be more relevant to agricultural robots, including Agriture, Agroamara, AMOR, Mobotware, SAFAR, and Stanley. Of these, four architectures, CARMEN, Agroamara, Mobotware, and SAFAR, had field trials for agricultural applications. However, open-source availability was limited and only Mobotware had been recently updated.

While different architectures may focus on different aspects of robotic systems, they tend to provide the means for (1) modularizing tasks for processes that are important to a functioning robot, (2) defining messaging systems and protocols for interprocess communication, and (3) defining operations that must occur across distributed processes. Several illustrative system architectures portray key features of architectural thinking that is needed for present and future intelligent agricultural machines.

In early efforts to promote architectural thinking about agricultural robots, Blackmore et al. (2002) proposed a conceptual system architecture for autonomous tractors that consisted of a set of objects or agents that have well-defined narrow interfaces between them. The two types of agents are processes and databases. A process carries out tasks to achieve a goal. Nine processes were defined and described: Coordinator, Supervisor, Mode Changer, Route Plan Generator, Detailed Route Plan Generator, Multiple Object Tracking, Object Classifier, Self-Awareness, and Hardware Abstraction Layer (Figure 5.14). Three databases were defined (Tractor, Implement, and GIS) and are used to store and retrieve data about the machine and its operational context.

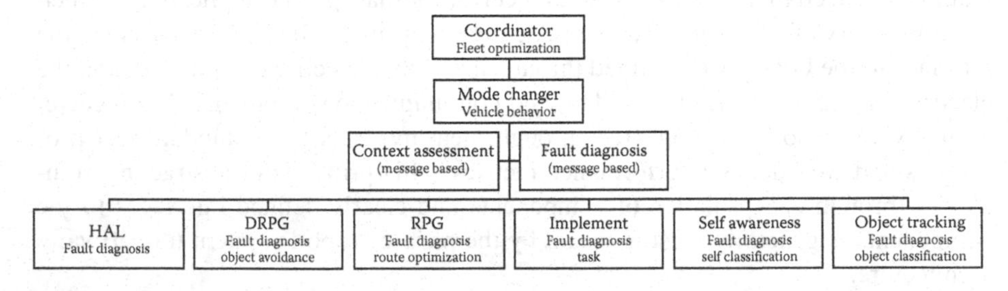

FIGURE 5.14 A conceptual system architecture for an autonomous tractor, which consists of 10 encapsulated processes, databases, and interprocess messaging. (From Blackmore, B.S., S. Fountas, and H. Have. 2002. Proposed system architecture to enable behavioral control of an autonomous tractor. In Zhang, Q. (ed.), *Automation Technology for Off-Road Equipment: Proceedings of the 2002 Conference*, St. Joseph, MI, USA: ASAE. With permission.)

Hierarchical relationships existed between the processes. The Coordinator process, for example, would reside on a computer in the farm office, and would facilitate the human farm manager to provide high-level operational directives to the robot (e.g., check nutrient status of corn in field 1 for 3 days) and provide robot status feedback to the farm manager. The Supervisor process would be hosted on the tractor, and would relay messages from the Coordinator to manage lower-level process. While there was no reported implementation in Blackmore et al. (2002), the thought behind the architecture specifically for a tractor was useful and similar structures are observed in later architectures for intelligent agricultural machines.

Torrie et al. (2002) described the Joint Architecture for Unmanned Ground Systems (JAUGS) Domain Model, which was in active development at that time, and highlighted some implementations of JAUGS in agriculture, including John Deere's Autonomous Orchard Tractor and Autonomous Gator. Torrie et al. stated that JAUGS was primarily a standard messaging architecture to enable components to communicate to one another in a standard manner. Thus, as long as different components (i.e., controllers, user interfaces, and sensors) comply with the standard, they are able to communicate without problems.

Later, JAUGS was changed to JAUS to be more generally applied to all types of unmanned vehicles. In 2005, JAUS transitioned to be a Society of Automotive Engineers standard developed under its aerospace standards division. Rowe and Wagner (2008) provided a clear description of the JAUS standard. The standard has two parts, the domain model that describes the goals for JAUS and the reference architecture that specifies an architecture framework, a message format, and a standard message set. They described how they used JAUS in an implementation that was their organization's entry into the DARPA 2007 Urban Challenge.

The Robotics Operating System (ROS; Open Source Robotics Foundation) is an open-source robotic operating system developed at the messaging layer to provide an interface for passing messages between processes running on different host computing platforms that make up the computer hardware architecture of a robot. ROS also provides a broad set of libraries and tools useful for robotics development. These resources have grown out of the experience of the ROS community. Libraries include (1) standard robot message definitions, (2) the transform library for managing coordinate transform data, (3) a robot description language for describing and modeling a robot, (4) means for collecting diagnostics about the state of the robot, and (5) packages for common robotics problem such as pose estimation, localization, and mobile navigation. One tool is for 3D visualization of sensor data types and another for developing graphical interfaces for a robot. ROS also includes integration with other open-source projects, including the robot simulator, Gazebo, the well-known computer vision library, OpenCV, pointcloudlibrary—for processing 3D data and depth images, and the motion planning library, MoveIt! (Quigley et al., 2009; ROS.org, 2015). While ROS is general and not specifically tailored to any specific application, it does have applicability to intelligent agricultural machines and can be used as a part of larger system architectures for agricultural machines.

Jensen et al. (2014) describe the development of a robotic software systems architecture called FroboMind intended to assist in the development of field robots for PA tasks (Figure 5.15). The authors make a case that, although there are

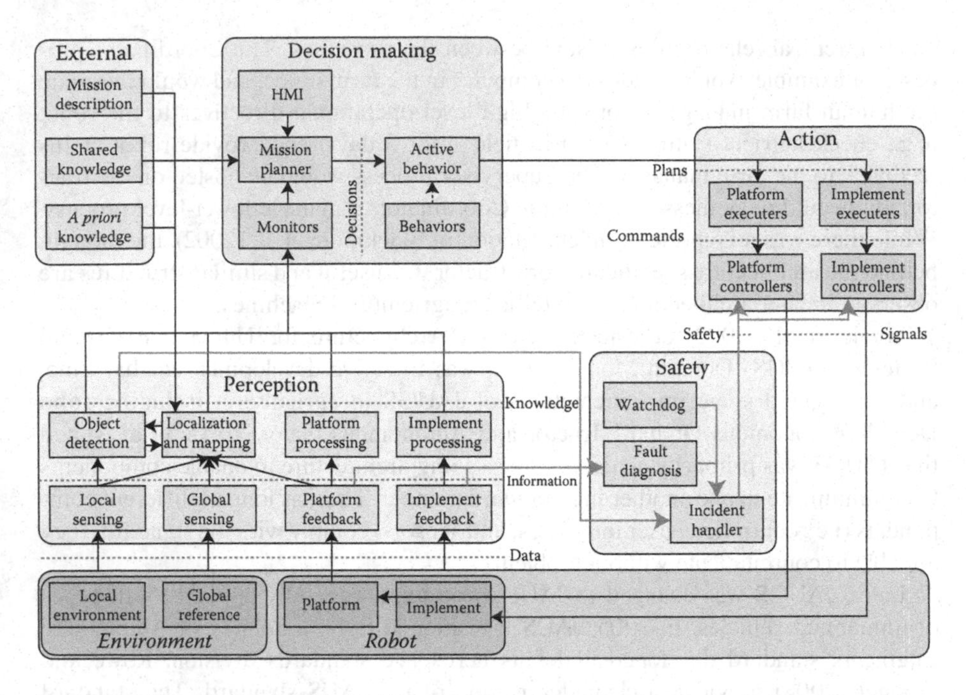

FIGURE 5.15 The FroboMind architecture layer consists of perception, decision making, and action layers along with a separate safety module. (From Jensen, K. et al., 2014. *Robotics*, 3:207–234. With permission.)

several examples of agricultural robots in the literature, the complexity of software required for autonomous systems in agriculture makes it progressively difficult for individual research groups to make progress, and greater collaboration is needed. They propose that a software architecture designed for field robots doing PA tasks will enable field experiments and more efficient reuse of existing work across projects.

FroboMind has a four-part structure, which from lowest to highest level include operating system, middleware, architecture, and components. The Linux operating system Ubuntu was chosen because of its large distribution and long-term support. ROS was used for the middleware, the software that connects software components or mediates between software applications, to define the internal communication structure between processes.

The FroboMind architecture level consists of four modules, which are perception (consisting of sensing and processing submodules), decision making (consisting of mission planning and behavior), action (consisting of executing and controlling), and safety modules, all of which are encompassing the layered framework introduced at the beginning of this chapter. At the component level, the software components, written in C++ or Python, are implemented as ROS packages. FroboMind is open source and has been used in the development of several agricultural robots. It is not a hard real-time system, but its "soft" real-time performance appears to be sufficient for agricultural robotics applications. Jensen et al. (2014) observed relatively rapid development of a new application using FroboMind in an experiment. A high level of software reuse was observed across several robotic implementations.

Robotics software system architectures provide services that are needed for future intelligent agricultural machine behavior. For small field robots, a common limitation is the low work rate associated with them. To overcome this limitation, multiple robots will need to operate collaboratively to meet agricultural timeliness requirements. Such an application requiring multiple vehicle coordination will thus need a distributed software architecture such as that provided by ROS. ROS enables the development of systems consisting of a number of processes running on different hosts, and potentially on different robots. These processes communicate on a peer-to-peer network and via wireless links such as IEEE 802.11.

With a multiple vehicle team focusing on completion of a field operation, vehicle coordination will be required. Thus, system architectures are needed to provide the middleware required for vehicle coordination. In this scheme, it is important that each robot can flexibly assume different roles. For example, one robot might need to become a follower of another robot, or next become a leader of all local followers, or be a follower of one robot and the leader of another. For this flexibility, the system architecture will need to provide the means for a role manager in each robot to assemble the correct processes and messages needed to act out the current role of that vehicle. To facilitate these interactions, publish–subscribe middleware can be used. This middleware enables processes to publish messages into the communication channel without directly sending them to specific receivers. Subscribers, or the receivers of messages, can filter the messages they need and ignore the rest (Matteucci, 2003). This capability, along with peer-to-peer networking, provides a kind of flexible network structure in which any robot can be a leader or follower and robots can come in and out of system, with an ongoing scheme for discovering which robots are available to complete the field operation and real-time knowledge of each robot's progress. While there may be a process providing overall coordination of the completion of the field operation, work by individual robots can be carried out in a flexible manner. Additionally, remote procedure calls, as enabled in ROS, for example, enable one process to call a function in another process, which also is a powerful tool in distributed systems like those associated with mobile field robots.

In summary, robotic software system architectures provide the means for handling complexity through well-defined processes and messaging, as well as higher-level features, such as those described above. These architectures also promote reusability, which enables research and development teams to build on one another's work and move toward more intelligence embedded into agricultural machines.

5.8 AUTONOMOUS VEHICLES AND FIELD ROBOTS

5.8.1 Small Field Robots in Research

Several examples of small field robots are covered in the literature. By studying the designs of these robots, insight can be gained into what machine forms might emerge as future field robots. Earlier agricultural field robots were typically designed for a specific purpose such as automated weeding or field data collection. More recently, field robots have been designed for more general field operations. The following are some examples of field robots.

Blasco et al. (2002) describe the development in Spain of a robot for weed control with a high-voltage electrode that eliminated weeds through electrical discharges. The electrode end-effector was positioned to the location of the weed plant by the use of a HEXA parallel linkage structure. Most of the work focused on machine vision perception for locating plants in images. In similar work, Astrand and Baerveldt (2002) reported on a robot developed in Sweden for the purpose of mechanical weeding in sugar beet. The robot had the capability to navigate along crop rows with a machine vision sensor and to distinguish between crop plants and weeds.

Bak and Jakobsen (2004) described the development of a 4WS/4WD robot designed to sense and map weed populations in row crops. The robot was guided along crop rows with a machine vision perception system. The robot's line tracking performance was reported. Bakker et al. (2010a,b) also reported on a small robot developed at Wageningen University in the Netherlands. A well-documented design process resulted in a 4WS/4WD robot powered by a 31.3-kW diesel engine and propelled by hydraulic transmission (Figure 5.16). The robot was designed to be a research platform for intrarow mechanical weeding for sugar beet. Development of a navigation controller for the robot was reported, but no work about weed detection and mechanical weeder control could be found.

The BoniRob is a field robot developed jointly by Hochscule Osnabrück, Germany and Amazone and Robert Bosch companies (Bangert et al., 2014; Figure 5.17). The BoniRob was developed for plant phenotyping, but is also a reusable platform for different application modules such as mechanical weed control and precision spraying.

FIGURE 5.16 A weeding robot developed at Wageningen University. (From Bakker, T., J. Bontsema, and J. Müller. 2010a. *Journal of Terramechanics*, 47:63–73. With permission.)

FIGURE 5.17 BoniRob autonomous field robot with a soil penetrometer application module. (From University of Applied Sciences Osnabrück, Germany. With permission.)

The modules are mechanically attached to a frame in the center of the robot, which also has a plug for electrical power and ethernet connection to the robot platform. The application module concept was inspired by the conventional tractor and implement paradigm. Langsenkamp (2014) reported on the BoniRob mechanical weed control actuator that presses individual weed plants into the ground.

The majority of the above robots had a similar machine form consisting of a relatively high clearance platform enabling movement over tall crop plants, as well as 4WS/4WD. This machine form is also found in other robots, including the Iowa State AgRover and the Embrapa Brazil Agribot (Godoy, 2012; Diaz, 2013). This platform design enables flexibility in the type of tools that can be used with the platform, application over a wide range of crop growth stages, and high maneuverability.

5.8.2 Other Robot Machine Form Approaches

More recent field robots have tended to take other approaches to machine form. One machine form is a low-clearance track vehicle with skid-steering. Field robots of this type emerged with the Armadillo robot designed and built as an electrically powered general-purpose tool carrier (Jensen et al., 2012). In the next generation, the Armadillo Scout was designed with main features of redesigned track modules and new battery technology (Nielsen et al., 2012; Griepentrog, 2012). Emerging from these efforts is the Grassbots project (Sørensen, 2014; Anon., 2014), which uses a similar machine form with a low-clearance chassis and tracks—thereby providing a low-weight platform for lowland grass harvesting. The project involves multiple university partners and enterprises who will lead commercialization efforts. The commercial Robotti field robot has a similar machine form, is electrically powered, and has been demonstrated as a tool carrier for the cultivation of row crops (Kongskilde, 2013; Grimstad, 2014).

Another approach has been to use commercially available machines into which machine intelligence is integrated to enable machine autonomy. This approach, building on proven machinery technology, has promise in the commercialization of intelligent agricultural machines. Noguchi (2013) presented several robotic vehicles that were built on small, conventional machinery platforms, including a wheeled tractor, a crawler tractor, a rice transplanter, and a small combine harvester. This approach used small, conventional agricultural machines typically found in Japan. Each machine was able to autonomously perform field operations for which it was designed. The wheeled tractor, for example, was able to carry out tillage, seeding, spraying, and harvesting autonomously when provided with a map. Similarly, the ASuBot was built on a Massey Ferguson 38-15 garden tractor (Jørgensen, 2011; Nielsen et al., 2011; Jensen et al., 2012). The Autonomous Mechanisation System (AMS; Blackmore et al., 2007a,b; Griepentrog, 2010; Jaeger-Hansen, 2013) was built on a 20-kW tractor (model Hakotrac 3000, Hako, Bad Oldesloe, Germany). These examples are similar to the early Demeter system built on a conventional windrower (model 2550 Speedrower, New Holland, Pennsylvania) using GPS and machine vision for localization. The system autonomously harvested over 40 ha each year in two cropping seasons (Pilarski et al., 2002).

5.8.3 Commercial Autonomous Machines

Commercial introductions of machine autonomy have generally been leader–follower pairs of machines (Hest, 2012; Posselius and Foster, 2012). Several of these examples are for on-the-go unloading of combine harvesters into a grain cart pulled by a tractor. For example, John Deere Machine Sync, described above, enables the combine to lead the tractor–grain cart follower (ASABE, 2013). The Case IH V2V system operates in a similar fashion (Case IH, 2011). The Kinze Manufacturing Autonomy project has demonstrated a fully autonomous tractor–grain cart product. With this system, combine operators call the autonomous tractor to follow the combine until the time of unloading when the tractor synchronizes with the combine. After unloading, the grain cart is driven to the side of the field to await further instructions. This system is currently in a multiple year test program (Kinze, 2014).

Fendt's entry into machine autonomy is a product called GuideConnect, which is a leader–follower system for two tractors performing field operations (Fendt, 2012). The leader tractor has an operator, and the following tractor is driverless. The follower tractor follows the course of the leader tractor at an operator-specified following distance and a lateral distance from the path of the leader tractor. At the end of the field, the follower tractor pauses while the leader tractor turns, and then the follower takes the same turning pattern.

These commercial introductions demonstrate that for specific situations, semiautonomous operations show potential to add value to farmers. Note that in each case, human operators are exercising supervision and are doing the high-level control and mission planning. That is, much of the high-level intelligence still resides in the human, and it can be amplified with machine intelligence to achieve higher productivity or efficiency goals.

5.9 SUMMARY AND FUTURE DIRECTIONS

The growing worldwide population requires increased food production from agriculture. However, the available land, water, and other production inputs are all limited. As such, about 70% of the additional food needs must come from efficiency-improving technologies (Simmons, 2011). PA is one of those technologies; machine automation is a key component in PA. Automation cannot only improve productivity, but also solve problems related to limited labor availability and high labor costs due to the aging farm population. Significant progress has been made since the early 2000s in automated guidance, variable-rate application, section control, machine coordination, and logistics support. The rapid adoption and impact of automation technology in agriculture cannot be understated. However, there is still a long way to go toward robotic farming. Many technologies, as required for developing intelligent agricultural machinery and discussed in the previous sections, are still in early development stages. Among those, the most challenging technologies are machine health awareness and safeguarding, mission planning, and implement monitoring. System architectures that will enable research teams to build upon one another's work are also critically needed.

Future intelligent machinery for production agriculture may take the form of multiple small robots. There are many advantages in using small autonomous robots in PA. With a small machine form, the soil compaction problem caused by large and heavy agricultural machines traveling over fields can be minimized. Research also indicated that utilizing small machines have the potential to reduce energy requirements for field operations (Toledo et al., 2014). In addition, vehicle safeguarding for a small machine can be implemented more easily than for a large machine, which makes the small-sized robot a more desired machine form for robotic farming. However, small autonomous robots have a competitive disadvantage over conventional machines because of their slow work rate.

Additionally, the right level of automation still needs to be sorted out in the market. Although fully automated machines are desired, some argue that such machines may not be commercialized in the foreseeable future due to safety concerns. Future intelligent machinery may still need human operators for some supervisory control, reacting to unexpected situations that could arise in fields, although the majority of machine functions will be automated. The leader–follower configuration for the combine harvester and tractor–grain cart combination for on-the-go grain tank unloading is an example. Development of this type of highly automated machine or semiautonomous machine will continue to be the focus of the agricultural equipment industry.

There are very few examples of successful commercialization of agricultural robots. The reasons for slow commercialization are mainly technical and economic. Technically, removing human operators from a machine performing a field operation means that the burden of the supervisory control of the machine must be placed on machine intelligence systems. To date, it has been difficult to achieve this level of intelligence. Economically, the high cost to equip the machine with the necessary intelligence is not feasible right now. Nevertheless, technology innovations in sensing and controls, precision guidance, machine communications, information management, and power electronics may eventually make robotic farming a reality.

REFERENCES

Adhikari, B. and M. Karkee. 2011. 3D reconstruction of apple trees for mechanical pruning. ASABE Paper No. 1111613, St. Joseph, MI, USA: ASABE.

Aerts, J.M., S. Van Buggenhout, E. Vranken, M. Lippens, J. Buyse, E. Decuypere, and D. Berckmans. 2003. Active control of the growth trajectory of broiler chickens based on online animal responses. *Poultry Science*, 82:1853–1862.

Aloimonos, Y. 1992. Purposive active vision. *CVGIP: Image Understanding*, 56:840–850.

Anon. 2014. *Grassbots til fremtidens intelligente landbrug*. Available at www.agromek.dk/presse/pressemeddelelser-fra-udstillerne.aspx?PubId=2405. Accessed on January 29, 2015.

ASABE. 2013. A salute to the winners. *Resource*, 20:4–17.

Ashraf, M.A., J. Takeda, H. Osada, and S. Chiba. 2003. Generalised steering strategy for vehicle navigation on sloping ground. *Biosystems Engineering*, 86:267–273.

Astrand, J.B. and A. Baerveldt. 2002. An agricultural mobile robot with vision-based perception for mechanical weed control. *Autonomous Robots*, 13:21–35.

Bak, T. and H. Jakobsen. 2004. Agricultural robotic platform with four wheel steering for weed detection. *Biosystems Engineering*, 87:125–136.

Bakker, T., J. Bontsema, and J. Müller. 2010a. Systematic design of an autonomous platform for robotic weeding. *Journal of Terramechanics*, 47:63–73.

Bakker, T., K. van Asselt, J. Bontsema, J. Müller, and G. van Straten. 2010b. A path following algorithm for mobile robots. *Autonomous Robots*, 29:85–97.

Bangert, W., A. Kielhorn, F. Rahe, A. Albert, P. Biber, F. Sellmann, S. Grzonka et al. 2014. Field-robot-based agriculture: "RemoteFarming.1" and "BoniRob-Apps". In *Proceedings of Agricultural Engineering, Landtechnik AgEng 2013. Components and Systems for better Solutions. VDI Conference*, Hannover, Germany, pp. 439–445.

Beder, C., B. Bartczak, and R. Koch. 2007. A comparison of PMD-cameras and stereo-vision for the task of surface reconstruction using patchlets. In *Proceedings of the 2007 IEEE Computer Society Conference on Computer Vision and Pattern Recognition*, Minneapolis, Minnesota, USA, pp. 1–8.

Belbachir, A.N. 2010. *Smart Cameras*. New York, NY, USA: Springer.

Berg, M.D., M.V. Kreveld, M. Overmars, and O. Schwarzkopf. 2000. *Computational Geometry*, 2nd edition. New York, NY, USA: Springer.

Billingsley, J. and M. Schoenfisch. 1997. The successful development of a vision guidance system for agriculture. *Computers and Electronics in Agriculture*, 16:147–163.

Blackmore, B.S., S. Fountas, and H. Have. 2002. Proposed system architecture to enable behavioral control of an autonomous tractor. In Zhang, Q. (ed.), *Automation Technology for Off-Road Equipment: Proceedings of the 2002 Conference*, St. Joseph, MI, USA: ASAE.

Blackmore, B.S., S. Fountas, S. Vougioukas, L. Tang, C.G. Sørensen, and R. Jørgensen. 2007a. Decomposition of agricultural tasks into robotic behaviours. *Agricultural Engineering International: The CIGR Ejournal*, manuscript PM 07 006. Vol. IX. October 2007.

Blackmore, B.S., H. Griepentrog, S. Fountas, and T. Gemtos. 2007b. A specification for an autonomous crop production mechanization system. *Agricultural Engineering International: The CIGR Ejournal*, manuscript PM 06 032. Vol. IX. September 2007.

Blasco, J., N. Aleixos, J.M. Roger, G. Rabatel, and E. Molto. 2002. Robotic weed control using machine vision. *Biosystems Engineering*, 83:149–157.

Bochtis, D.D., S.G. Vougioukas, and H.W. Griepentrog. 2009. A mission planner for an autonomous tractor. *Transactions of the ASABE*, 52:1429–1440.

Burks, T., D. Bulanon, K.S. You, Z. Ni, and A. Sundararajan. 2013. Orchard and vineyard production automation. In Zhang, Q. and Pierce, F.J. (eds.), *Agricultural Automation: Fundamentals and Practices*. Boca Raton, FL, USA: CRC Press, pp. 149–204.

Butler, D., L. Holloway, and C. Bear. 2012. The impact of technological change in dairy farming: Robotic milking systems and the changing role of the stockperson. *Journal of the Royal Agricultural Society of England*, 173:1–6.

Case, IH. 2011. *New Case IH Technologies Earn Gold and Silver Honors in Global Innovation Competition*. Available at www.caseih.com/en_us/PressRoom/News/Pages/2011-01-11-TECHNOLOGIES-EARN-GOLD-AND-SILVER.aspx. Accessed on January 29, 2015.

Cheng, J., J. Yi, and D. Zhao. 2007. Design of a sliding mode controller for a trajectory tracking problem of marine vessels. *Control Theory and Applications*, 1:233–237.

Cho, S.I. and N.H. Ki. 1999. Autonomous speed sprayer guidance using machine vision and fuzzy logic. *Transactions of the ASAE*, 42:1137–1143.

Choset, H. and P. Pignon. 1997. Coverage path planning: The boustrophedon cellular decomposition. In *Proceedings of International Conference on Field and Service Robotics*, Canberra, Australia.

Continental AG. 2013. *No Blind Spot: Continental Proviu ASL360 360-Degree Camera System for Retrofitting*. Available at www.continental-corporation.com/www/pressportal_com_en /themes/press_releases/3_automotive_group/cvam/press_releases/pr_2013_10_10_proviu_asl360_en.html. Accessed on January 24, 2015.

Craessaerts, G., J. De Baerdemaeker, and W. Saeys. 2010. Fault diagnostic systems for agricultural machinery. *Biosystems Engineering*, 106:26–36.

de Koning, K. and J. Rodenburg. 2004. Automatic milking: State of the art in Europe and North America. In Meijering, A., Hogeveen, H., and De Koning, C.J.A.M. (eds.), *Automatic Milking, A Better Understanding*. The Netherlands: Wageningen Academic Publishers, pp. 27–37.

Deere. 2015. *John Deere Machine Sync*. Available at www.deere.com/en_US/products/equipment/ag_management_solutions/guidance/machine_sync/machine_sync.page. Accessed on January 22, 2015.

Diaz, J.F.A., H.B. Guerrero, A.J. Tiberti, R.A. Tabile, G.T. Tangerino, C.J. Torres, R.V. Souza et al., 2013. Agribot-mobile robot to support of agricultural precision activities. In *Proceedings of the 22nd International Congress of Mechanical Engineering (COBEM 2013)*, November 3–7, Ribeirão Preto, SP, Brazil, pp. 6243–6254.

Edan, Y., S. Han, and N. Kondo. 2009. Automation in agriculture. In: Nof, S.Y. (Ed.), *Springer Handbook of Automation*. Springer, Berlin, Heidelberg, pp. 1095–1128.

Elkaim, G., M. O'Conner, T. Bell, and B. Parkinson. 1997. System identification and robust control of farm vehicles using CDGPS. *Proceedings of ION GPS*, 10:1415–1426.

Fabret, S., P. Soueres, M. Taix, and L. Cordessed. 2001. Farmwork path planning for field coverage with minimum overlapping. In *Proceedings of 2001 8th IEEE International Conference*, IEEE, Piscataway, NJ, USA, pp. 691–694.

Fendt. 2012. *GuideConnect: Two Tractors–One Driver.* Available at bit.ly/LDAz1N. Accessed on January 29, 2015.

Ferentinos, K.P., K.G. Arvanitis, H.J. Tantau, and N. Sigrimis. 2006. Special aspects of IT for greenhouse cultivation. In Munack, A. (ed.), *CIGR Handbook of Agricultural Engineering, Vol. VI Information Technology*, St. Joseph, MI, USA: ASABE, pp. 294–312.

Frost, A.R., D.J. Parsons, K.F. Stacey, A.P. Robertson, S.K. Welch, D. Filmer, and A. Fothergill. 2003. Progress towards the development of an integrated management system for broiler chicken production. *Computers and Electronics in Agriculture*, 39:227–240.

Gerrish, J.B., G.C. Stockman, L. Mann, and G. Hu. 1985. Image processing for path-finding in agricultural field operations. ASAE Paper No. 853037, St. Joseph, MI, USA: ASABE.

Godoy, E.P., G.T. Tangerino, R.A. Tabile, R.Y. Inamasu, and A.J.V. Porto. 2012. Networked control system for the guidance of a four-wheel steering agricultural robotic platform. *Journal of Control Science and Engineering*, 2012:10, Article ID 368503.

Gokturk, S. and C. Tomasi. 2004. 3D head tracking based on recognition and interpolation using a time-of-flight depth sensor. In *Proceedings of the 2004 IEEE Computer Society Conference on Computer Vision and Pattern Recognition (CVPR)*, Washington D.C., USA, pp. 211–217.

Gomez-Gil, J., J.C. Ryu, S. Alonso-Garcia, and S.K. Agrawal. 2011. Development and validation of globally asymptotically stable control laws for automatic tractor guidance. *Applied Engineering in Agriculture*, 27:1099–1108.

Gould, R.G. 1959. The LASER, light amplification by stimulated emission of radiation. In Franken, P.A. and Sands, R.H. (eds.), *The Ann Arbor Conference on Optical Pumping*, Ann Arbor, MI, USA: The University of Michigan.

Gray, K. 2002. Obstacle detection sensor technology. In Zhang, Q. (ed.), *Automation Technology for Off-Road Equipment: Proceedings of the 2002 Conference*, St. Joseph, MI, USA: ASAE, pp. 442–450.

Grest, D. and R. Koch. 2007. Single view motion tracking by depth and silhouette information. In Ersboll, B.K. and Pedersen, K.S. (eds.), *Scandinavian Conference on Image Analysis (SCIA 2007)*, LNCS 4522, Berlin, Heidelberg, Germany: Springer-Verlag, pp. 719–729.

Griepentrog, H.W., B.S. Blackmore, and S.G. Vougioukas. 2006. Positioning and navigation. In Munack A. (ed.), *CIGR Handbook of Agricultural Engineering, Vol. VI Information Technology*, St. Joseph, MI, USA: ASABE, pp. 195–204.

Griepentrog, H.W., C.L. Jæger-Hansen, and K. Dühring. 2012. Electric agricultural robot with multi-layer-control. In *Proceedings of the International Conference of Agricultural Engineering*, Valencia, Spain.

Griepentrog, H.W., A. Ruckelshausen, R.N. Jørgensen, and I. Lund. 2010. Autonomous systems for plant protection. In *Precision Crop Protection—The Challenge and Use of Heterogeneity*. The Netherlands: Springer, pp. 323–334.

Grimstad, L. 2014. *Powertrain, Steering and Control Components for the NMBU Agricultural Mobile Robotic Platform*. MS thesis. Ås, Akershus, Norway: Norwegian University of Life Sciences.

Grovum, M.A. and G.C. Zoerb. 1970. An automatic guidance system for farm tractors. *Transactions of the ASAE*, 13:565–576.

Guo, L., Q. Zhang, and S. Han. 2002. Agricultural machinery safety alert system using ultrasonic sensors. *Journal of Agricultural Safety and Health*, 8:385–396.

Hagras, H., V. Callaghan, M. Colley, and M. Carr-West. 1999. A fuzzy-genetic based embedded-agent approach to leaning and control in agricultural autonomous vehicles. In *Proceedings of the IEEE International Conference Robotics and Automation*, Detroit, MI, USA, pp. 1005–1010.

Hagras, H., M. Colley, V. Callaghan, and M. Carr-West. 2002. Online learning and adaptation of autonomous mobile robots for sustainable agriculture. *Autonomous Robots*, 13:37–52.

Heidarbeigi, K., H. Ahmadi, and M. Omid. 2009. Fault diagnosis of Massey Ferguson gearbox using power spectral density. *Journal of Agricultural Technology*, 5:1–6.

Heidarbeigi, K., H. Ahmadi, and M. Omid. 2010. Adaptive vibration condition monitoring techniques for local tooth damage in gearbox. *Modern Applied Science*, 4:104–110.

Hest, D. 2012. New driverless tractor, grain cart systems coming this year. *Farm Industry News*. Available at farmindustrynews.com/precision-guidance/new-driverless-tractor-grain-cart-systems-coming-year?page=1. Accessed on January 29, 2015.

Huang, T.S. 1996. Computer vision: Evolution and promise. In Vandoni, C.E. (ed.), *19th CERN School of Computing*, Geneva, Switzerland: CERN, pp. 21–25.

Iida, M. and T.F. Burks. 2002. Ultrasonic sensor development for automatic steering control of orchard tractor. In Zhang, Q. (ed.), *Automation Technology for Off-Road Equipment: Proceedings of the 2002 Conference*, St. Joseph, MI, USA: ASAE, pp. 221–229.

Jaeger-Hansen, C., K. Jensen, M. Larsen, S. Hundevadt, H.W. Griepentrog, and R.N. Jørgensen. 2013. *Evaluating the Portability of the Frobomind Robot Software Architecture to New Autonomous Platform.* Available at http://orbit.dtu.dk/files/89360726/Jaeger2013a.pdf. Accessed on January 29, 2015.

Jain, L.C., A. Quteishat, and C.P. Lim. 2007. Intelligent machines: An introduction. In Chahl, J.S., Mizutani, A., and Sato-Ilic, M. (eds.), *Innovations in Intelligent Machines 1*, Germany: Springer-Verlag, pp. 1–9.

Jarvis, J.F. and E. Grant. 2014. *Intelligent Machine.* Available at www.accessscience.com/content.intelligent-machine/348225. Accessed on November 24, 2014.

Jensen, K., A. Bøgild, S. Nielsen, M. Christiansen, and R. Jørgensen. 2012. Frobomind, proposing a conceptual architecture for agricultural field robot navigation. *CIGR 2012 Conference Paper*, Spain.

Jensen, K., M. Larsen, S.H. Nielsen, L.B. Larsen, K.S. Olsen, and R.N. Jørgensen. 2014. Towards an open software platform for field robots in precision agriculture. *Robotics*, 3:207–234.

Jin, J. and L. Tang. 2006. *Optimal Path Planning for Arable Farming, ASABE Paper Number 061158*, ASABE, St. Joseph, MI, USA.

Jin, J. and L. Tang. 2010. Optimal coverage path planning for arable farming on 2D surfaces. *Transactions of the ASABE*, 53:283–295.

Jin, J. and L. Tang. 2011. Coverage path planning on three-dimensional terrain for arable farming. *Journal of Field Robotics*, 28:424–440.

Jørgensen, R.N. 2011. *Robots Using ROS: ASuBot.* Available at orgprints.org/20038/. Accessed on January 29, 2015.

Karkee, M. and B.L. Steward. 2010. Study of the open and closed loop characteristics of a tractor and a single axle towed implements systems. *Journal of Terramechanics*, 47:379–393.

Khodabakhshian, R. 2013. Maintenance management of tractors and agricultural machinery: Preventive maintenance systems. *Agricultural Engineering International: CIGR Journal*, 15:147–159.

Kim, D. and T. Oh. 1999. Tracking control of a two-wheeled mobile robot using input–output linearization. *Control Engineering Practice*, 7:369–373.

Kinze. 2014. *Kinze Adds New Features to Its Autonomous Harvest System.* Available at http://www.kinze.com/article.aspx?id=341. Accessed on January 29, 2015.

Kise, M., Q. Zhang, and F. Rovira-Más. 2005. A stereovision-based crop row detection method for tractor-automated guidance. *Biosystems Engineering*, 90:357–367.

Klose, R., J. Penlington, and A. Ruckelshausen. 2009. Usability study of 3D time-of-flight cameras for automatic plant phenotyping. In *Image Analysis for Agricultural Products and Processes*, Potsdam, Germany: Leibniz Institut fur Agrartechnik, pp. 93–105.

Kongskilde. 2013. *New Automated Agricultural Platform—Kongskilde Vibro Crop Robotti.* Available at www.kongskilde.com/ro/en/News/Year%202013/09-09-2013%20-%20New%20automated%20agricultural%20platform%20-%20Kongskilde%20Vibro%20Crop%20Robotti. Accessed on January 29, 2015.

Kramer, J. and M. Scheutz. 2007. Development environments for autonomous mobile robots: A survey. *Autonomous Robots*, 22:101–132.

Kshetri, S., B.L. Steward, and S.J. Birrell. 2014. Dielectric spectroscopic sensor for particle contaminant detection in hydraulic fluids. ASABE Paper Number 1914249, St. Joseph, MI, USA: ASABE.

Langsenkamp, F., F. Sellmann, M. Kohlbrecher, A. Kielhorn, W. Strothmann, A. Michaels, A. Ruckelshausen, and D. Trautz. 2014. *Tube Stamp for Mechanical Intra-Row Individual Plant Weed Control.* Available at my.hs-osnabrueck.de/ecs/fileadmin/groups/156/Veroeffentlichungen/2014-CIGR_2014_Tube_Stamp_for_mechanical_intra-row_individual_Plant_Weed_Control.pdf. Accessed on January 28, 2014.

Larsen, W.E., G.A. Nielsen, and D.A. Tyler. 1994. Precision navigation with GPS. *Computers and Electronics in Agriculture*, 11:85–95.

LaRue, J. 2014. Management considerations for variable rate irrigation. In *Proceedings of the 26th Annual Central Plains Irrigation Conference*, February 25–26, 2014, Colby, KS, USA, pp.110–114.

Lely. 2014. *Celebrating 20,000 Steps of Progression in Dairy Farming*. Available at www.lely.com/en/home/media-centre/news-en-events/news/celebrating-20000-steps-of-progression-in-dairy-farming. Accessed on January 20, 2015.

Lin, T., L. Weng, and A. Tsai. 2014. Object tracking and collision avoidance using particle filter and vector field histogram methods. ASABE Paper No. 141906189, St. Joseph, MI, USA: ASABE.

Luck, J.D., T.S. Stombaugh, and S.A. Shearer. 2011a. *Basics of Automatic Section Control for Agricultural Sprayers*. Available at www2.ca.uky.edu/agc/pubs/aen/aen102/aen102.pdf. Accessed on January 31, 2015.

Luck, J.D., R.S. Zandonadi, B.D. Luck, and S.A. Shearer. 2010. Reducing pesticide over-application with map-based automatic boom section control on agricultural sprayers. *Transactions of the ASABE*, 53:685–690.

Luck, J.D., R.S. Zandonadi, and S.A. Shearer. 2011b. A case study to evaluate field shape factors for estimating overlap errors with manual and automatic section control. *Transactions of the ASABE*, 54:1237–1243.

Ma, R. 2005. DEM generation and building detection from Lidar data. *Photogrammetric Engineering and Remote Sensing*, 71:847–854.

Marchant, J.A. 1996. Tracking of row structure in three crops using image analysis. *Computers and Electronics in Agriculture*, 15:161–179.

Marchant, J.A. and R. Brivot. 1995. Real time tracking of plant rows using a Hough transform. *Real Time Imaging*, 1:363–375.

Marr, D. 1982. *Vision: A Computational Investigation into the Human Representation and Processing of Visual Information.* San Francisco, CA, USA: W. H. Freeman.

Matteucci, M. 2003. *Publish/Subscribe Middleware for Robotics: Requirements and State of the Art.* Technical Report, Dipartimento di Elettronica e Informazione, Politecnico di Milano, Milano, Italy.

Mobileye. 2015. *Lane Departure Warning*. Available at www.mobileye.com/technology/applications/lane-detection/lane-departure-warning. Accessed on January 31, 2015.

Nielsen, S.H., A. Bøgild, K. Jensen, and K.K. Bertelsen. 2011. Implementations of Frobomind using the robot operating system framework. *Automation and System Technology in Plant Production*, 7:10–14.

Nielsen, S.H., K. Jensen, A. Bøgild, O.J. Jørgensen, N.J. Jacobsen, C.L.D. Jæger, and R.N. Jørgensen. 2012. A Low Cost, modular robotics tool carrier for precision agriculture research. In *Proceedings of 11th International Conference on Precision Agriculture*. Indianapolis, Indiana, USA.

Noguchi, N. 2013. Agricultural vehicle robot. In Zhang, Q. and Pierce, F.J. (eds.), *Agricultural Automation: Fundamentals and Practices*. Boca Raton, FL, USA: CRC Press, pp. 15–40.

Noh, K.M. and D.C. Erbach. 1993. Self-tuning controller for farm tractor guidance. *Transactions of the ASAE*, 36:1583–1594.

O'Connor, M., G. Elkaim, T. Bell, and B. Parkinson. 1996. Automatic steering of a farm vehicle using GPS. In *Proceedings of the Third International Conference on Precision Agriculture*, Minneapolis, MN, USA.

Oksanen, T. and A. Visala. 2007. Path planning algorithms for agricultural machines. *Agricultural Engineering International: The CIGR Ejournal*. 4.

Oksanen, T. and A. Visala. 2009. Coverage path planning algorithms for agricultural field machines. *Journal of Field Robotics*, 26:651–668.

Parish, R.L. and C.E. Goering. 1970. Developing an automatic steering system for a hydrostatic vehicle. *Transactions of the ASAE*, 13:523–527.

Pedersen, S.M., S. Fountas, H. Have, and B.S. Blackmore. 2006. Agricultural robots–System analysis and economic feasibility. *Precision Agriculture*, 7:295–308.

Pilarski, T., M. Happold, H. Pangels, M. Ollis, K. Fitzpatrick, and A. Stentz. 2002. The Demeter system for automated harvesting. *Autonomous Robots*, 13:9–20.

Posselius, J. and C. Foster. 2012. Autonomous self-propelled units: What is ready today and to come in the near future. In *Proceedings of the 2012 Meeting of the Club of Bologna*. Available at www.clubofbologna.org/ew/documents/KNR_S1_1_Posselius.pdf. Accessed on January 29, 2015.

Purswell, J.L. and R.S. Gates. 2013. Automation in animal housing and production systems. In Zhang, Q. and F.J. Pierce (eds.), *Agricultural Automation: Fundamentals and Practices*, Boca Raton, FL, USA: CRC Press, pp. 205–230.

Quigley, M., K. Conley, B. Gerkey, J. Faust, T. Foote, J. Leibs, R. Wheeler, and A.Y. Ng. 2009. ROS: An open-source robot operating system. *ICRA Workshop on Open Source Software*, 3:5.

Reid, J.F. 2004. Mobile intelligent equipment for off-road environments. In Zhang, Q., Iida, M., and Mizushima, A. (eds.), *Automation Technology for Off-Road Equipment: Proceedings of the 7–8 October 2004 Conference in Kyoto, Japan*, St. Joseph, MI, USA: ASAE.

Reid, J.F., S.W. Searcy, and R.J. Babowicz. 1985. Determining a guidance directrix in row crop images. ASAE Paper No. 853549, St. Joseph, MI, USA: ASABE.

Ringbeck, T. and B. Hagebeuker. 2007. *A 3D Time of Flight Camera for Object Detection, Optical 3-D Measurement Techniques, ETH Zürich*. Available at www.ifm.com/obj/o1d_paper-pmd.pdf. Accessed on January 31, 2015.

ROS. 2015. *ROS Core Components*. Available at www.ros.org/core-components. Accessed on January 1, 2015.

Rovira-Más, F., Q. Wang, and Q. Zhang. 2007. Design of stereo perception systems for automation of off-road vehicles. ASABE Paper No. 073126, St. Joseph, MI, USA: ASABE.

Rowe, S. and C. Wagner. 2008. *An Introduction to the Joint Architecture for Unmanned Systems (JAUS)*. Technical Report. Ann Arbor, MI, USA: Cybernet Systems Corporation. Available at www.openskies.net/papers/07F-SIW-089%20Introduction%20to%20JAUS.pdf. Accessed on January 31, 2015.

Ryo, T., N. Noguchi, and A. Mizushima. 2004. Automatic guidance with a laser scanner for a robot tractor in an orchard. In Zhang, Q., Iida, M., and Mizushima, A. (eds.), *Automation Technology for Off-Road Equipment: Proceedings of the 7–8 October 2004 Conference in Kyoto, Japan*, St. Joseph, MI, USA: ASAE, pp. 369–373.

Rzevski, G. 2003. On conceptual design of intelligent mechatronic systems. *Mechatronics*, 13:1029–1044.

Schacter, D.L., D.T. Gilbert, and D.M. Wegner. 2011. *Psychology*, 2nd Edition. New York, NY, USA: Worth Publishers.

Schwarte, R., H.G. Heinol, B. Buxbaum, Z.P. Xu, T. Ringbeck, Z.G. Zhang, W. Tai, K. Hartmann, W. Kleuver, and X.M. Luan. 1998. Novel 3D vision systems based on layout optimized PMD structures. *Technisches Messen*, 65:264–271.

Sharda, A., J.P. Fulton, T.P. McDonald, and C.J. Brodbeck. 2011. Real-time nozzle flow uniformity when using automatic section control on agricultural sprayers. *Computers and Electronics in Agriculture*, 79:169–179.

Sharda, A., J.P. Fulton, T.P. McDonald, W.C. Zech, M.J. Darr, and C.J. Brodbeck. 2010. Real-time pressure and flow dynamics due to boom section and individual nozzle control on agricultural sprayers. *Transactions of the ASABE*, 53:1363–1371.

Simmons, J. 2011. Making safe, affordable and abundant food a global reality. *Presented at the Range Beef Cow Symposium XXII*, November 29, 30, and December 1, 2011, Mitchell, Nebraska. Available at http://digitalcommons.unl.edu/rangebeefcowsymp/300/. Accessed on January 30, 2015.

Smith, L.A., R.L. Schafer, and R.E. Young. 1985. Control algorithms for tractor-implement guidance. *Transactions of the ASAE*, 28:415–419.

Sørensen, C.A.G. 2014. *Grassbots*. Available at eng.au.dk/en/research/research-projects/mechanical-and-materials-engineering-research-projects/grassbots/. Accessed on January 29, 2015.

Stombaugh, T.S., E.R. Benson, and J.W. Hummel, 1999. Guidance control of agricultural vehicles at high field speeds. *Transactions of the ASAE*, 42:537–544.

Thuilot, B., C. Cariou, P. Martinet, and M. Berduca. 2002. Automatic guidance of a farm tractor relying on a single CP-DGPS. *Autonomous Robots*, 13:53–71.

Tillett, N.D. 1991. Automatic guidance sensors for agricultural field machines: A review. *Journal of Agricultural Engineering Research*, 50:167–187.

Toledo, O.M., B.L. Steward, L. Tang, and J. Gai. 2014. Techno-economic analysis of future precision field robots. ASABE Paper No. 141903313, St. Joseph, MI, USA: ASABE.

Torrie, M.W., D.L. Cripps, and J.P. Swensen. 2002. Joint Architecture for Unmanned Ground Systems (JAUGS) applied to autonomous agricultural vehicles. In Zhang, Q. (ed.), *Automation Technology for Off-Road Equipment: Proceedings of the 2002 Conference*, St. Joseph, MI, USA: ASAE, pp. 1–12.

TRW. 2015. *Camera Technology.* Available at www.trw.com/electronic_systems/sensor_technologies/camera. Accessed on January 31, 2015.

Tu, X. 2013. *Robust Navigation Control and Headland Turning Optimization of Agricultural Vehicles.* Graduate thesis and dissertation. Iowa State University, Ames, IA, USA.

Tu, X., S. Du, L. Tang, H. Xin, and B. Wood. 2011. A real-time automated system for monitoring individual feed intake and body weight of group housed turkeys. *Computers and Electronics in Agriculture*, 75:313–320.

van Straten, G. and L.G. van Willigenburg. 2006. Control and optimization. In Munack, A. (ed.), *CIGR Handbook of Agricultural Engineering, Vol. VI Information Technology*, St. Joseph, MI, USA: ASABE, pp. 124–138.

Wang, Q. and Q. Zhang. 2007. Stereo vision based trajectory tracking for agricultural vehicles. ASABE Paper No. 071016, St. Joseph, MI, USA: ASABE.

Wang, S. and P. Yang. 2005. Nonlinear modeling and analysis of vehicle planar motion dynamics. In *Proceedings of IEEE International Conference on Mechatronics*, Taipei, Taiwan, pp. 90–95.

Wei, J., F. Rovira-Más, J.F. Reid, and S. Han. 2005. Obstacle detection using stereo vision to enhance safety of autonomous machines. *Transactions of the ASAE* 48:2389–2397.

Wilson, J.N. 2000. Guidance of agricultural vehicles—A historical perspective. *Computers and Electronics in Agriculture*, 25:3–9.

Xu, Z., R. Schwarte, H. Heinol, B. Buxbaum, and T. Ringbeck. 1998. Smart pixels-photonic mixer device (PMD): New system concept of a 3D-imaging camera-on-a-chip. In *Proceedings: Fifth International Conference on Mechatronics and Machine Vision in Practice*, Nanjing, China, pp. 259–264.

Yang, S.X. and C. Luo. 2004. A neural network approach to complete coverage path planning. *IEEE Transactions on Systems, Man, and Cybernetics—Part B: Cybernetics*, 34:718–725.

Zhang, Q. and H. Qiu. 2004. A dynamic path search algorithm for tractor automatic navigation. *Transactions of the ASAE*, 47:639–646.

Zhou, J., M. Zhang, G. Liu, and S. Li. 2008. Fuzzy control for automatic steering and line tracking of agricultural robot. In *Proceedings of the International Conference on Computer Science and Software Engineering*, Wuhan, Hubei, China, pp. 1094–1097.

6 Precision Agriculture in Large-Scale Mechanized Farming

Chenghai Yang, Ruixiu Sui, and Won Suk Lee

CONTENTS

6.1 INTRODUCTION AND CURRENT STATUS

Research activities in precision agriculture started with the development of yield monitors, grid soil sampling, soil sensors, positioning systems, and variable-rate technology (VRT) at universities and research institutes in the United States and Europe in the late 1980s. By the early 1990s, grain yield monitors and variable-rate controllers became commercially available. With advances in global positioning systems (GPS), geographic information systems (GIS), remote sensing, and sensor technology, the agricultural community has experienced and witnessed the development and application of various precision agriculture technologies since the mid-1990s.

The central concept of precision agriculture is to identify within-field variability and to manage that variability. More specifically, precision agriculture uses a suite of electronic sensors and spatial information technologies (i.e., GPS, GIS, and remote sensing) to map within-field soil and crop growth variability and to optimize farming inputs (e.g., fertilizers, pesticides, seeds, and water) to the specific conditions for each area of a field with the aim to improve farm input efficiency, increase farm profits, and reduce environmental impacts. To automatically implement the concept of precision agriculture, the following four main steps are generally involved: (1) measuring spatial variability; (2) analyzing data and making site-specific recommendations; (3) implementing the variable-rate application (VRA) of farm inputs; and (4) evaluating the economic and environmental benefits.

A broader view of precision agriculture would include more than VRA. It is more about helping farmers better manage their operations and correct inadvertent errors using sensing and control to automate and more precisely carry out field operations. For example, automatic guidance, boom section control, and planter monitoring are all examples of precision agriculture technology, but really have nothing to do with VRA.

Precision agriculture has the potential to improve the use efficiency of farm inputs, increase farm profits, reduce adverse environmental impacts, and improve sustainability. These benefits are important for both producers and the general public and will affect the pace of adoption of precision agriculture. For a new farming practice to be widely adopted in production agriculture, the practice must yield an economic profit except for regulatory requirements. Precision agriculture requires additional costs associated with new equipment and data collection and analysis. If the initial investment for equipment is high, actual economic returns of VRA will be low or even negative for the first few years. Some costs associated with data collection for a field can be accurately determined, while other costs for new equipment and data analysis are difficult to estimate for each field. Nevertheless, it is certain that these costs will go down if the same equipment and data analysis software or services are used for multiple fields over multiple years.

Despite technological advances and potential benefits, the adoption of precision agriculture technologies has been slower than envisioned in the United States as well as in other parts of the world. Using the U.S. Department of Agriculture's Agricultural Resource Management Survey (ARMS) data collected between 1996 and 2009, Schimmelpfennig and Ebel (2011) examined trends in the adoption of four key information technologies, including yield monitors, VRA technologies, guidance systems, and GPS-based soil maps, in the production of major field crops. While yield monitoring was used on over 40% of U.S. grain crop acres, the adoption rates for VRT were only 12% for corn, 8% for soybeans, and 14% for winter wheat. The use of GPS-based soil maps declined to about 15%, while guidance systems were adopted on 15%–35% of nationally planted acres for corn, soybeans, and winter wheat. Some of the factors that could contribute to the low and mixed adoption rates include the lack of farm operator education, technical sophistication, and farm management acumen.

Using the 2010 ARMS data of corn producers, Ebel and Schimmelpfennig (2012) attempted to understand the lower VRT adoption rates and to see if the adoption is sequential. Based on 1445 observations from the survey, the adoption rates were 39% for yield monitors and 17% for VRT for corn in 2010. Only 45% of the yield monitor users actually created yield maps. About 40% of those who used yield monitors and created yield maps also made use of VRT. In comparison, only 10% of those who used yield monitors, but did not create yield maps, performed variable-rate fertilizer application.

In 2013, *CropLife* magazine and the Center for Food and Agricultural Business at Purdue University conducted the 16th survey of crop input dealers and their use of precision technology (Holland et al., 2013). About 51% of the respondents offered controller application of single-nutrient fertilizer, while 47% offered a multinutrient fertilizer option. Variable-rate pesticide application decreased from 22% in 2011 to 16% in 2013, while variable-rate lime application made a minor gain from 45% in 2011 to 47% in 2013. Variable-rate seeding increased to 32% in 2013 from 24% in 2011. Precision agronomic services, such as soil sampling with GPS and GIS field mapping, were offered by 66% of the respondents, an increase from 59% in the 2011 survey. GPS guidance systems with manual (light bar) and automatic (auto-steer) control were offered by 65% and 61%, respectively, of responding dealerships. GPS-enabled sprayer boom sections (53%), satellite/aerial imagery (39%), field mapping with GIS for billing purposes (32%), and GPS for logistics (21%) all made gains from the 2011 survey. The use of telemetry for field-to-home office communications jumped from 7% in 2011 to 15% in 2013. Chlorophyll/greenness sensors also increased to 7% from 4% previously. Soil electrical conductivity mapping (12%) and other vehicle-mounted soil sensors for mapping (3%) were similar to the 2011 results.

When asked about their propensity to invest in precision technology in the future in the 2013 survey, the responding dealerships indicated that investment would continue to grow. About 81% of the respondents said they plan to allocate funds to precision technology, a slight increase from the 80% of the respondents investing in precision technology in 2011. Although most crop input dealers currently offer one or more precision agriculture technologies to their customers, the adoption rates of these technologies by farmers are still relatively low as shown by the ARMS data (Schimmelpfennig and Ebel, 2011). The adoption pattern has also been uneven geographically. The survey results show that precision technologies are clearly more popular in the Midwest than in other parts in the United States.

Many studies in the United States have shown that VRAs of farm inputs are superior to uniform rate application in terms of efficient input use, but evidence of profitability has been mixed (Swinton and Lowenberg-DeBoer, 1998; Bullock and Lowenberg-DeBoer, 2007). A principal cause of the nonprofitability of VRT has been that farmers have had insufficient information about crop yield response to managed inputs, field characteristics, and weather (Bullock et al., 2009). Review articles and book chapters on the adoption and profitability of VRT and other precision agriculture technologies around the world can be found in the literature (Griffin and Lowenberg-DeBoer, 2005; Srinivasan, 2006; Bramley, 2009; Oerke et al., 2010; Robertson et al., 2011; Zhang and Pierce, 2013).

This chapter will discuss and illustrate with examples how precision agriculture technologies have been used for precision fertilizer application, water management, crop pest management, and specialty crop production in large-scale mechanized farming in the United States. Some of the challenges and research needs will also be discussed.

6.2 PRECISION FERTILIZER APPLICATION

Fertilizer application is an essential and critical practice in agricultural production systems. Fertilization is one of the greatest cost inputs in crop production. Taking fertilizer nitrogen (N) as an example, it has been the largest increase in the use of agricultural inputs during the past few decades (Johnston, 2000). In conventional fertilization management systems, fertilizer N is uniformly applied across a field. Uniform N fertilizer rate across entire fields can result in over- and underapplications of N because crop responses to N fertilization are often variable within individual fields (Vetch et al., 1995) and plants in some parts of the field may need more N while plants in other parts may require less. Therefore, on some parts of the field, more N should be applied or much less to none on other parts of the field (Raun and Johnson, 1999). Either underuse or overuse of N fertilizer can create negative effect on desired growth pattern of plants and cause decrease of yield and quality (Fernandez et al., 1996; Gerik et al., 1998). Additionally, overfertilization with N will increase production costs while increasing the potential for negative environmental impact (Bakhsh et al., 2002; USEPA, 2003). Owing to substantially increased environmental concerns and rising N prices, there is an urgent need of innovative technologies and systems that can apply the fertilizer more precisely so as to increase fertilizer use efficiency, maximize farm profit, and minimize environmental impacts.

VRA technologies can be divided into two main categories: map-based and sensor-based. Map-based VRA uses predetermined prescription maps in VRA operation while sensor-based VRA uses real-time information from various sensors to perform VRA "on-the-go." No GPS receiver might be needed for the sensor-based VRA.

6.2.1 Equipment for Fertilizer VRA

The practice of map-based variable-rate fertilizer application requires hardware and software, including a prescription map, a fertilizer applicator equipped with a VRA controller and relevant software, and a GPS receiver (Figure 6.1). In field operations, as the applicator travels, the GPS receiver determines the location of the applicator in the field. Based on the spatial information from the GPS receiver and the data from the prescription map, the VRA controller generates an electrical signal to control a mechanical actuator to apply fertilizer at a desired rate to that specific location in the field. Although most of the equipment for variable-rate fertilization was map-based, sensor-based fertilizer VRA systems have been developed in recent years. Instead of

FIGURE 6.1 A map-based variable-rate liquid fertilizer applicator. Prescription maps can be uploaded and displayed in the controller installed inside the cab.

using the application rates given in the prescription map in the map-based system, the sensor-based VRA applicator uses multiple sensors and data acquisition and processing systems to collect and interpret real-time information of plant health conditions, determine the application rate using predetermined algorithm, and control the actuator to apply the desired amount of fertilizer as the applicator travels across the field (Figure 6.2).

Fertilizers can be applied as solids and liquid. Both solid and liquid fertilizer applicators with VRA capability are commercially available for agricultural production. Most VRA liquid fertilizer applicators use servo valves, flow meters, and speed sensors to directly control the flow of the liquid fertilizer to achieve a desired application rate. As the applicator moves across the field, the VRA controller is constantly updated with the applicator location information provided by the GPS receiver and desired application rate at the location, and then adjusts the flow rate of the liquid

FIGURE 6.2 Illustration of a sensor-based variable-rate application system including sensors, controls, and the actuator.

fertilizer to match the desired rate by controlling the servo valve opening based on the inputs from the speed sensor and flow meter and the swath width of the applicator (Yang, 2000; Grisso et al., 2011). Some other control methods such as chemical injection control and pulse width modulation (PWM) flow control are also available for VRA fertilization, especially for the top-dressing (Sui and Thomasson, 2006; Taylor and Fulton, 2010; Bora et al., 2011).

6.2.2 Prescription Map

Map-based variable-rate fertilization requires prescription maps. A prescription map provides the information to the VRA controller of the applicator to apply how much fertilizer at each specific location within the field. The prescription map for variable-rate fertilizer should include spatial coordinates of each location and the fertilizer application rate associated with each location within the field. Normally, a prescription map can be created using GIS or other software. One or multiple inputs, including soil properties, crop yield, plant health conditions, and field topography, are often used to establish the management zone and calculate nutrient application rate. This input information can be obtained through various means, including soil sampling, soil survey maps, soil electric conductivity mapping, yield monitors, remote sensing imagery, and ground-based plant health sensing (Sui and Thomasson, 2006). No prescription map is needed for sensor-based variable-rate fertilization. Sensor-based applicators use sensors to collect the information for determining the nutrient application rate in real time *in situ*. The information from the sensors such as soil properties and plant characteristics is processed for the rate control on-the-go. Though it does not involve a prescription map, a sensor-based system must have an algorithm programmed into the controller so that the controller can calculate the application rate using various sensor measurements as the applicator moves through the field. So far, the majority of variable-rate fertilizer applicators utilize prescription maps. A few sensor-based VRA systems are commercially available for fertilization application. Optical sensors for plant canopy reflectance measurement and electrical or electromagnetic sensors for soil electrical conductivity measurement are the most popular sensing devices used for sensor-based VRA.

6.2.3 Variable-Rate Fertilization Practice

Variable-rate fertilization has become a common practice in crop production for many producers in the United States and Europe. This technology applies plant nutrients based on plant needs in each location within a field rather than the average of the field. Adoption of this technology would increase fertilizer use efficiency resulting in improvement of crop yield and farming profitability and reduction of environment impacts. On-farm studies have been conducted to evaluate variable-rate fertilization technologies and economic feasibility in field operation (Carr et al., 1991; Mulla et al., 1992; Wibawa et al., 1993). The following are a few examples.

Yang et al. (2000) investigated the differences in yield and economic return between uniform and variable-rate N and phosphorus (P) fertilizer applications for grain sorghum for 2 years in Texas, USA. Three 14-ha fields were used with three fertilizer application strategies (uniform rate N, uniform rate N and P, and variable-rate N and P) in the study. Soil samples were collected from the fields and analyzed for soil properties, including soil texture, organic matter, electrical conductivity (EC), and nutrient contents. N and P application rates were calculated based on the yield goal and soil properties. Prescription maps for N and P application were created and the fertilizers were applied accordingly using a variable-rate fertilizer applicator. The sorghum was harvested with a yield monitor. Yield data analysis indicated that the variable-rate treatment resulted in significantly higher yields than the uniform N and P treatment for both years. Yield data also showed that yield in the area with variable-rate treatment was distributed more evenly than that with the two uniform rate treatments. The variable-rate treatment had positive relative economic returns of $27/ha in the first year and $23/ha in the second year over the uniform N and P treatment. However, if the costs for use of the VRA technology were considered, these returns would be lower.

Koch et al. (2004) evaluated the economic feasibility of variable-rate N application in two continuous corn cropping system fields in Colorado for two growing seasons. One field was 18.5 ha under furrow irrigation, and the other was 58 ha under center pivot irrigation. Commercial software was used to generate site-specific management zones (SSMZ) on the fields with GIS data layers, including bare soil aerial imagery, field topography, and the farmer's past crop and soil management experience. Four N management strategies were compared, which were uniform N application with a constant yield goal, variable-rate N application based on grid soil sampling with a constant yield goal, variable-rate N application based on SSMZ using a constant yield goal (SSMZ-CYG), and variable-rate N application based on SSMZ using variable yield goal (SSMZ-VYG). Nitrogen application rate for each strategy was determined using an algorithm, which included yield goal, soil nitrate residual, and soil organic content as components in the calculating formula. After analysis of the yields, costs, and net returns, they found that the SSMZ-VYG N management strategy used 6%–46% less total N fertilizer and produced $18.21 to $29.57/ha more net return when compared with the uniform N management. They also pointed out that it would be a more profitable way for producers to add VRA systems onto existing uniform N fertilizer applicators for variable-rate N fertilization.

The impact of variable-rate fertilization on surface and underground water quality was evaluated in wild blueberry fields in central Nova Scotia, Canada (Saleem et al., 2013a,b). Uniform and variable-rate fertilizer application treatments were used in these studies. The fertilizer was applied in the fields according to the experimental design. Surface runoff and subsurface water samples associated with different treatments were collected from different zones. The runoff water samples were analyzed for water quality, including total phosphorus (TP), dissolved reactive phosphorus (DRP), particular phosphorus (PP), and inorganic N concentrations. After every heavy rainfall, the subsurface water samples were collected and analyzed for nitrate N and ammonium N. Compared to the uniform fertilizer application treatment, the variable-rate treatment significantly reduced the TP, DRP, PP, and inorganic N losses in the runoff, but did not affect the yield with 40% less fertilizer applied. The variable-rate treatment significantly decreased nitrate and ammonium loading in subsurface water as compared with the uniform treatment.

To make sensor-based fertilization, a real-time sensing and control system is required to determine the fertilizer needs and apply the desired rate at each specific location across the field. Sensors are the key components in the sensor-based fertilizer application system. Several sensors are commercially available for VRA use. Crop Circle and GreenSeeker crop sensors are commonly used for plant canopy reflectance measurement. The Crop Circle sensor is made by Holland Scientific Inc. (Lincoln, NE). It uses modulated LEDs as a light source and is able to measure the reflectance in three bands. The GreenSeeker sensor, manufactured by NTech Industries (Ukiah, CA), also uses modulated LEDs as light sources and measures the light reflected from the plant canopy in two bands. Plant canopy reflectance in different bands measured by these sensors has often been used to calculate vegetation indices, which indicate plant characteristics. The most commonly used index is normalized difference vegetation index (NDVI). NDVI is calculated by dividing the difference between the reflectances at near-infrared (NIR) and red bands by the sum of the reflectances at the two bands; that is, $NDVI = (NIR - Red)/(NIR + Red)$. Other vegetation indices, including reflectance band ratios and individual band reflectance, have also been employed for crop management.

Sui et al. (1989, 2005) and Sui and Thomasson (2006) reported the development of a ground-based sensing system for determining N status in cotton plants in real time *in situ*. The system includes an active optical sensor, an ultrasonic sensor, and a data acquisition unit (DAQ). The optical sensor was able to measure crop canopy reflectance in four wavebands, including a blue band (400–500 nm), green band (520–570 nm), red band (610–710 nm), and NIR band (750–1100 nm). The ultrasonic sensor determined plant height. The DAQ was an intelligent device, which simultaneously collected and processed data from the optical sensor, ultrasonic sensor, and spatial information from a GPS receiver on-the-go. Spectral reflectance and plant height data were compared to laboratory measurements of cotton plant leaf N content and used to train an artificial neural network (ANN) for predicting N status in cotton plants. The trained ANN was able to predict N status of the cotton plants at 90% accuracy when N status was divided into two categories, deficient and nondeficient. The results suggested that this real-time crop sensing system had promising potential for sensor-based fertilizer application.

For successful use of a sensor-based variable-rate fertilization system, an understanding of how the sensor works and what it actually measures is required. The knowledge of relationships between sensor measurements and plant needs is also necessary. Taking crop canopy sensor and soil EC sensor as examples, the crop canopy sensor is positioned above the crop canopy and measures crop canopy reflectance in specific spectral bands, and the soil EC sensor measures soil electrical conductivity in different depths of the soil. The reflectance data and soil EC data can be processed and used for various applications based on the relationship between these data and other variables of interest. For example, Khalilian et al. (2008, 2011) used NDVI and soil EC data to develop an algorithm for N application in cotton. The results indicated the potential for using midseason specific plant NDVI data for VRA of N for cotton. Similar studies and results in cotton have been reported by Taylor et al. (2007), Scharf et al. (2008), and Sharma et al. (2008).

Scharf et al. (2011) conducted an on-farm study on sensor-based variable-rate N applications in corn. In their research, multiple crop reflectance sensors were installed on the N application equipment to measure plant canopy reflectance to control side-dress N rate. Using the sensor's outputs in the visible and NIR channels, a parameter referred to as relative visible/NIR was calculated by dividing the ratio of the reflectance at the visible band to the reflectance at the NIR band in the target area by the ratio of the reflectance at the visible band to the reflectance at the NIR band in the high-N area. The N application rate was determined using this relative visible/NIR parameter with an algorithm for N use in corn. According to the rate, N was applied site-specifically using the sensor-based variable-rate fertilizer applicator. Fifty-five on-farm demonstrations were made in 5 years. The results showed that the sensor-based N application increased partial profit by $42/ha and yield by 110 kg/ha while reducing N use by 16 kg/ha as compared to the producer's N rate. It was obvious that the sensor-based variable-rate N application for corn produced economic and environmental benefits.

Raun et al. (2001) found that grain yield potential of winter wheat could be predicted using canopy spectral reflectance characteristics. In their work, canopy spectral reflectances at the red and NIR bands were measured by utilizing an optical instrument. NDVI was calculated from the reflectance measurements. Yield potential was estimated using NDVI measurements and cumulative growing degree-days. Raun et al. (2002) reported that when compared with uniform rate N application, the variable-rate N fertilization in wheat could increase N use efficiency by more than 15% using the N fertilization optimization algorithm, which was developed with the canopy optical reflectance measurements. Biermacher et al. (2009) conducted on-farm evaluation of the profitability of a sensor-based variable-rate N application system in comparison with the conventional uniform N application method. The variable-rate system used optical sensors to detect plant canopy spectral reflectance in the field. The optical reflectance measurements were used by the system's controller through the algorithm to calculate the N needs on each 0.37 m^2 grid in real time *in situ* and applied the N fertilizer according to the needs on-the-go. The experiments were conducted in wheat crops for three growing seasons across nine locations in Oklahoma, USA. They found that the mean net returns with the sensor-based variable-rate N application system were not statistically different from the net returns using the conventional uniform N application methods. The sensor-based VRA system used by Biermacher et al. (2009) was modified in the size of sensing grid and the algorithm for determining the on-the-go N application rate in wheat. Using the sensing method and the algorithm used in the modified VRA system, Boyer et al. (2011) did a similar study and obtained similar results.

6.2.4 Challenges and Research Needs

Variable-rate fertilization technology has been developed for more than 20 years. Currently, various types of equipment and control devices for variable-rate fertilizer application are available on the market. Some producers have used variable-rate fertilizer application for crop production. However, the adoption of this technology has been slow. There are many reasons for the slow adoption. The main reason is probably that the lack of sufficient evidence to demonstrate that variable-rate fertilization can significantly increase net returns in crop production. Some research has shown that adoption of variable-rate fertilization could provide positive economic returns, but results vary depending on how the technology is used and what the specific field conditions are. The uncertainty of the profitability and the cost of implementation are some of the producers' concerns about investment in variable-rate fertilization technology. To address these concerns to accelerate the adoption of this technology, further research needs to be conducted in the following aspects: (1) development of low-cost variable-rate components (i.e., sensors, controllers, and actuators) and integration of these components into/with existing fertilizer application equipment; (2) development of low-cost and easy-to-use tools to determine site-specific fertilizer application rates and create prescription maps so that field profitability can be enhanced more effectively with minimum inputs; and (3) long-term systematic studies on the economic feasibility and environmental impacts of variable-rate fertilization.

6.3 PRECISION CROP PEST MANAGEMENT

Pesticides are widely used to control a variety of crop pests to minimize yield loss and quality reduction in crop production. The three major types of crop pests targeted by pest application are insects, weeds, and diseases. Pesticides (i.e., insecticides, herbicides, and fungicides) are commonly applied by ground-driven sprayers or aerial applicators. Traditional uniform pesticide application has been commonly used for pest control, though VRA may be more appropriate for the management of certain patchy weeds and diseases. Since insects are mobile within and across fields, traditional uniform application may be appropriate. This section will discuss the approaches to variable-rate pest management and some research and commercial activities on the use of VRT for the control of crop weeds and diseases.

6.3.1 Variable-Rate Pesticide Application Methods

The map-based and sensor-based methods also apply to variable-rate pesticide applications. Both methods require the application rates to be determined based on site-specific conditions, though only the map-based method requires a prescription map. The selection of the application methods depends on whether the pest can be identified and correct application rates determined quickly on-the-go.

Pest distribution and density over space and time are not uniform and can be affected by a variety of factors. Some crop pests occur randomly within a field, while others tend to occur in similar patterns spatially and temporally. For example, some weeds often occur in aggregated patches of varying size or in stripes along the direction of cultivation (Thornton et al., 1990; Gerhards and Christensen, 2006; Christensen et al., 2009). If pest occurrences are consistent in density and locations over years, maps from previous years can be used to regulate pesticide applications in subsequent years. Otherwise, either sensor-based VRA can be used for real-time pest control or map-based control can be used within the season if pest maps can be quickly generated using ground-based sensors or remote sensing imagery.

The key to the implementation of variable-rate pesticide application is to generate pest maps or to detect pest presence on-the-go. Several techniques such as grid sampling, ground-based spectral sensors, and remote sensing can be used to document the distribution and intensity of weeds or diseases within fields during the growing season. The maps derived from these methods can be used for both within-season and postseason control. Yield maps from yield monitors can also be used to map the distribution of a pest if that pest causes a significant yield loss. For real-time control, ground-based soil organic matter sensors can be used with VRA preemergence herbicides because the amount of soil organic matter influences the effectiveness of some herbicides, often mentioned on the label. Ground-based spectral sensors or digital cameras can be used to detect the presence of weeds or fungi for VRA of herbicides or fungicides.

6.3.2 Variable-Rate Controllers

VRA will not affect the basic functions of existing pesticide applicators. The required changes will be necessary to accommodate the addition of a task computer, a GPS receiver, a controller, sensors, and valves. There are different types of control systems on the market that are adaptable to precision application, including flow-based control of a tank mix, chemical injection control, and chemical injection control with carrier control (Humburg, 2003). The flow-based control of a tank mix is the simplest of the three and it combines a flow meter, a ground speed sensor (or GPS speed), and a controllable valve (servo valve), with an electronic controller to apply the desired rate of the tank mix. These systems can make rate changes across the boom quickly. However, the changes in flow rate directly affect the pressure to be delivered to the spray nozzles. This can result in large changes in droplet size in the spray when the commanded flow rate is outside the best operating range for the nozzles.

Direct injection of the chemical into a stream of the carrier (water) uses the controller and a chemical pump to manage the rate of chemical injection rather than the flow rate of a tank mix. The flow rate of the carrier is usually constant, and the injection rate is varied to accommodate changes in the commanded application rate. However, the principal limitation of chemical injection systems without carrier control is the transport delay from the injection point to the application nozzles (Tompkins et al., 1990; Sudduth et al., 1995).

Chemical injection with carrier control can overcome some of the limitations and have many of the advantages of both of the earlier systems, but it requires that the control system change both the chemical injection rate and the water carrier rate to respond to application rate changes. One control loop manages the injection pump, while a second controller operates a servo valve to provide a matching flow of water. Chemical injection with carrier control will result in less application error than chemical injection without carrier control because carrier control minimizes the concentration variations to within dynamic response differences between the two subsystems, thus reducing the effect of transport delays (Steward and Humburg, 2000). Nevertheless, the range of carrier control is limited to the workable pressure range of the nozzles. With the advent of PWM nozzles (Giles et al., 1996) and variable orifice nozzles (Bui, 2005), the range of flow rates has been expanded while minimizing changes in droplet size.

Several agricultural equipment vendors, including Raven Industries, Micro-Trak, Mid-Tech, and DICKEY-john, started to provide these variable-rate controllers in the 1990s. Many of the controllers have been adapted to existing applicators for variable-rate pesticide applications. Meanwhile, numerous research activities have continued on the development and evaluation of variable-rate pesticide application systems (Al-Gaadi and Ayers, 1999; Han et al., 2001; Dammer et al., 2009; Liu et al., 2014). More advanced variable control systems have been developed in the last 20 years and are available from more vendors.

6.3.3 Variable-Rate Pest Management Activities

6.3.3.1 Herbicide Application

Reductions in herbicide use achieved with site-specific applications depend on the distribution and density of weeds in the field. In an evaluation of site-specific postemergence weed control of broadleaf and grass weeds in corn based on grid sampling-derived prescription maps, Williams et al. (2000) showed a reduction of herbicide use by 11.5%–98.0% compared with conventional herbicide spraying. Throp and Tian (2004) used a weed map developed from remote sensing imagery for variable-rate herbicide treatments in a soybean field. Of the four herbicide treatments, the VRA performed the best when considering both weed kill effectiveness and herbicide use efficiency. Koller and Lanini (2005) evaluated variable-rate herbicide applications based on weed infestation maps developed from the previous year using grid sampling. Their results showed that when either the weed seedling map or the mature weed map was used, weed control in terms of subsequent weed cover was comparable to uniform herbicide application, while the total amount of herbicide applied was reduced by 39% for the seedling map and 24% for the mature weed map. Nolte et al. (2011) reported that site-specific placement of herbicide in a field with nonuniform soil textures reduced levels of seedling injury by 30% in regions of the experimental field where soil texture was classified as light or sandy, while application rates were 35% lower than the standard rate with no significant differences in weed control.

Detection of weeds in agricultural crops using airborne and satellite imagery has been a challenge due to the similarity in spectral reflectance between weed and crop plants and the complex interaction among crop, weed, and soil background. Therefore, other research has focused on the use of ground-based sensors for weed detection and variable-rate herbicide application. Shearer and Jones (1991) used photoelectric sensing to detect weeds between crop rows and to activate a spray nozzle on a ground-based spray system. Testing in soybeans showed a 15% reduction in herbicide usage with no compromise in weed control. Hanks and Beck (1998) investigated two commercial photoelectric sensor-based systems, the Detectspray Model S-50 and the WeedSeeker Model PhD 1620, and field testing of the two systems showed a 63%–85% reduction in herbicide use with no significant loss in weed control.

Photoelectric sensors can be easily incorporated into variable-rate herbicide application systems, but this type of sensor cannot distinguish weeds from crop plants and is generally limited to the detection of significant weed cover between crop rows. Therefore, ground-based weed sensors using machine vision technology have been investigated. In machine vision, images are collected with ground-based cameras and then classified to distinguish weeds from crop and soil background using image processing techniques. Although most of the studies in this area have focused on the evaluation of different color indices, morphological or texture parameters, and complex algorithms for segmentation of weeds under naturally variable lighting conditions, only a few machine vision-based weed sensing systems have been tested for real-time herbicide application. Giles and Slaughter (1997) developed a machine vision-guided precision band sprayer for small plant foliar spraying. The system reduced application rates by 66%–80% and increased spray deposition efficiency on the target plants by 2.5–3.7 times greater than conventional broadcast spraying. Tian (2002) integrated a real-time machine vision sensing system and individual nozzle controlling devices with a commercial map-driven herbicide sprayer to create an intelligent weed sensing and spraying system. Field experiments showed that the integrated system operated with a 91% overall delivery accuracy and that potential herbicide savings ranged from 52% for one single threshold to 71% for four threshold levels under normal weed conditions.

6.3.3.2 Fungicide Application

Depending on types of pathogens and their spatial patterns, disease management plans differ greatly. The pattern of infection may be random for wind-dispersed or seed-borne diseases. For soilborne diseases, primary disease patterns may reflect previous disease occurrence in the field. If patterns of diseases or disease risk are predictable and stable between years, site-specific fungicide application can be implemented using the previous year's or current year's maps. On the other hand, if disease patterns vary from year to year, only the current year's maps or sensor-based real-time application can be employed. Bjerre et al. (2006) provided an overview of site-specific management of crop leaf pathogens based on canopy spectral reflectance and remote sensing imagery with VRT. Similar to weed management, grid sampling, ground-based sensors, and airborne and satellite imagery can be used to map the distribution and severity of various diseases.

Any pest that causes sufficient plant stress to distort the reflectance characteristics of crop foliage is a candidate for detection by means of remote sensing. Airborne and spaceborne imagery has been widely used to detect and map a large number of crop diseases, but early detection remains difficult to impossible. By the time disease symptoms can be detected on the remote sensing imagery, damage could have already been done to the crop in many cases. For some diseases, this may be early enough to take control measures to minimize the damage; for others, it may be too late to correct the problem within the season. However, remote sensing imagery obtained in the current season can be useful for the management of recurring diseases, such as soilborne fungi, in the following seasons. Another challenge for remote sensing detection of crop diseases is that multiple biotic and abiotic conditions may coexist and produce similar effects on the color, geometry, or vigor of the upper crop foliage. Crop diseases and insects and some soil problems can all cause morphological (wilting or stunting) and physiological (chlorosis, darkening, or dehydration) changes in a crop. If only one dominant disease occurs or if multiple diseases or stresses with distinctive symptoms are present, remote sensing imagery will be able to discriminate the infected areas; otherwise, discrimination of the diseases may be possible with additional knowledge of the dynamic behaviors of the diseases or other stresses and relevant information of the specific soil and crop conditions. Moreover, high spatial resolution multispectral and/or hyperspectral imagery taken at multiple times may be necessary. As remote sensing imagery with finer spatial and temporal resolution is becoming more available and less expensive, it will present a great opportunity for both growers and researchers to more effectively use this data source for the detection of crop diseases. Many crop diseases have been identified as good candidates for remote sensing, but practical procedures for farming operations are still lacking. Efforts need to be devoted to the development of operational methodologies for detecting and mapping these candidate diseases. Meanwhile, more research is needed to evaluate more advanced imaging systems and image processing techniques for distinguishing the diseases that are difficult to detect or occur with other stresses.

Although remote sensing has been successfully used to detect many diseases, very few ground-based disease detection sensors are available for real-time site-specific fungicide application. However, plant cover sensors that may indirectly indicate potential disease occurrence have been used to regulate fungicide application before the onset of disease symptoms. Dammer and Ehlert (2006) evaluated a plant cover sensor (CROP-Meter) for real-time VRA of fungicides against cereal diseases and they achieved a savings of 7%–38% over 11 fields without negative influences on yield and disease occurrence. The sensor's signal was correlated with the leaf area index (LAI), which was then used to regulate the application rate based on the assumption that there is a tendency for higher disease occurrence in dense canopies for some diseases such as powdery mildew. However, this relatively simple method of controlling fungicide application does not take into account the differences in disease distribution. Dammer et al. (2009) incorporated a decision support system into the previous CROP-Meter-based variable-rate fungicide application system. The decision support system was used to create a map of management zones considering infection probabilities for fungal diseases using weather data and field-specific data. The application rates within these management zones were further adjusted based on the CROP-Meter measurements of local vegetation differences. Compared with conventional uniform spraying, variable-rate treatment with the plant cover sensor resulted in fungicide savings from 13.9% to 32.6% with a decision support map and from 11.1% to 20.3% without the decision support map.

Higher levels of biomass (thicker rice) are more likely to have higher incidences of sheath blight. Baker and Meggs (2006) compared two fungicides (Quadris and Stratego) at variable rates by NDVI zone derived from multispectral imagery and conventional blanket application. Savings averaged $2.59/ha ($1.05/ac) for Stratego in rice, $25.62/ha ($10.37/ac) for Quadris in rice, and $8.48/ha ($3.43/ac) for Quadris in soybeans after all costs for flying and imagery were factored. Follow up with the farmers on the rice fields with variable-rate fungicide applications indicated yields to be at or above 5-year averages, with good milling quality.

Fungicide application can be regulated according to canopy density as previously discussed or according to disease incidence. When a critical threshold for a disease is exceeded, fungicides have to be sprayed immediately because some pathogens can quickly spread throughout the crop canopy. For some diseases, such as rusts and powdery mildew, early detection and variable-rate treatment may be appropriate. For some diseases that tend to spread quickly, treatment decisions based solely on disease observation may not be sufficient. In this case, uniform application may be more appropriate.

Some diseases tend to occur in similar spatial patterns within fields over years. Site-specific treatment can be performed before the onset of the disease based on previous years' infection maps. One such disease is cotton root rot that is caused by the soilborne fungus *Phymatotrichopsis omnivore*. Cotton root rot is a serious and destructive disease that has affected cotton production in the southwestern and south central United States for over a century. Recent research has shown that the commercial fungicide Topguard (flutriafol) is able to control the disease (Isakeit et al., 2010). Yang et al. (2005, 2012) monitored and mapped the progression of cotton root rot within and across growing seasons in south and central Texas using airborne multispectral imagery, as infected plants had higher red reflectance and lower NIR reflectance compared to noninfected plants. The imagery from 2010 to 2014 along with the image data collected from 2000 to 2002 has demonstrated that cotton root rot tends to occur in the same general areas within fields over recurring years, though other factors such as weather and cultural practices may affect its initiation and severity. This recurrent pattern of cotton root rot incidence should provide the producer with greater confidence to use aerial imagery for making site-specific treatment decisions.

Figure 6.3 shows two color-infrared images of a 105-ha (260-ac) cotton field near Edroy, Texas, over a 10-year period. The estimated percent infection areas were 17.0% in 2001 and 17.5% in 2011 under natural conditions. The overall infection patterns between 2001 and 2011 were similar, though there were changes in the locations of infected areas. A change detection analysis showed that 9.0% of the field was infected in both years, while 8.0% of the field was infected only in 2001 and 8.5% only in 2011 in addition to the common infection areas. Thus, a total of 25.5% of the field was infected in either 2001 or 2011. To accommodate the potential variation of the infection, if a 5-m buffer is created around the infected areas on the overlaid map, about 40% of the field should be treated. Treatment plans for this disease should be simply on/off application and multiple levels of treatment are not necessary. Considering the cost of the fungicide flutriafol at $124/ha ($50/ac) for the recommended rate of 2.4 L/ha (32 oz/ac), site-specific treatment will reduce fungicide use by 60% and the savings from fungicide reduction will be $74/ha ($30/ac) per year for this particular field. Research is currently undergoing to demonstrate to cotton growers in Texas how to adapt variable-rate controllers to their existing applicators for site-specific flutriafol application.

6.3.4 Challenges and Research Needs

Field results have demonstrated that variable-rate pesticide application is appropriate and potentially profitable for managing crop weeds and fungal diseases. Although

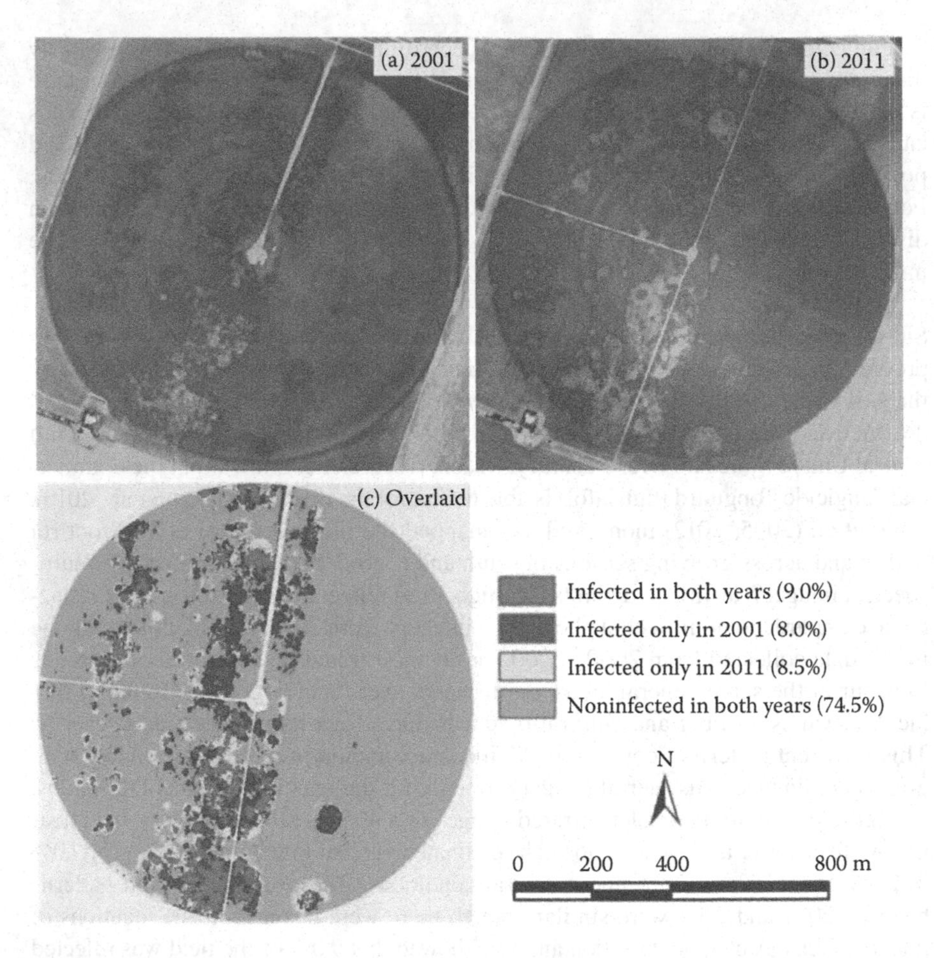

FIGURE 6.3 **(See color insert.)** Color-infrared images taken in (a) 2001 and (b) 2011 and (c) overlaid classification map for an irrigated cotton field infected with root rot near Edroy, Texas. (Adapted from Yang, C., C.J. Fernandez, and J.H. Everitt. 2005. *Transactions of the ASAE*, 48(4):1619–1626; Yang, C. et al. 2012. *Proceedings of the Beltwide Cotton Conferences*, Memphis, TN: National Cotton Council of America, pp. 475–480.)

the individual technologies required to implement site-specific pesticide application are available, it is still a challenge to integrate these technologies for either a weed or a disease management system. How to accurately and reliably detect weeds and diseases or to predict their risk and pressure for either real-time sensor-based application or map-based application remains to be a major challenge. Producers do not want to see critical weeds and diseases left untreated in their fields. Treatment decisions have to take into account the potential expansion of the targeted pest. Not all the weeds or diseases are good candidates for VRA and traditional uniform application remains very effective for many crop pests. More research is needed to identify relevant weeds and diseases for which site-specific management will be both technically feasible and economically profitable in the long term.

6.4 PRECISION WATER MANAGEMENT

Irrigation plays a key role in agriculture production. Irrigated lands produce approximately 40% of the world's total food on 17% of its cropped lands (Fereres and Connor, 2004). In the United States, irrigated agriculture is a major consumer of freshwater, accounting for 80% of the nation's consumptive water use. Irrigation is essential for crop production in arid and semiarid regions. However, in recent years, the acreage of irrigated land has increased rapidly in humid regions due to the uncertainty in the amount and timing of precipitation. For example, crop producers in the Mid-South region of the United States, which has approximately 1300 mm annual precipitation, have become increasingly reliant on supplemental irrigation to ensure adequate yields and reduce risks of production losses due to water stress during the crop growing season. Increasing groundwater withdrawals are resulting in a decline in the aquifer levels across the region. Global agricultural production is facing a serious shortage of water. Improved irrigation technologies are needed to increase water use efficiency for sustainable use of water resources.

There are various irrigation methods, including surface irrigation, sprinkler irrigation, and microirrigation. A sprinkler irrigation system utilizes sprinkler nozzles to distribute irrigation water under pressure. Compared to surface irrigation methods, sprinkler irrigation can significantly improve irrigation efficiency. Majority of the sprinkler irrigation systems in production agriculture are center pivot systems.

As described in the previous chapters, soil and plant characteristics in agricultural fields can vary considerably within a field. It is desirable to treat the plants based on the plant needs at each specific location of the field. Precision agriculture technologies allow farmers to make site-specific adjustment of production inputs for optimal profit. Variable-rate irrigation (VRI) can be used for optimizing irrigation water use efficiency.

6.4.1 VRI Concept

VRI technologies were designed to site-specifically apply irrigation water at variable rates within the field to adjust the temporal and spatial variability in soil and plant characteristics. VRI is normally implemented on self-propelled center pivot and linear move sprinkler irrigation systems. Similar to other VRA systems in precision agriculture, VRI practices require hardware and software. VRI hardware requirements include a GPS receiver to determine the spatial position of the irrigation system and an intelligent electronic device to control individual sprinklers or groups of sprinklers to deliver the desired amount of irrigation water on each specific location within the field according to VRI prescription. The software required includes the algorithms to calculate the water application rates and the computer programs to create VRI prescription maps.

Currently, two primary control methods are used to realize VRI: speed control and duty-cycle control (LaRue and Evans, 2012). The speed control method changes the travel speed of the sprinkler irrigation system to vary the water application depth. As the other operation parameters of the irrigation system remain constant, the water application depth is inversely proportional to the travel speed of the system in the field. This means the higher the travel speed, the lower the water application depth. Although the speed control method is easy to implement and inexpensive, it is only able to vary the application rate in the travel direction of the irrigation system, not along the lateral pipeline, resulting in difficulty to develop VRI randomly shaped management zones to address the variability of soil and plant characteristics across the field. The duty-cycle control method changes the duty cycle of individual sprinklers or groups of sprinklers installed along the lateral pipeline. As the VRI system moves in a constant speed, the GPS receiver determines the system's position in the field. Then, using a preloaded VRI prescription map or the real-time information collected from the field, the VRI controller adjusts the on/off time of the sprinklers to achieve the desired water application rate. The duty-cycle control method is capable of varying the irrigation rate in the system's travel direction and along the lateral pipeline, which offers the flexibility in the development of the management zone.

6.4.2 VRI Control

Studies on the development of controls for VRI using the sprinkler irrigation systems have been conducted since the 1990s. A few of these studies are reviewed as follows. Fraisse et al. (1992) evaluated the feasibility to apply different irrigation water depths using solenoid valves to control the flow to each sprinkler head or set of sprinkler heads. Performance of the low sprinkler heads and solenoid valves subjected to rapid pulsing was tested in the lab. Their results indicated that water distribution pattern of the sprinkler heads was not significantly affected at a pulsing frequency of 1 cycle/min or higher, and the electrical solenoid valves could be employed to vary water application depth with self-propelled sprinkler irrigation systems (Fraisse et al., 1995a,b). McCann and Stark (1993) patented a method and apparatus for site-specific application of irrigation water and chemicals using center-pivot or linear-moving sprinkler irrigation systems. Evans et al. (1996) discussed site-specific applications of irrigation water and chemicals using self-propelled irrigation systems coupled with climatic data, soil properties, and plant growth conditions. Camp et al. (1997) and Omary et al. (1997) reported their work on the development of a site-specific center pivot irrigation system for precision management of water and nutrients. They added three parallel manifolds in each segment along the main pipeline of a commercial center pivot system. The manifolds and nozzles were sized to provide 1×, 2×, and 4× nominal application rate. Using a programmable logic controller, the three manifolds could be operated individually or in various combinations to provide eight application rates. Comparing the measured water delivery to the designed parameters, this VRI system was able to deliver water to the control zones at rates very close to the design (Stone et al., 2006).

A prototype center pivot VRI system was developed by researchers at the University of Georgia and the Farmscan group (Perth, Western Australia). The VRI system changed irrigation water application rates by cycling sprinklers on and off and by varying the center pivot travel speed (Perry et al., 2003). Performance of this prototype VRI system was evaluated. The results showed uniformity coefficients of the system, with application rates of 20%, 80%, and 50%, were 86%, 94%, and 95%, respectively. King et al. (1998) patented a variable-flow-rate sprinkler head. In their invention, the sprinkler nozzle size was reduced by inserting a retractable concentric pin into the nozzle bore. The flow rate of the nozzle could be varied by cycling the concentric pin in and out of the nozzle bore using a linear actuator such as an electric solenoid. Following this approach, King and Kincaid (2004) constructed and tested prototypes of the variable-flow-rate sprinkler for site-specific irrigation management. The lab test results showed that cycling insertion of the concentric pin in the sprinkler nozzle bore was able to vary a time-averaged flow rate over a range of 36%–100%. The prototypes of this variable-flow-rate sprinkler were also tested in field on a three-span linear-move irrigation system to evaluate the water application uniformity (King et al., 2005). It was found that the application uniformity was equal to or greater than 90% with the tested application rates. Han et al. (2009) developed a variable-rate lateral irrigation system. To vary the irrigation water application rate, they used the pulsing system described by Perry et al. (2003) to control the duty cycle of individual sprinklers or groups of sprinklers and a speed control system to change the travel speed of the lateral irrigation system.

Evans et al. (2010) reported their work on the development of a site-specific irrigation system. The site-specific system was tested on a linear move sprinkler system. Results indicated that the site-specific system was capable of switching between midelevation spray application (MESA) and low-energy precision application (LEPA) irrigation methods and varying water application depths according to the defined location in the field. Pierce et al. (2006) and Chávez et al. (2010a,b) developed and tested a remote irrigation monitoring and control system (RIMCS) for continuous move systems. The RIMCS was installed on a linear move irrigation system for site-specific irrigation. Coupled with a GPS receiver, a single board computer (SBC) with wireless Ethernet was employed to control sprinkler nozzles through solenoids to vary irrigation water application rates according to irrigation prescription maps. The SBC could wirelessly communicate with a remote server. The RIMCS was also able to monitor the irrigation system performance and soil and crop conditions through wireless sensor networks.

Sprinkler irrigation systems equipped with VRI controllers are now commercially available. Figure 6.4 shows a center pivot VRI system manufactured by Valmont Industries in 2011. The VRI zone control package of this system includes five VRI zone control units, a GPS receiver, and computer software. Each VRI zone control unit controls the duty cycle of the sprinklers in two independent zones by turning electric solenoid valves on and off to achieve desired application depths in individual zones. The GPS receiver determines the pivot's position in the field for the identification of control zones in real time. The VRI prescriptions could be created using the software and wirelessly uploaded to the system's control panel. The performance status of the system could be remotely monitored using a smart device such as a smart phone (Sui and Fisher, 2014).

6.4.3 Application Uniformity Test

Some work has been reported on the evaluation of commercial VRI system performance. Perry et al. (2004) and Dukes and Perry (2006) tested the uniformity of center pivot and linear move VRI systems in Georgia and Florida, USA. They reported

FIGURE 6.4 A center pivot variable-rate irrigation system running at the research farm of USDA-ARS Crop Production Systems Research Unit in Stoneville, Mississippi, USA.

that both the sprinkler cycling rate and the travel speed of the VRI systems had no significant effect on the uniformity of irrigation water application. O'Shaughnessy et al. (2013) tested the uniformity of two center pivot VRI systems in a windy location in Texas, USA with five application rates ranging from 30% to 100%. The test results showed that uniformity coefficients at different application rates varied from 84.4% to 90.8% with an average of 88.8%. Sui and Fisher (2014) evaluated the uniformity of the center pivot VRI system shown in Figure 6.4. They found that the average coefficient of uniformity over the application rates of 30%, 50%, 70%, and 100% was 84.3%. The effect of application rate on the uniformity was significant, with higher application rates providing higher uniformity. The uniformity of a control zone could be influenced by the overlap of sprinkler coverage between the adjacent control zones.

6.4.4 Challenges and Research Needs

Even though research efforts on VRI have been made for more than 20 years, adoption of this technology by producers for agriculture production has been quite limited. There are about 175,000 center pivot and linear move sprinkler systems in the United States. However, there are only about 200 sprinkler systems that have zone control VRI capacities in the United States, and about 500 speed control VRI systems around the world (Evans et al., 2013).

The slow adoption of VRI technologies could be due to a number of reasons. The main reason could be that the economic benefits gained by VRI technologies are still not very clear to producers and irrigation industries. There is very little scientific information documenting that the VRI technologies conserve water or energy on a field scale for crop production (Evans and King, 2012). More research is needed to demonstrate that the adoption of VRI technology for water, and nutrient management will increase farming profit and protect the environment. Evans et al. (2013) pointed out the needs and tools required to encourage adoption of VRI technologies, including (1) low-cost, reliable variable equipment such as frequency pump motors, solenoid valves, and pressure regulators; (2) guidelines and tools to assist consultants and producers to develop management zones and write VRI prescriptions; (3) guidelines for placement of sensor networks and use of the information from the sensors installed across the fields; (4) easy-to-use basic, generalized decision support systems for VRI in both humid and arid regions; and (5) technical assistant training on VRI technologies for producers, consultants, and other relevant personnel.

6.5 PRECISION TECHNOLOGIES FOR SPECIALTY CROPS

Specialty crops are defined in law as "fruits and vegetables, tree nuts, dried fruits, and horticulture and nursery crops, including floriculture" (USDA, 2014). Plants commonly considered fruits and tree nuts include apple, avocado, banana, blueberry, citrus, cherry, coconut, coffee, cranberry, grape (including raisin), kiwi, mango, nectarine, olive, papaya, peach, pear, pecan, persimmon, pineapple, pistachio, raspberry, strawberry, and walnut. Some of the common specialty vegetable crops include artichoke, asparagus, bean, beet, broccoli, Brussels sprouts, cabbage, carrot, celery, chives, cucumber, eggplant, garlic, horseradish, leek, lettuce, melon, mushroom, pea, onion, pepper, potato, pumpkin, radish, spinach, squash, sweet corn, sweet potato, taro, tomato, and watermelon (USDA, 2014).

Among these crops, the following are grown in large-scale production in the United States: apple, blueberry, citrus, cranberry, grape, peach, pecan, pistachio, strawberry, walnut, lettuce, potato, sweet corn, and tomato. However, compared to traditional grain crops, specialty crop productions are relatively small in scale; therefore, precision technologies for specialty crops have not been well developed and are still in developmental stages. Crop management practices are also different, and thus more creative approaches and solutions are needed. This section reviews some precision technologies used for precision fertilization, water management, and crop pest management for specialty crop production. Some other related research activities with application examples are also discussed.

6.5.1 Precision Fertilizer and Pesticides Application

To correctly prescribe proper fertilizer amounts for tree crops, it is vital to quantify the canopy volume. In the early 2000s, Tumbo et al. (2002) conducted a study to compare manual canopy volume measurements with those by ultrasonic and laser sensors for citrus production. They reported that the laser sensor yielded better results than those by the ultrasonic sensor due to a higher resolution, and was also faster in acquiring data. However, both sensors showed good potential in automatic canopy measurements. Then, Zaman et al. (2005) measured citrus tree canopy sizes using ultrasonic sensors with a differential global positioning system (DGPS) receiver, and used them for creating prescription maps for variable-rate nitrogen fertilizer applications. They used a commercial granular fertilizer spreader (MidTech Legacy 6000) for variable-rate nitrogen application, and reported cost savings of 38%–40% compared to uniform applications. Also, Miller et al. (2005) conducted a field trial for a commercial VRT controller with a spinner disk pull-type granular fertilizer spreader in a 16-ha citrus grove during the 2002–2004 fruit seasons. They tested the spreader with a prescription map and photocell-based canopy sensing. The total amount of applied granules was compared between actual applied and weighed amounts, and they found an average absolute difference of 8%. Also, they compared actual application rates with target rates, and found a good agreement between them with a coefficient of determination of 0.98.

Further, Schumann et al. (2006) investigated the performance of a variable-rate fertilizer spreader in a citrus grove, where tree sizes measured from an ultrasonic sensor were used to determine application rates. Six different nitrogen rates were applied variably throughout the grove, while actual application rates were calculated using gear tooth speed sensors by monitoring the conveyer chain speed. They found that there were time lags in the system due to the spreader's response time, the DGPS receiver latency, and the time to read the prescription map, and those lags made the system inappropriate for single-tree variable-rate fertilization. They reported an average on–off response time of the spreader was less than 3 s, and an average time for changing rates was between 2 and 5 s. Zaman and Schumann (2006) conducted a study to find out important soil properties affecting citrus tree growth, and to implement variable-rate soil amendment applications by dividing the grove into different management zones using the identified soil properties. They reported that management zones could be well divided by the NDVI and soil organic matter contents for implementing VRA of elemental iron and dolomite.

Instead of measuring canopy volume for citrus, another study was conducted using spectral characteristics to determine fertilization application rates, instead of analyzing leaf samples for nitrogen concentration in a laboratory. Min et al. (2008) developed a hyperspectral citrus leaf nitrogen sensing system using detector arrays in 680–950 nm and 1400–2500 nm, linear variable filters, a halogen light source, a longpass filter, and data acquisition cards. The detector arrays showed a very good linearity between integration time and voltage outputs ($r > 0.99$), and stabilities within ±0.1% and ±0.5% for the two sensors. They reported a root mean square difference (RMSD) of 1.69 g/kg in predicting citrus leaf nitrogen contents.

For apple production, Sharda et al. (2014) investigated spray coverage of various emitters using a solid set canopy delivery (SSCD) system for tree crops, which can be used for providing precise pesticide applications and reducing pesticide drift. Six different emitters were studied with four mounting configurations (upper side and underside of leaves). Based on spray depositions on water-sensitive paper cards, they found that an 80° hollow cone emitter produced the best coverages of 58% for the upper side and 21% of the underside of leaves.

In another study for blueberry production, Esau et al. (2014) developed a prototype variable-rate sprayer for spot application of herbicides in wild blueberry production. The prototype consisted of digital color cameras, a variable-rate controller, solenoid valve nozzles, a height sensor, and a pocket PC, which were mounted on a tractor. Based on 97 weed patches mapped in a trial field, they reported that weed areas were reduced from 28% to 3% after herbicide application, and that the sprayer performed well with a 69% herbicide savings.

6.5.2 Water Management

For tree and vine crops, microsprinkler irrigation is commonly used in the United States. Boman et al. (2012) described the current status of microsprinkler irrigation in the United States, which is commonly used for better freeze protection and more savings in water, energy, and fertigation than other irrigation methods. Especially, microsprinkler irrigation can be used for automatic irrigation along with real-time moisture sensors.

Torre-Neto et al. (2001) demonstrated an automated microsprinkler irrigation system for citrus for VRI using tensiometers, temperature sensors, and RJ-485 communication standard. They reported that the system performed well with lower power consumption and potential significant water savings. Further Torre-Neto et al. (2005) presented hardware implementation of wireless sensor and actuator nodes, field stations, a base station for automated and spatially variable irrigation in a six hectares grove for citrus production in Brazil. Parsons et al. (2010) utilized capacitance moisture sensors to implement an automatic irrigation system for citrus. They installed the sensors at five different depths in an orange grove in Florida to trigger microsprinkler irrigation, which could be adjusted by soil type, season, and grower's preference. In Israel, a similar system was also developed for apple production using tensiometers (Meron et al., 2001).

For nectarine production, Coates et al. (2006) developed a variable-rate microsprinkler irrigation system for precise irrigation and automatic detection of faulty drip lines and damaged emitters. Based on monitoring line pressure, the system automatically turned off the microsprinklers when there were any drip line damages or breaks. The system was able to variably apply water across the orchard.

For cranberry production, Pelletier et al. (2013) conducted a study to determine the relationship between cranberry yield and soil water potential to determine the irrigation threshold. They tried three different water treatments (wet, dry, and control) and found that the dry treatment saved irrigation water by 21%–93% compared to the control, and that yield was affected depending on soil water potential thresholds.

For detecting water stress in almond, walnut, and grape, Dhillon et al. (2014) developed a mobile sensor system (Figure 6.5) to predict water status using an infrared thermometer and sensors for measuring microclimatic conditions (photosynthetically active radiation, air temperature and humidity, and wind speed). Using leaf temperature and the microclimatic conditions, they conducted a stepwise regression and a canonical discriminant analysis to predict water stressed and unstressed leaves, and reported misclassification errors ranging from 1.6% to 9.6%, indicating the feasibility of using the developed sensor system for managing irrigation.

As an alternative irrigation method, Lamm et al. (2012) overviewed subsurface drip irrigation (SDI), which has been used since the 1960s. They suggested the use of an RTK GPS receiver to accurately place drip lines. One of the advantages of the SDI is that it enables irrigation of tree crops such as walnuts and almonds without wetting

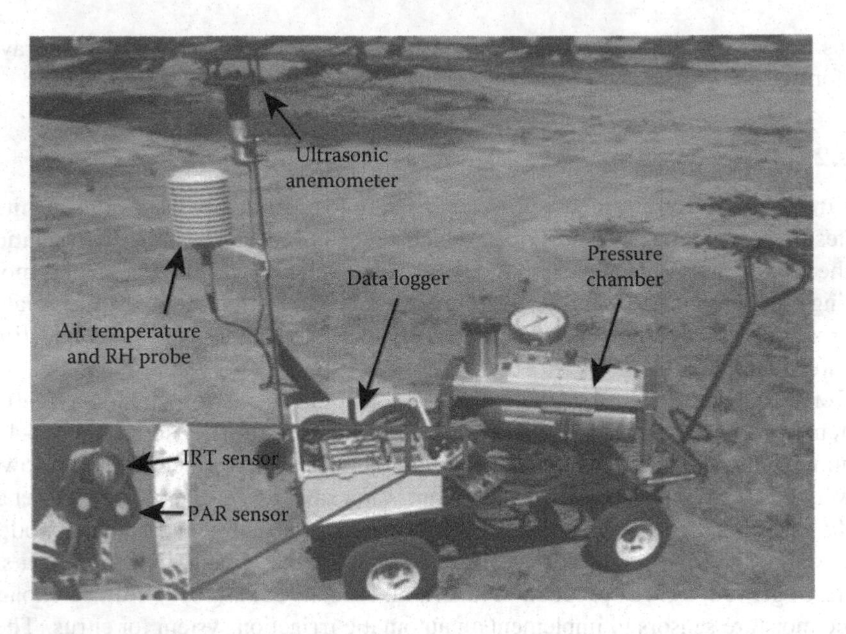

FIGURE 6.5 Mobile sensor suite for measuring leaf temperature and microclimatic conditions. The pressure chamber was used to measure midday stem water potential for validation of the results. (From Dhillon, R. et al. 2014. *Transactions of the ASABE*, 57(1):297–304. With permission.)

the harvested nuts while they are being dried on the ground. They discussed challenges in design, installation, operation, management, cropping, and maintenance. For specialty crop production, SDI can provide uninterrupted management practices, including multiple harvests, spraying, mowing, and tilling, and reduce weed germination.

6.5.3 Crop Pest Management

There have been many studies for specialty crop disease detection. For tomato, Zhang et al. (2005) explored the feasibility of utilizing airborne multispectral imaging to detect tomato late blight disease. They developed the following five vegetation indices using red (R) and NIR bands, and also spectra collected by a handheld spectrometer from the field: R, NIR, NIR/R, NIR-R, and NDVI. With cluster analysis and classification process, they were able to identify the diseased plants with an average accuracy of 87%.

Multiple studies were conducted to detect the Huanglongbing (HLB) disease, also known as citrus greening. The HLB was first found in the south Florida in 2005, and its insect vector (Asian citrus psyllid or ACP) was found in 1998. In California, the ACP was first found in 2008, and HLB was found in southern California in 2012 (UC IPM Online, 2014). Typically, ground inspection is conducted to identify disease symptomatic canopies along with a polymerase chain reaction (PCR) analysis; however, ground scouting is very labor intensive, time consuming, and costly. Thus, an aerial detection would be a good alternative. Kumar et al. (2012) acquired airborne hyperspectral and multispectral images of HLB-infected citrus groves in Florida, and identified the infected canopies using various analysis algorithms, including image-derived spectral library, spectral angle mapping (SAM), mixture tuned matched filtering (MTMF), and linear spectral unmixing. They reported that MTMF yielded the best detection accuracy of 80%, while SAM using multispectral images produced an accuracy of 87%. Also, Li et al. (2012) conducted a study to detect HLB-infected canopies using airborne hyperspectral and multispectral images. They utilized red edge position (REP) to distinguish infected citrus canopies and reported that REP worked better for indoor spectral data than those for outdoor measurements. Various detection algorithms were implemented such as parallelepiped, minimum distance, Mahalanobis distance, SAM, spectral information divergence (SID), spectral feature fitting, and MTMF. Figure 6.6 illustrates the disease density maps obtained using these methods. They reported that detection accuracies were more than 60% for most methods and up to 95% for SID; however, simpler methods (minimum distance and Mahalanobis distance) yielded more consistent results among all datasets.

Further, Li et al. (2014) proposed a new method, extended spectral angle mapping (ESAM), for identifying HLB-diseased citrus canopies using airborne hyperspectral imaging. The method consisted of the Savitzky–Golay smoothing, support vector machine classifier, vertex component analysis to find pure end members, SAM, and REP for removing false-positives. They reported that the ESAM method yielded a

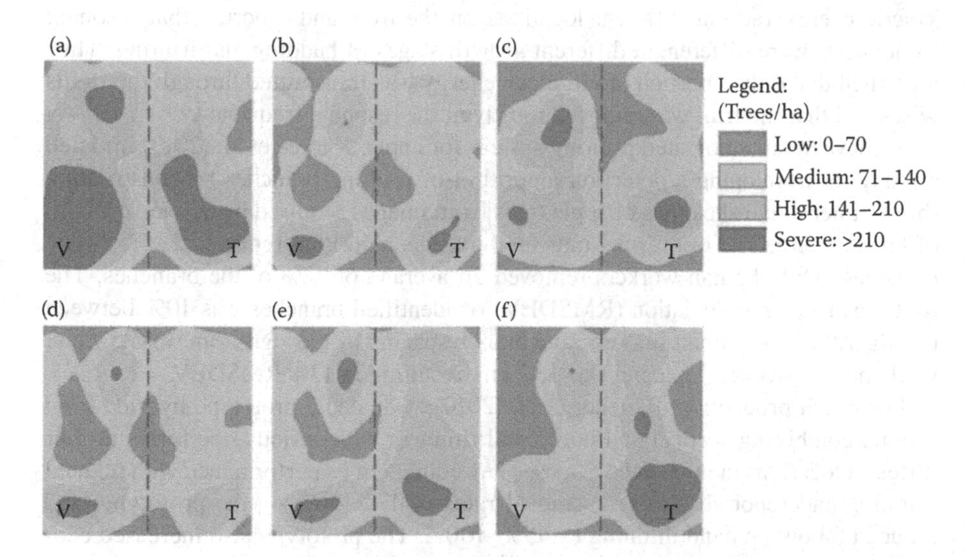

FIGURE 6.6 (See color insert.) HLB disease density maps in a citrus grove, obtained using various detection algorithms. The dashed line in the middle of each map indicates the boundary between training (T) and validation (V) sets. (a) Scouted infected trees, (b) MinDist result, (c) MahaDist result, (d) SAM result, (e) SID result, and (f) MTMF result. (From Li, X. et al. 2012. *Computers and Electronics in Agriculture*, 83:32–46. With permission.)

correct detection accuracy of 86%, while tree maturity status affected the disease detection accuracy.

Instead of using standard manned aircraft, Garcia-Ruiz et al. (2013) utilized a low-altitude unmanned aerial vehicle (UAV) to detect HLB disease for citrus. With a high spatial resolution, images acquired by the UAV performed better than those acquired by an aircraft when six spectral band images and seven vegetation indices were used for stepwise regression analysis. Accuracies and false-negatives were 67%–85% and 7%–32% for UAV images, and 61%–74% and 28%–45% for aerial images.

Besides citrus, tomato is also one of the most consumed specialty crops in the United States. Jones et al. (2010) investigated the spectral signature of tomato bacterial leaf spot disease in the ultraviolet, visible, and NIR regions, and analyzed reflectance measurements of healthy and diseased samples using partial least squares regression, correlation coefficients, and stepwise multiple linear regression. They identified important wavelengths to distinguish diseased tomato leaves, developed disease prediction models, and reported an RMSD of 4.9% in predicting disease severity in percent using the best prediction model.

6.5.4 Other Related Studies for Specialty Crops

Toward the development of an efficient mechanical harvester for cherry, Du et al. (2013) investigated vibratory energy requirements and harvest efficiency using kinetic energy on limbs of fruiting branches, and conducted experiments on upright fruiting offshoot trees for mechanical harvesting. They identified constant resonant frequencies of 8–10 Hz within the upright offshoots. They calculated a relative kinetic energy ratio at different locations on the tree, and reported that resonant frequencies were different at different growth stages of budding and fruiting. They identified that only a portion of vibratory energy was transmitted through the trellis wires and that the fruit was removed mostly at the resonant frequency.

To develop an automated pruning system for apple, Karkee et al. (2013) initiated a study for developing a detection algorithm of pruning branches by constructing three-dimensional skeletons of apple trees from images acquired by a time-of-flight-of-light camera. The test results indicated that the algorithm removed 19.5% of the branches, while human workers removed an average of 22% of the branches. The root mean squared deviation (RMSDEV) of identified branches was 10% between the algorithm and human workers, and branch spacing by the algorithm was 35.7 cm, while human workers' spacing was 33.7 cm, resulting in 13% RMSDEV.

For peach production, Baugher et al. (2010) evaluated a prototype hybrid string thinner combining vertical and horizontal thinners with various tree forms at four different locations in the United States. They compared its performance with manual thinning and reported blossom removal rates of 17%–56% by the prototype, and reduced follow-up hand thinning by 19%–100%. The prototype also increased economic savings of $236–$1490 per acre.

For lettuce production, commercial thinners are currently available (Agmechtronix, 2014; Vision Robotics Corp., 2014). For crops at a seedling stage, machine vision is used to detect seedlings to be removed, and herbicide is sprayed to kill them. The system can be attached to a three-point hitch of a tractor, and the operation speed is 3–4 miles/h. The system can provide advantages of savings over labor-intensive hand weeding and increase yield.

6.5.5 Challenges and Research Needs

As illustrated above, currently there are not many precision technologies available for specialty crop production due to lack of low-cost and affordable commercial sensing systems. Many sensing systems are still in the research stages, and it is relatively difficult to find sponsors for the specialty crop market due to its relatively small scale compared with other traditional grain and field crop markets. To address the needs of specialty crop industries, the U.S. government established the specialty crop research initiative (SCRI) program and provided $215 million for research and extension projects during 2008–2012 to encourage specialty crop research and commercialization of developed tools and technologies. More integrated research will be needed among many disciplines, including growers, agricultural engineers, agronomists, horticulturalists, and soil scientists, along with specialty crop growers to identify current challenges, and to develop more creative solutions with more investment and support from the general public.

6.6 SUMMARY AND FUTURE DEVELOPMENT

Significant progress has been made in precision agriculture technologies over the last two decades, despite the fact that the integration and adoption of these technologies have been relatively slow. Precision agriculture as a farming strategy is gradually changing the way farmers manage their fields. Some technologies developed for precision agriculture have become standard practices in production agriculture. Yield monitors and guidance systems are the two most widely used precision agriculture technologies by individual farmers today. Other technologies such as GPS-based soil sampling, real-time crop and soil sensors, and remote sensing have been used by some producers, crop consultants, and agricultural dealers for VRAs of fertilizers, herbicides, fungicides, water, seeds, and lime. Although research and field operations have demonstrated the feasibility and potential economic benefits of VRAs of various farming inputs, it remains a major challenge for the farmer to integrate all the technologies into his or her routine farming practice. Moreover, the farmer is still not completely convinced that VRT will change his or her bottom line and will be suitable for all the crops or fields.

Survey results indicate that most agricultural dealers provide one or more precision agriculture technologies, including VRT, for their customers. Owing to the sophistication, initial costs, and time commitment of these technologies, it is more appropriate for the dealers to provide precision agriculture services until more user-friendly and integrated systems are available. More research is needed to develop improved site-specific recommendation algorithms for sensor-based and map-based applications. It is important to keep the algorithms or decision rules simple enough for practical applications. VRA can be both profitable and environmentally beneficial for fields with large variability in crop yield and soil nutrients as well as for fields with patched weeds and isolated disease infections. However, VRA may not be suitable or necessary for fields with little soil and crop growth variability. More research is needed to develop criteria and guidelines to identify such fields or particular field operations (i.e., fertilization, pesticide application, irrigation, and seeding) for VRAs. VRA has great potential for high-value specialty crops such as vegetables and fruits, because of their high production cost, susceptibility to pests, and intensive labor requirements. More emphasis should be put on the development and application of precision agriculture technologies for specialty crop management.

Precision agriculture is an evolving technology. Although it is important to continue to develop new techniques and enhance existing technologies, more research should be devoted to the integration, application, and adoption of these technologies so that more and more farmers and dealers will be able to adapt them to current practices. Precision agriculture involves a great deal of technologies and requires additional investments of money and time, but it can be practiced at different levels depending on the specific field and crop conditions and the resources and technology services available to the farmer. If practiced properly, precision agriculture can improve farm input efficiency, increase farm profitability and minimize adverse environmental impacts, thus improving the long-term sustainability of production agriculture. After all, precision agriculture is becoming an indispensable component of agricultural production systems.

DISCLAIMER

The mention of a commercial product is solely for the purpose of providing specific information and should not be construed as a product endorsement by the authors or the institutions with which the authors are affiliated.

REFERENCES

Agmechtronix. 2014. *Row Crop Thinner*. Available at http://www.agmechtronix.com/Products/ RCT/. Accessed on July 18, 2014.

Al-Gaadi, K. and P. Ayers. 1999. Integrating GIS and GPS into a spatially variable rate herbicide application system. *Applied Engineering in Agriculture*, 15(4):255–262.

Baker, W.H. and B.A. Meggs. 2006. Using multispectral imagery to make variable rate applications in rice. In Norman, R.J., Meullenet J.F., and Moldenhauer, K.A.K. (eds.), *Research Series 540: B.R. Wells Rice Research Series 2005*. Fayetteville, AR: Arkansas Agricultural Experiment Station, pp. 102–106.

Bakhsh, A., R.S. Kanwar, T.B. Bailey, C.A. Cambardella, D.L. Karlen, and T.S. Colvin. 2002. Cropping system effects on NO_3–N loss with subsurface drainage water. *Transactions of the ASAE*, 45:1789–1797.

Baugher, T.A., J. Schupp, K. Ellis, J. Remcheck, E. Winzeler, R. Duncan, S. Johnson et al. 2010. String blossom thinner designed for variable tree forms increases crop load management efficiency in trials in four United States peach-growing regions. *HortTechnology*, 20(2):409–414.

Biermacher, J.T., F.M. Epplin, B.W. Brorsen, J.B. Solie, and W.R. Raun. 2009. Economic feasibility of site-specific optical sensing for managing nitrogen fertilizer for growing wheat. *Precision Agriculture*, 10:213–230.

Bjerre, K.D., L.N. Jørgensen, and J.E. Olesen. 2006. Site-specific management of crop diseases. In Srinivasan, A. (ed.), *Handbook of Precision Agriculture: Principles and Applications*. Binghamton, NY: The Haworth Press, Inc., pp. 185–205.

Boman, B., B. Sanden, R.T. Peters, and L. Parsons. 2012. Current status of microsprinkler irrigation in the United States. *Applied Engineering in Agriculture*, 28:359–366.

Bora, G.C., M.D. Schrock, D.L. Oard, and J. Grimm. 2011. Performance of pulse width modulated single to multifold outlet system for variable-rate anhydrous ammonia application. *Transactions of the ASABE*, 54(2):397–402.

Boyer, C.N., B.W. Brorsen, J.B. Solie, and W.R. Raun. 2011. Profitability of variable rate nitrogen application in wheat production. *Precision Agriculture*, 12:473–487.

Bramley, R.G.V. 2009. Lessons from nearly 20 years of precision agriculture research, development, and adoption as a guide to its appropriate application. *Crop & Pasture Science*, 60:197–217.

Bui, Q.D. 2005. Varitarget—A new nozzle with variable flow rate and droplet optimization. *ASAE*, St. Joseph, MI: ASAE, Paper No. 051125.

Bullock, D.S. and J. Lowenberg-DeBoer. 2007. Using spatial analysis to study the values of variable rate technology and information. *Journal of Agricultural Economics*, 58(3):517–535.

Bullock, D.S., M.L. Ruffo, D.G. Bullock, and G.A. Bollero. 2009. The value of variable rate technology: An information-theoretic approach. *American Journal of Agricultural Economics*, 91(1):209–223.

Camp, C.R., E.J. Sader, D.E. Evans, L.J. Usrey, and M. Omary. 1997. Modified center pivot system for precision management of water and nutrients. *Applied Engineering in Agriculture*, 14(1):23–31.

Carr, P.M., G.R. Carlson, J.S. Jacobsen, G.A. Nielsen, and E.O. Skogley. 1991. Farming soils, not fields: A strategy for increasing fertilizer profitability. *Journal of Production Agriculture*, 4:57–61.

Chávez, J.L., F.J. Pierce, T.V. Elliott, and R.G. Evans. 2010a. A remote irrigation monitoring and control system (RIMCS) for continuous-move systems: Part A. Description and development. *Precision Agriculture*, 11(1):1–10.

Chávez, J.L., F.J. Pierce, T.V. Elliott, R.G. Evans, Y. Kim, and W.M. Iversen. 2010b. A remote irrigation monitoring and control system (RIMCS) for continuous-move systems: Part B. Field testing and results. *Precision Agriculture*, 11(1):11–26.

Christensen, S., H.T. Søgaard, P. Kudsk, M. Nørremark, I. Lund, E.S. Nadimi, and R. Jørgensen. 2009. Site-specific weed control technologies. *Weed Research*, 49:233–241.

Coates, R.W., M.J. Delwiche, and P.H. Brown. 2006. Control of individual microsprinklers and fault detection strategies. *Precision Agriculture*, 7:85–99.

Dammer, K.-H. and D. Ehlert. 2006. Variable-rate fungicide spraying in cereals using a plant cover sensor. *Precision Agriculture*, 7:137–148.

Dammer, K.-H., H. Thole, T. Volk, and B. Hau. 2009. Variable-rate fungicide spraying in real time by combining a plant cover sensor and a decision support system. *Precision Agriculture*, 10:431–442.

Dhillon, R., V. Udompetaikul, F. Rojo, J. Roach, S. Upadhyaya, D. Slaughter, B. Lampinen, and K. Shackel. 2014. Detection of plant water stress using leaf temperature and microclimatic measurements in almond, walnut, and grape crops. *Transactions of the ASABE*, 57(1):297–304.

Du, X., D. Chen, Q. Zhang, P.A. Scharf, and M.D. Whiting. 2013. Response of UFO (upright fruiting offshoots) on cherry trees to mechanical harvest by dynamic vibration excitation. *Transactions of the ASABE*, 56(2):345–354.

Dukes, M.D. and C.D. Perry. 2006. Uniformity testing of variable-rate center pivot irrigation control system. *Precision Agriculture*, 7(2):205–218.

Ebel, R. and D. Schimmelpfennig. 2012. Production cost and the sequential adoption of precision technology. In *Proceedings of the Agricultural & Applied Economics Association's Annual Meeting*, Seattle, WA, August 12–14, 2012.

Esau, T.J., Q.U. Zaman, Y.K. Chang, D. Groulx, A.W. Schumann, and A.A. Farooque. 2014. Prototype variable rate sprayer for spot-application of agro chemicals in wild blueberry. *Applied Engineering in Agriculture,* 30(5):717–725.

Evans, R.G., S. Han, S.M. Schneider, and M.W. Kroeger. 1996. Precision center-pivot irrigation for efficient use of water and nitrogen. In Robert, P.C., Rust, R.H., and Larson, W.E. (eds.), *Proceedings of the 3rd International Conference on Precision Agriculture*, Madison, WI: ASA-CSSA-SSSA, pp. 75–84.

Evans, R.G., W.M. Iversen, W.B. Stevens, and J.D. Jabro. 2010. Development of combined site-specific MESA and LEPA methods on a linear move sprinkler irrigation system. *Applied Engineering in Agriculture*, 26(5):883–895.

Evans, R.G. and B.A. King. 2012. Site-specific sprinkler irrigation in a water-limited future. *Transactions of the ASABE*, 55(2):493–504.

Evans, R.G., J. LaRue, K.C. Stone, and B.A. King. 2013. Adoption of site-specific variable rate sprinkler irrigation systems. *Irrigation Science*, 31:871–887.

Fernandez, C.J., K.J. McInnes, and J.T. Cothren. 1996. Water status and leaf area production in water—and nitrogen-stressed cotton. *Crop Science*, 36:1224–1233.

Fereres, E. and D.J. Connor. 2004. Sustainable water management in agriculture. In Cabrera, E. and Cobacho R. (eds.), *Challenges of the New Water Policies for the XXI Century*, Lisse, The Netherlands: A. A. Balkema, 157–170.

Fraisse, C.W., H.R. Duke, and D.F. Heermann. 1995a. Laboratory evaluation of variable water application with pulse irrigation. *Transactions of the ASAE*, 38(5):1363–1369.

Fraisse, C.W., D.F. Heermann, and H.R. Duke. 1992. Modified linear-move system for experimental water application. In Jeyen, J., Mwendera, E., and Badji, M. (eds.), *Proceedings of the Advances in Planning, Design, and Management of Irrigation Systems as Related to Sustainable Land Use*, Leuven, Belgium: Center for Irrigation Engineering, 1, pp. 367–376.

Fraisse, C.W., D.F. Heermann, and H.R. Duke. 1995b. Simulation of variable water application with linear-move irrigation systems. *Transactions of the ASAE*, 38(5):1371–1376.

Garcia-Ruiz, F., S. Sankaran, J.M. Maja, W.S. Lee, J. Rasmussen, and R. Ehsani. 2013. Comparison of two aerial imaging platforms for identification of Huanglongbing-infected citrus trees. *Computers and Electronics in Agriculture*, 91:106–115.

Gerhards, R. and S. Christensen. 2006. Site-specific weed management. In Srinivasan, A. (ed.), *Handbook of Precision Agriculture: Principles and Applications.* Binghamton, NY: The Haworth Press, Inc., pp. 185–205.

Gerik, T.J., D.M. Oosterhuis, and H.A. Torbert. 1998. Managing cotton nitrogen supply. *Advances in Agronomy*, 64:115–147.

Giles, D.K., G.W. Henderson, and K. Funk. 1996. Digital control of flow rate and spray droplet size from agricultural nozzles for precision chemical application. In Roberts, P.C., Rust R.H., and Larson, W.E. (eds.), *Proceedings of the 3rd International Conference on Precision Agriculture*, Madison, WI: ASA/CSSA/SSSA, pp. 729–738.

Giles, D.K. and D.C. Slaughter. 1997. Precision band sprayer with machine-vision guidance and adjustable yaw nozzles. *Transactions of the ASAE*, 40(1):29–36.

Griffin, T.W. and J. Lowenberg-DeBoer. 2005. Worldwide adoption and profitability of precision agriculture. *Revista de Politica Agricola*, 14:20–38.

Grisso, R.B., M. Alley, W. Thomason, D. Holshouser, and G.T. Roberson. 2011. *Precision Farming Tools: Variable-Rate Application*. Blacksburg, VA: Virginia Cooperative Extension, pp. 442–505.

Han, S., L.L. Hendrickson, B. Ni, and Q. Zhang. 2001. Modification and testing of a commercial sprayer with PWM solenoids for precision spraying. *Applied Engineering in Agriculture*, 17(5):591–594.

Han, Y.J., A. Khalilian, T.O. Owino, H.J. Farahani, and S. Moor. 2009. Development of Clemson variable-rate lateral irrigation system. *Computers and Electronics in Agriculture*, 68:108–113.

Hanks, J.E. and J.L. Beck. 1998. Sensor-controlled hooded sprayer for row crops. *Weed Technology*, 12:308–314.

Holland, J.K., B. Erickson, and D.A. Widmar. 2013. *Precision Agricultural Services Dealership Survey Results*. Willoughby, OH: CropLife Magazine and West Lafayette, IN: The Center for Food and Agricultural Business, Purdue University.

Humburg, D. 2003. *Site-Specific Management Guidelines: Variable Rate Equipment Technology for Weed Control.* Brooking, SD: Department of Agricultural and Biosystems Engineering, South Dakota State University, pp. SSMG-7.

Isakeit, T., R.R. Minzenmayer, A. Abrameit, G. Moore, and J.D. Scasta. 2010. Control of phymatotrichopsis root rot of cotton with flutriafol. In *Proceedings of the Beltwide Cotton Conferences*, Memphis, TN: National Cotton Council of America, pp. 200–203.

Johnston, A.E. 2000. Efficient use of nutrients in agricultural production systems. *Communication in Soil Science and Plant Analysis*, 31:1599–1620.

Jones, C.D., J.B. Jones, and W.S. Lee. 2010. Diagnosis of bacterial spot of tomato using spectral signatures. *Computers and Electronics in Agriculture*, 74(2):329–335.

Karkee, M., B. Adhikari, S. Amatya, and Q. Zhang. 2013. Identification of pruning branches in tall spindle apple trees for automated pruning. *Computers and Electronics in Agriculture*, 103:127–135.

Khalilian, A., W. Henderson, Y. Han, W. Porter, and E. Barnes. 2011. Sensor based nitrogen management for cotton production in coastal plain. In *Proceedings of the Beltwide Cotton Conferences*, Memphis, TN: National Cotton Council of America, pp. 531–537.

Khalilian, A., W. Henderson, Y. Han, and P.J. Wiatrak. 2008. Improving nitrogen use efficiency in cotton through optical sensing. In *Proceedings of the Beltwide Cotton Conferences*, Memphis, TN: National Cotton Council of America, pp. 583–587.

King, B.A., G.L. Foster, D.C. Kincaid, and R.B. Wood. 1998. Variable flow sprinkler head. U.S. Patent No. 5785246.

King, B.A. and D.C. Kincaid. 2004. A variable flow rate sprinkler for site-specific irrigation management. *Applied Engineering in Agriculture*, 20(6):765–770.

King, B.A., R.W. Wall, D.C. Kincaid, and D.T. Westermann. 2005. Field testing of a variable rate sprinkler and control system for site-specific water and nutrient application. *Applied Engineering in Agriculture*, 21(5):847–853.

Koch, B., R. Khosla, W.M. Frasier, D.G. Westfall, and D. Inman. 2004. Economic feasibility of variable-rate nitrogen application utilizing site-specific management zones. *Agronomy Journal*, 96:1572–1580.

Koller, M. and W.T. Lanini. 2005. Site-specific herbicide applications based on weed maps provide effective control. *California Agriculture*, 59(3):182–187.

Kumar, A., W.S. Lee, R. Ehsani, L.G. Albrigo, C. Yang, and R.L. Mangan. 2012. Citrus greening disease detection using aerial hyperspectral and multispectral imaging techniques. *Journal of Applied Remote Sensing*, 6:063542.

Lamm, F.R., J.P. Bordovsky, L.J. Schwankl, G.L. Grabow, J. Enciso-Medina, R.T. Peters, P.D. Colaizzi. 2012. Subsurface drip irrigation: Status of the technology in 2010. *Transactions of the ASABE*, 55:483–491.

LaRue, J. and R. Evans. 2012. Considerations for variable rate irrigation. In *Proceedings of the 24th Annual Central Plains Irrigation Conference*, Colby, Kansas, February 21–22, 2012, pp. 111–116.

Li, H., W.S. Lee, K. Wang, R. Ehsani, and C. Yang. 2014. Extended spectral angle mapping (ESAM) for citrus greening disease detection using airborne hyperspectral imaging. *Precision Agriculture*, 15(2):162–183.

Li, X., W.S. Lee, M. Li, R. Ehsani, A. Mishra, C. Yang, and R. Mangan. 2012. Spectral difference analysis and airborne imaging classification for citrus greening infected trees. *Computers and Electronics in Agriculture*, 83:32–46.

Liu, H., H. Zhu, Y. Shen, Y. Chen, and H.E. Ozkan. 2014. Development of digital flow control system for multi-channel variable-rate sprayers. *Transactions of the ASABE*, 57(1):273–281.

McCann, I.R. and J.C. Stark. 1993. Method and apparatus for variable application of irrigation water and chemicals. U.S. Patent No. 5,246,164.

Meron, M., R. Hallel, M. Peres, B. Bravdo, and R. Wallach. 2001. Tensiometer actuated automatic micro irrigation of apples. *ISHS Acta Horticulturae*, 562: 63–69.

Miller, W.M., A.W. Schumann, J.D. Whitney, and S. Buchanon. 2005. Variable rate applications of granular fertilizer for citrus test plots. *Applied Engineering in Agriculture*, 21(5):795–801.

Min, M., W.S. Lee, T.F. Burks, J.D. Jordan, A.W. Schumann, J.K. Schueller, and H. Xie. 2008. Design of a hyperspectral nitrogen sensing system for citrus. *Computers and Electronics in Agriculture*, 63(2):215–226.

Mulla, D.J., A.U. Bhatti, M.W. Hammond, and J.A. Benson. 1992. A comparison of winter wheat yield and quality under uniform versus spatially variable fertilizer management. *Agriculture, Ecosystems & Environment*, 38:301–311.

Nolte, K.D., M.C. Siemens, and P. Andrade-Sanchez. 2011. *Integrating Variable Rate Technologies for Soil-Applied Herbicides in Arizona Vegetable Production.* Publication No. AZ1538. Tucson, AZ: The University of Arizona Cooperative Extension.

Oerke, E.-C., R. Gerhards, G. Menz, and R.A. Sikora. 2010. *Precision Crop Protection—The Challenge and Use of Heterogeneity*. Heidelberg, Germany: Springer.

Omary, M., C.R. Camp, and E.J. Sadler. 1997. Center pivot irrigation system modification to provide variable water application depths. *Applied Engineering in Agriculture*, 13(2):235–239.

O'Shaughnessy, S.A., Y.F. Urrego, S.R. Evett, P.D. Colaizzi, and T.A. Howell. 2013. Assessing application uniformity of a variable rate irrigation system in a windy location. *Applied Engineering in Agriculture*, 29(4):497–510.

Parsons, L.R., W. Bandaranayake, J. Holeton, and D. Lankford. 2010. Automatic citrus irrigation in Florida using capacitance probes. In *Proceedings of the Third International Symposium on Soil Water Measurement Using Capacitance, Impedance and TDT*, Murcia, Spain. April 7–9, 2010.

Pelletier, V., J. Gallichand, and J. Caron. 2013. Effect of soil water potential threshold for irrigation on cranberry yield and water productivity. *Transactions of the ASABE*, 56(6):1325–1332.

Perry, C., M.D. Dukes, and K.A. Harrison. 2004. Effects of variable rate sprinkler cycling on irrigation uniformity. ASAE Paper No. 04-1117. St. Joseph, MI: ASABE.

Perry, C., S. Pocknee, and O. Hansen. 2003. A variable rate pivot irrigation control system. In Stafford J. and Werner A. (eds.), *Proceedings of the Fourth European Conference on Precision Agriculture*, pp. 539–544.

Pierce, F.J., J.L. Chavez, T.V. Elliott, G.R. Matthews, R.G. Evans, and Y. Kim. 2006. A remote real-time continuous-move irrigation control and monitoring system. ASABE Paper No. 062162. St. Joseph, MI: ASABE.

Raun, W.R. and G.V. Johnson. 1999. Improving nitrogen use efficiency for cereal production. *Agronomy Journal*, 91:357–363.

Raun, W.R., J.B. Solie, G.V. Johnson, M.L. Stone, E.V. Lukina, W.E. Thomason, and J.S. Schepers. 2001. In-season prediction of potential grain yield in winter wheat using canopy reflectance. *Agronomy Journal*, 93:131–138.

Raun, W.R., J.B. Solie, G.V. Johnson, M.L. Stone, R.W. Mullen, K.W. Freeman, W.E. Thomason, and E.V. Lukina 2002. Improving nitrogen use efficiency in cereal grain production with optical sensing and variable rate application. *Agronomy Journal*, 94:815–820.

Robertson, M.J., R.S. Llewellyn, R. Mandel, R. Lawes, R.G.V. Bramley, L. Swift, N. Metz, and C. O'Callaghan. 2011. Adoption of variable rate technology in the Australian grains industry: Status, issues and prospects. *Precision Agriculture*, 13:181–199.

Saleem, S.R., Q.U. Zaman, A.W. Schumann, A. Madani, Y.K. Chang, and A.A. Farooque. 2013a. Impact of variable rate fertilization on nutrient losses in surface runoff for wild blueberry fields. *Applied Engineering in Agriculture*, 30(2):179–185.

Saleem, S.R., Q.U. Zaman, A.W. Schumann, A. Madani, A.A. Farooque, and D.C. Percival. 2013b. Impact of variable rate fertilization on subsurface water contamination in wild blueberry cropping system. *Applied Engineering in Agriculture*, 29(2):225–232.

Scharf, P., L.F. Oliveira, E.D. Vories, G. Stevens, D. Dunn, and K.A. Sudduth. 2008. Managing N with sensors: Some practical issue. In *Proceedings of the Beltwide Cotton Conferences*, Memphis, TN: National Cotton Council of America, pp. 1585–1588.

Scharf, P.C., D.K. Shannon, H.L. Palm, K.A. Sudduth, S.T. Drummond, N.R. Kitchen, L.J. Mueller,V.C. Hubbard, and L.F. Oliveira. 2011. Sensor-based nitrogen applications out-performed producer-chosen rates for corn in on-farm demonstrations. *Agronomy Journal*, 103:1683–1691.

Schimmelpfennig, D. and R. Ebel. 2011. *On the Doorstep of the Information Age: Recent Adoption of Precision Agriculture.* Publication No. EIB-80. Washington, DC: U.S. Department of Agriculture, Economic Research Service.

Schumann, A.W., W.M. Miller, Q.U. Zaman, K.H. Hostler, S. Buchanon, and S. Cugati. 2006. Variable rate granular fertilization of citrus groves: Spreader performance with single-tree prescription zones. *Applied Engineering in Agriculture*, 22(1):19–24.

Sharda, A., M. Karkee, Q. Zhang, I. Ewlanow, U. Adameit, and J. Brunner. 2014. Effect of emitter type and mounting configuration on spray coverage for solid set canopy delivery system. *Computers and Electronics in Agriculture*, 112:184–192.

Sharma, A., G. Dilawari, S. Osborne, J.C. Banks, R. Taylor, and P. Weckler. 2008. On-the-go sensor system for cotton management. In *Proceedings of the Beltwide Cotton Conferences*, Memphis, TN: National Cotton Council of America, pp. 588–593.

Shearer, S.A. and P.T. Jones. 1991. Selective application of post-emergence herbicides using photoelectrics. *Transactions of the ASAE*, 34(4):1661–1666.

Srinivasan, A. 2006. *Handbook of Precision Agriculture: Principles and Applications.* Binghamton, NY: The Haworth Press, Inc.

Steward, B.L. and D.S. Humburg. 2000. Modeling the Raven SCS-700 chemical injection system with carrier control with sprayer simulation. *Transactions of the ASAE*, 43(2):231–245.

Stone, K.C., E.J. Sadler, J.A. Millen, D.E. Evans, and C.R. Camp. 2006. Water flow rates from a site-specific irrigation system. *Applied Engineering in Agriculture*, 22(1):73–78.

Sudduth, K.A., S.C. Borgelt, and J. Hou. 1995. Performance of a chemical injection sprayer system. *Applied Engineering in Agriculture*, 11(3):343–348.

Sui, R. and D.K. Fisher. 2014. Field test of a center pivot irrigation system. *Applied Engineering in Agriculture*, 31(1):83–88. DOI: 10.13031/aea.31.10539.

Sui, R. and J.A. Thomasson. 2006. Ground-based sensing system for cotton nitrogen status determination. *Transactions of the ASABE*, 49(6):1983–1991.

Sui, R., J.B. Wilkerson, W.E. Hart, L.R. Wilhelm, and D.D. Howard. 2005. Multi-spectral sensor for detection of nitrogen status in cotton. *Applied Engineering in Agriculture*, 21(2):167–172.

Sui, R., J.B. Wilkerson, L.R. Wilhelm, and F.D. Tompkins. 1989. A Microcomputer-based morphometer for bush-type plants. *Computer and Electronics in Agriculture*, 4:43–58.

Swinton, S.M. and J. Lowenberg-DeBoer. 1998. Evaluating the profitability of site-specific farming. *Journal of Production Agriculture*, 11:439–446.

Taylor, R., J.C. Banks, S. Osborne, T. Sharp, J. Solie, and B. Raun. 2007. In-season cotton management using real time sensors. In *Proceedings of the Beltwide Cotton Conferences*, Memphis, TN: National Cotton Council of America, pp. 908–912.

Taylor, R. and J. Fulton. 2010. *Sensor-Based Variable Rate Application for Cotton*. http://www.cottoninc.com/fiber/AgriculturalDisciplines/Engineering/Precision-Crop-Management-for-Cotton/Sensor-Based-Variable-Rate-App/Sensor-Based-App-Oct-2010.pdf. Accessed on November 3, 2014.

Thornton, P.K., R.H. Fawcett, J.B. Dent, and T.J. Perkins. 1990. Spatial weed distribution and economic thresholds for weed control. *Crop Protection*, 9:337–342.

Thorp, K.R. and L.F. Tian. 2004. Performance study of variable rate herbicide application based on remote sensing imagery. *Biosystems Engineering*, 88(1):35–47.

Tian, L. 2002. Sensor-based precision chemical application system. *Computers and Electronics in Agriculture*, 36(23):133–149.

Tompkins, F.D., K.D. Howard, C.R. Mote, and R.S. Freeland. 1990. Boom flow characteristics with direct chemical injection. *Transactions of the ASAE*, 33(3):737–743.

Torre-Neto, A., R.A. Ferrarezi, D.E. Razera, E. Speranza, W.C. Lopes, T.P.F.S. Lima, L.M. Rabello, and C.M.P. Vaz. 2005. Wireless sensor network for variable rate irrigation in citrus. *Information and Technology for Sustainable Fruit and Vegetable Production*. FRUTIC 05, Montpellier, France, September 12–16, 2005.

Torre-Neto, A., J.K. Schueller, and D.Z. Haman. 2001. Automated system for variable rate microsprinkler irrigation in citrus: A demonstration unit. In *Proceedings of the Third European Conference on Precision Agriculture*, Montpellier, France, June 18–20, 2001. CDROM, pp. 725–730.

Tumbo, S.D., M. Salyani, J.D. Whitney, T.A. Wheaton, and W.M. Miller. 2002. Sensors for measurements of citrus canopy volume. *Applied Engineering in Agriculture*, 18(3):367–372.

UC IPM Online. 2014. *Asian Citrus Psyllid and Huanglongbing Disease*. Available at http://www.ipm.ucdavis.edu/PMG/PESTNOTES/pn74155.html. Accessed on November 3, 2014.

USDA. 2014. *Definition of Specialty Crop*. Available at www.nifa.usda.gov/funding/scri/ scri.html. Accessed on November 3, 2014.

USEPA. 2003. National management measures to control nonpoint pollution from agriculture. EPA 841-B-03-004. Washington, DC: USEPA.

Vetch, J.A., G.L. Malzer, P.C. Robert, and D.R. Huggins. 1995. Nitrogen specific management by soil condition: Managing fertilizer nitrogen in corn. In Robert, P.C., Rust, R.H., and Larson, W.E. (eds.), *Site Specific Management for Agricultural Systems*. Madison WI: ASA/CSSA/ SSSA.

Vision Robotics Corp. 2014. *Application—Lettuce Thinner*. Available at http://www.vision-robotics.com/vrc/index.php?option=com_phocagallery&view=category&id=3:lettuce-thinner&Itemid=26. Accessed on November 3, 2014.

Wibawa, W.D., D.L. Dludlu, L.J. Swenson, D.G. Hopkins, and W.C. Dahnke. 1993. Variable fertilizer application based on yield goal, soil fertility, and soil map unit. *Journal of Production Agriculture*, 6:255–261.

Williams, M.M., R. Gerhards, and D.A. Mortensen. 2000. Two-year weed seedling population responses to a post-emergent method of site-specific weed management. *Precision Agriculture*, 2:247–63.

Yang, C. 2000. A variable rate applicator for controlling rates of two liquid fertilizers. *Applied Engineering in Agriculture*, 17(3):409–417.

Yang, C., J.H. Everitt, and J.M. Bradford. 2000. Comparison of uniform and variable rate nitrogen and phosphorus fertilizer applications for grain sorghum. *Transactions of the ASAE*, 44(2):201–209.

Yang, C., C.J. Fernandez, and J.H. Everitt. 2005. Mapping *Phymatotrichum* root rot of cotton using airborne three-band digital imagery. *Transactions of the ASAE*, 48(4):1619–1626.

Yang, C., G.N. Odvody, C.J. Fernandez, J.A. Landivar, R.R. Minzenmayer, and R.L. Nichols. 2012. Monitoring cotton root rot progression within and across growing seasons using remote sensing. In *Proceedings of the Beltwide Cotton Conferences*, Memphis, TN: National Cotton Council of America, pp. 475–480.

Zhang, M., Z. Qin, and X. Liu. 2005. Remote sensed spectral imagery to detect late blight in field tomatoes. *Precision Agriculture*, 6:489–508.

Zaman, Q. and A.W. Schumann. 2006. Nitrogen management zones for citrus based on variation in soil properties and tree performance. *Precision Agriculture*, 7:45–663.

Zaman, Q., A.W. Schumann, and W.M. Miller. 2005. Variable rate nitrogen application in Florida citrus based on ultrasonically-sensed tree size. *Applied Engineering in Agriculture*, 21(3):331–335.

Zhang, Q. and F.J. Pierce. 2013. *Agricultural Automation: Fundamentals and Practices*. Boca Raton, FL: CRC Press.

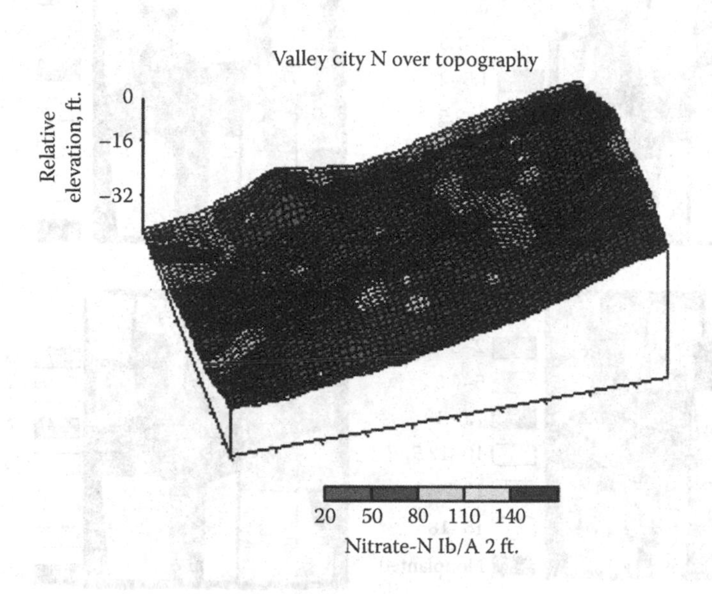

FIGURE 1.5 Residual soil nitrate from Valley City, North Dakota, over the landscape.

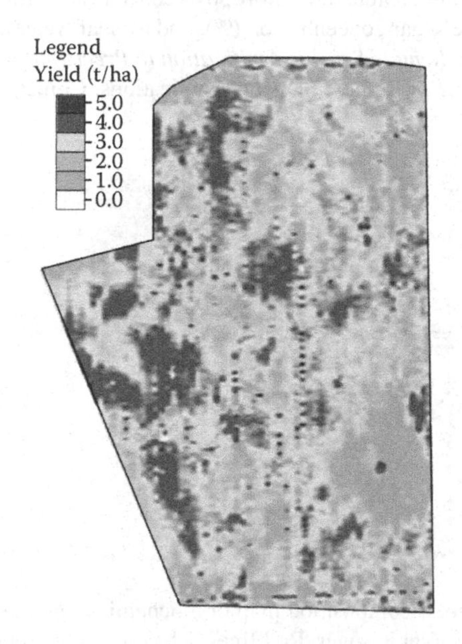

FIGURE 2.4 Typical combine-derived yield map.

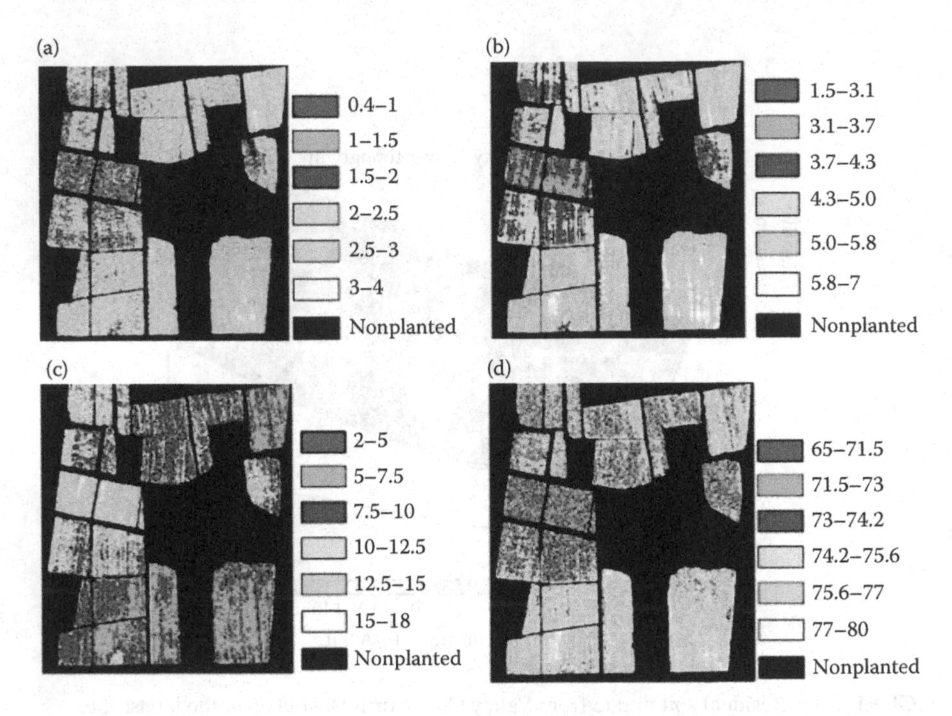

FIGURE 3.9 Map plots of biochemical parameters, including chlorophyll, total nitrogen, soluble sugar, and leaf water content. (a) Chlorophyll concentration (mg g^{-1}), (b) nitrogen concentration (%), (c) soluble sugar concentration (%), and (d) leaf water content (%). (From Liu, L.Y. 2002. *Hyperspectral Remote Sensing Application in Precision Agriculture*. Postdoctoral research report of Institute of Remote Sensing Applications, Chinese Academy of Sciences. With permission.)

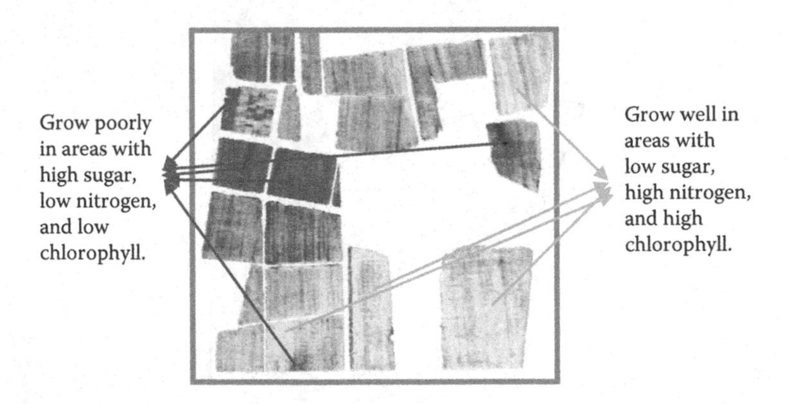

FIGURE 3.10 Pseudocolor composition map of biochemical parameters, including chlorophyll, total nitrogen, and soluble sugar. Red lines: The crops grow poorly in areas with high sugar, low nitrogen, and low chlorophyll. Green lines: The crops grow well in the areas with low sugar, high nitrogen, and high chlorophyll. (From Liu, L.Y. 2002. *Hyperspectral Remote Sensing Application in Precision Agriculture*. Postdoctoral research report of Institute of Remote Sensing Applications, Chinese Academy of Sciences. With permission.)

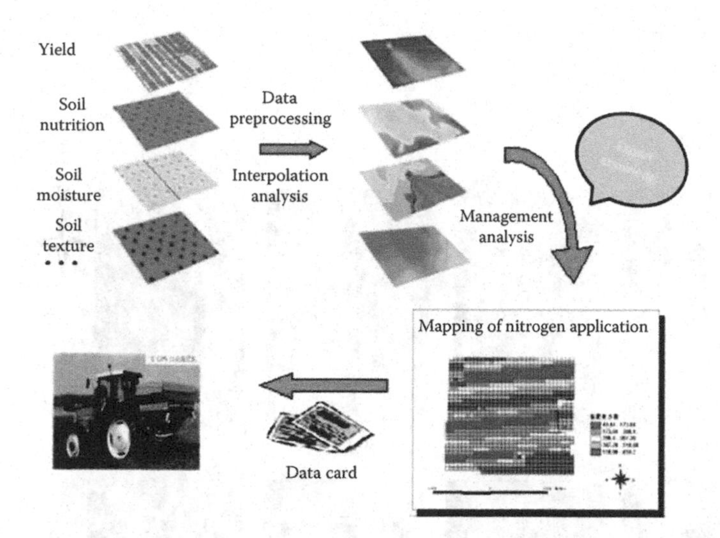

FIGURE 3.16 Key links in decision making for precision agriculture management and prescription generation. (From Chen, L.P. et al. 2002. *Transactions of the CSAE*, 18(2):1145–1148. With permission.)

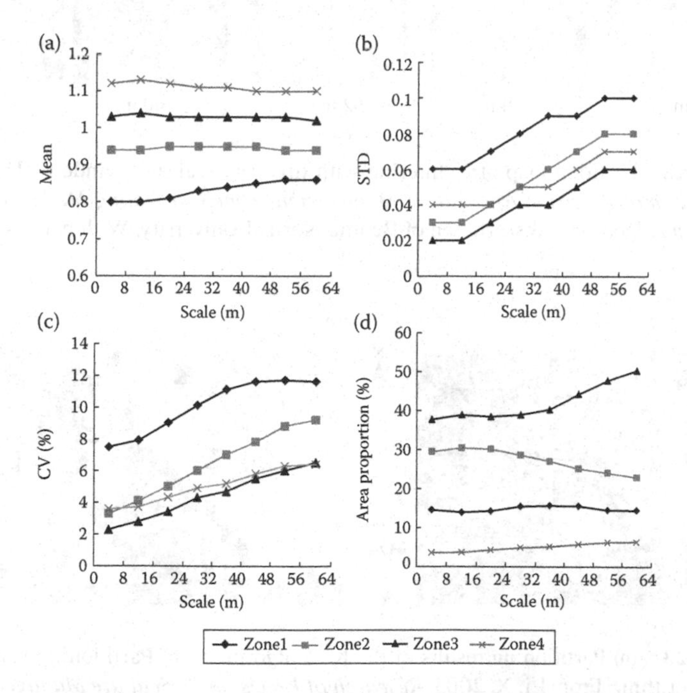

FIGURE 3.20 Changes in the mean, standard deviation, coefficient of variation, and proportion of area with scale for each partition. (a) Illustrates mean value changes with scale increasing for each partition; (b) illustrates standard deviation value changes with scale increasing for each partition; (c) illustrates coefficient of variation value changes with scale increasing for each partition; (d) illustrates area proportion (%) changes with scale increasing for each partition. (From Li, X. 2005. *Research of Precision Agriculture Management Zone Generating Methods Based on '3S' Technique.* Doctorate dissertation of Beijing Normal University. With permission.)

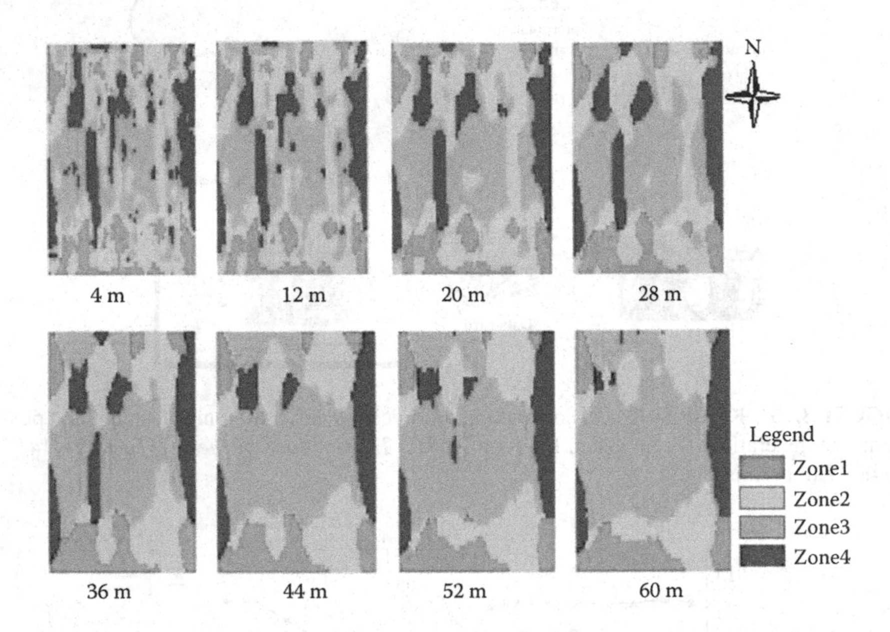

FIGURE 3.22 Partition map after filtering with different scales of window. (From Li, X. 2005. *Research of Precision Agriculture Management Zone Generating Methods Based on '3S' Technique*. Doctorate dissertation of Beijing Normal University. With permission.)

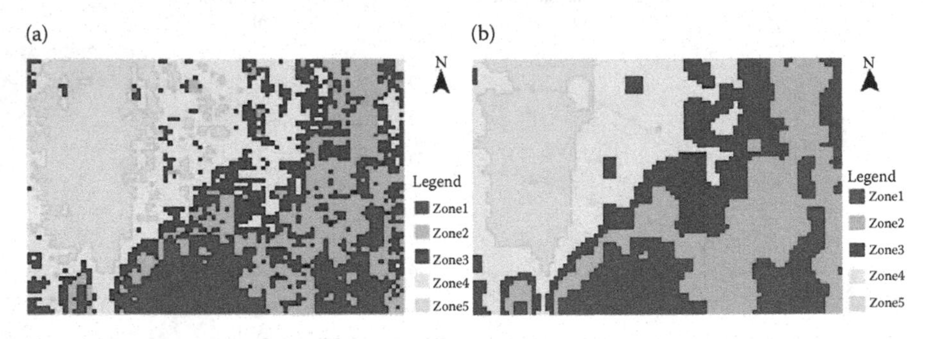

FIGURE 3.23 (a) Partitioning results of the K-M algorithm. (b) Partitioning results of the SC-KM algorithm. (From Li, X. 2005. *Research of Precision Agriculture Management Zone Generating Methods Based on '3S' Technique*. Doctorate dissertation of Beijing Normal University. With permission.)

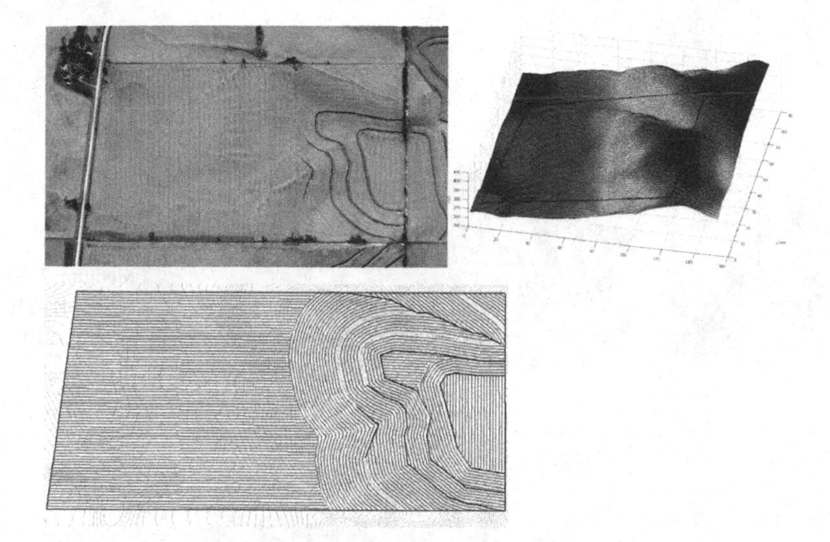

FIGURE 5.7 Examples of an optimized 3D coverage path planning algorithm for a 3D terrain where terraces and valleys exist. (From Jin, J. and L. Tang. 2011. *Journal of Field Robotics*, 28:424–440. With permission.)

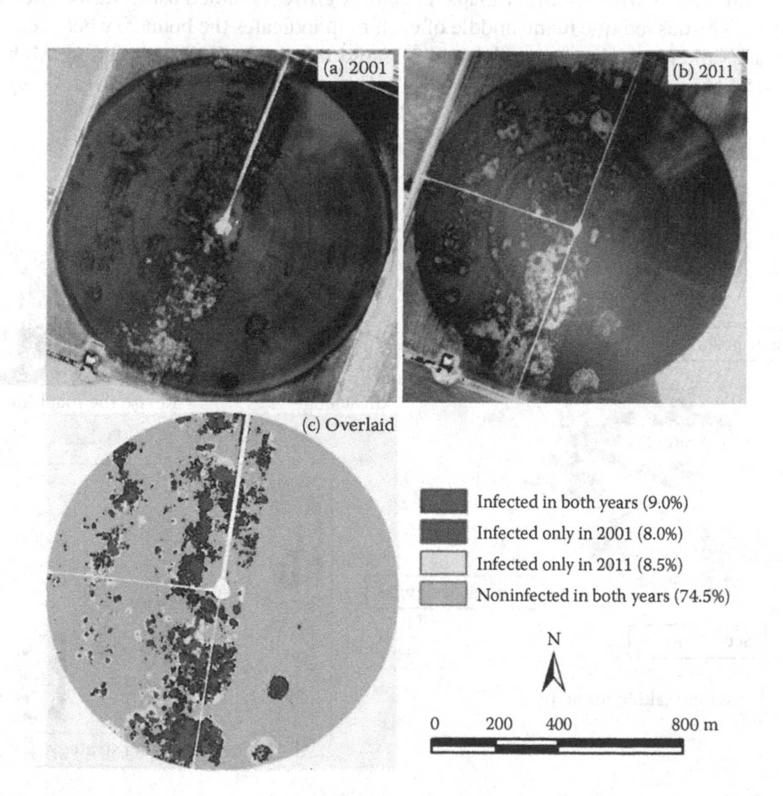

FIGURE 6.3 Color-infrared images taken in (a) 2001 and (b) 2011 and (c) overlaid classification map for an irrigated cotton field infected with root rot near Edroy, Texas. (Adapted from Yang, C., C.J. Fernandez, and J.H. Everitt. 2005. *Transactions of the ASAE*, 48(4):1619–1626; Yang, C. et al. 2012. *Proceedings of the Beltwide Cotton Conferences*, Memphis, TN: National Cotton Council of America, pp. 475–480.)

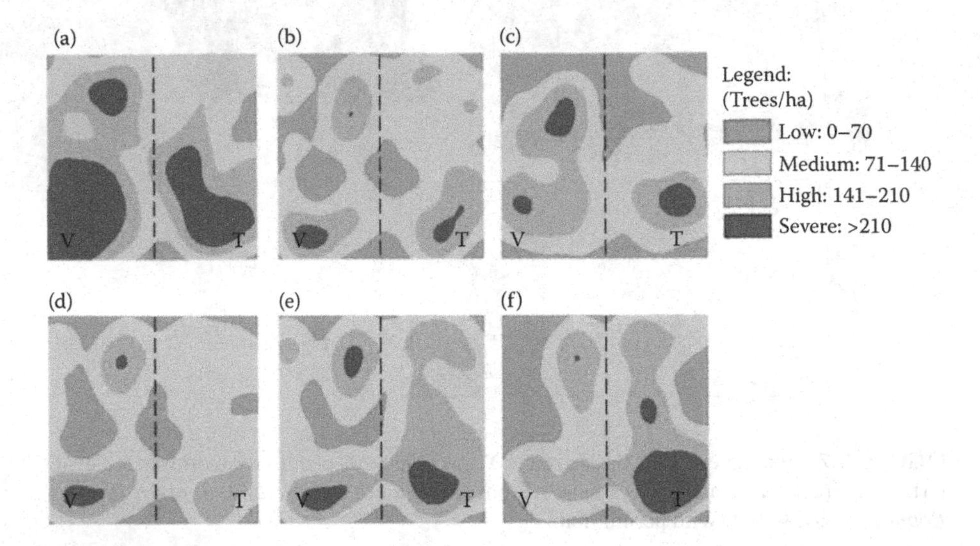

FIGURE 6.6 HLB disease density maps in a citrus grove, obtained using various detection algorithms. The dashed line in the middle of each map indicates the boundary between training (T) and validation (V) sets. (a) Scouted infected trees, (b) MinDist result, (c) MahaDist result, (d) SAM result, (e) SID result, and (f) MTMF result. (From Li, X. et al. 2012. *Computers and Electronics in Agriculture*, 83:32–46. With permission.)

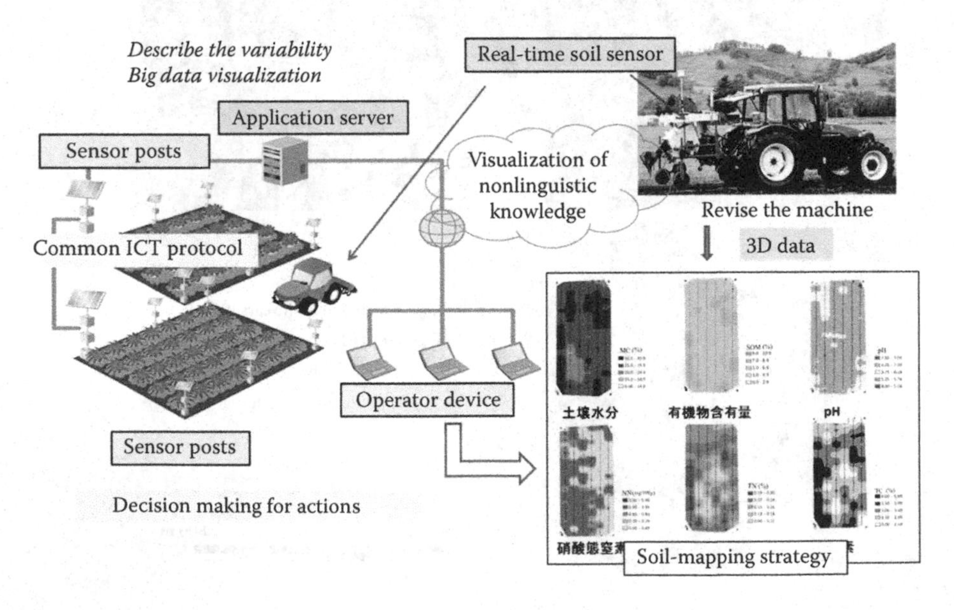

FIGURE 7.10 Precision restoring agriculture in Fukushima toward traceable management against rumor damage.

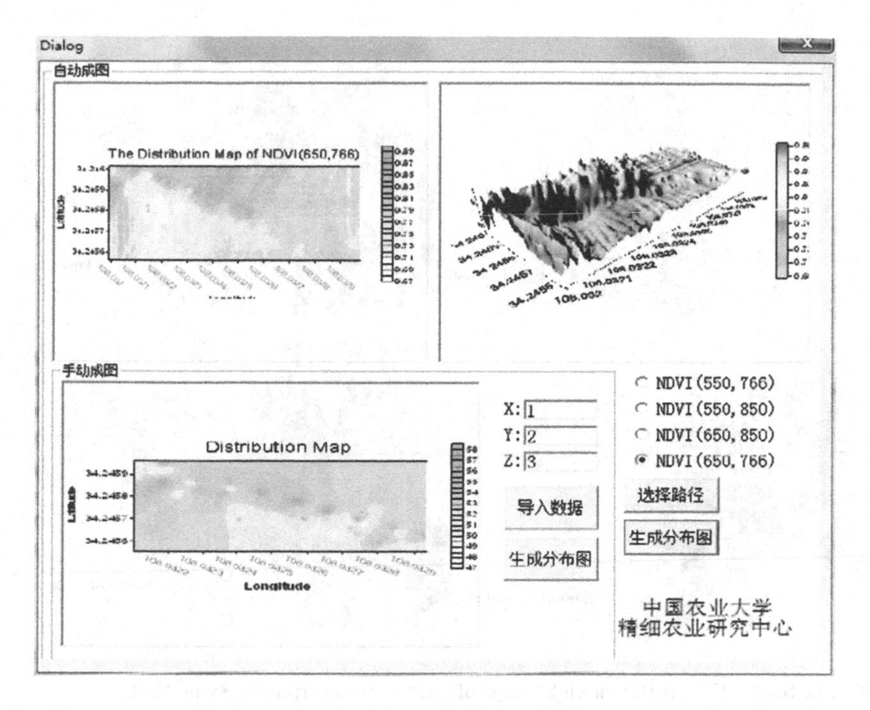

FIGURE 8.19 Distribution of chlorophyll content of wheat.

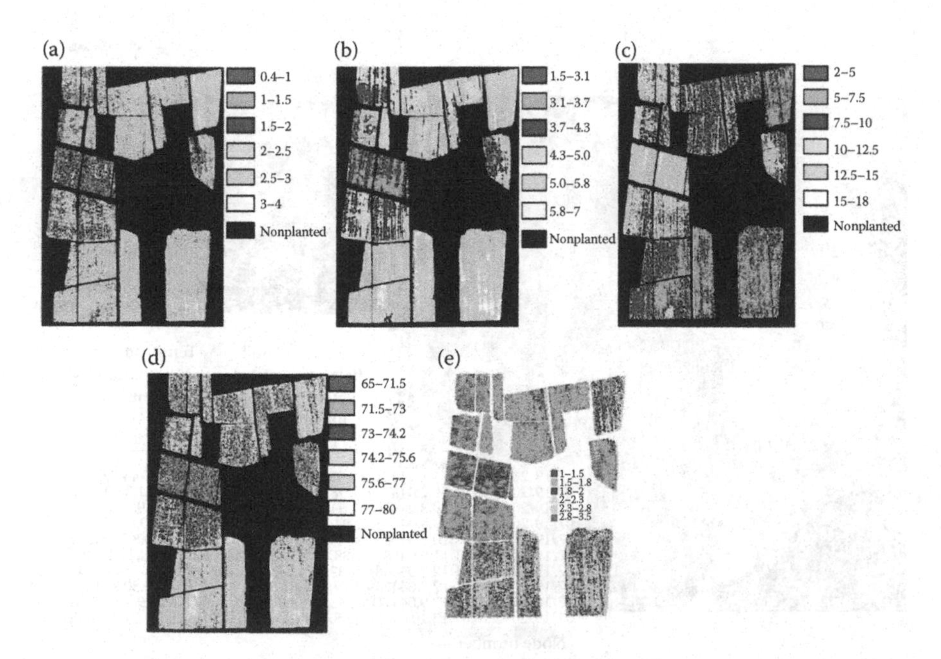

FIGURE 8.39 Maps of chlorophyll, total nitrogen, soluble sugar, and water of leaf biochemical parameter. (a) Chlorophyll content (mg/g), (b) nitrogen content (%), (c) soluble sugar content (%), (d) water content (%), and (e) LAI.

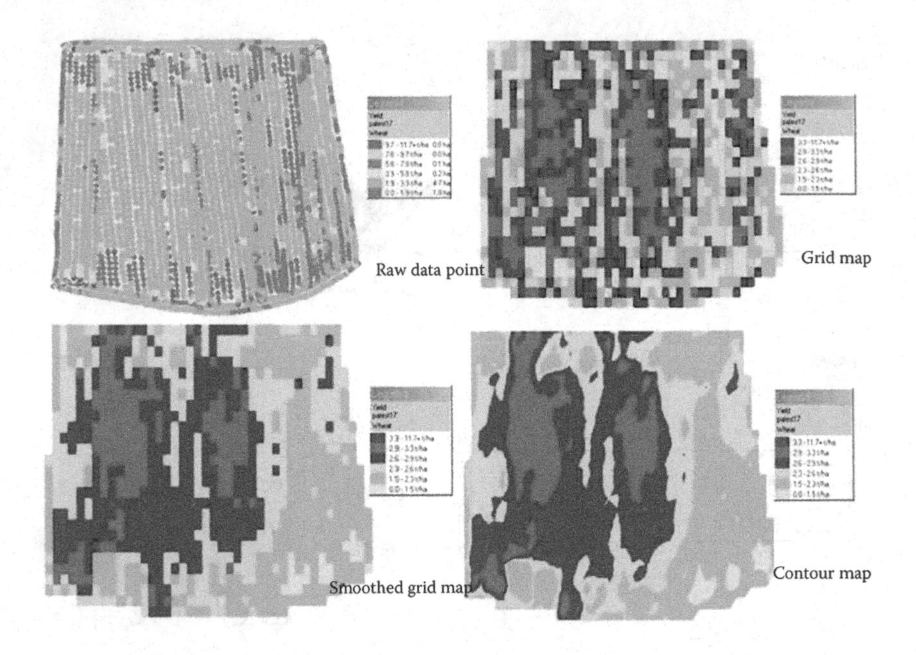

FIGURE 8.40 Four different yield maps of winter wheat from the same field.

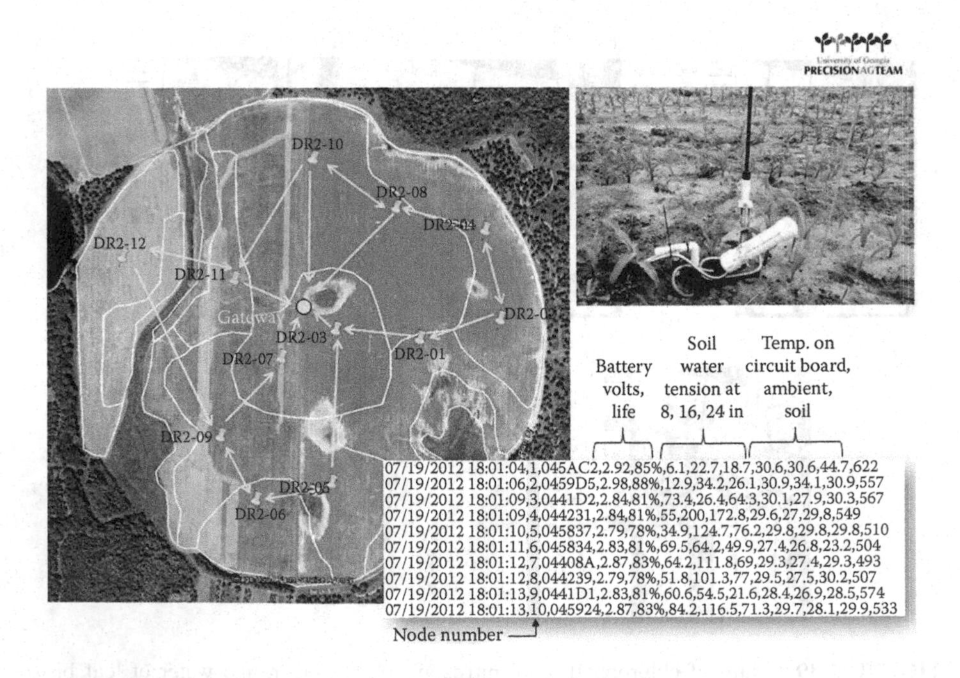

FIGURE 10.13 Irrigation soil sensor network. (From Vellidis, G. 2015. Irrigation sensor network. Personal communication on teaching material, January 14, 2015. With permission.)

7 A Systems Approach to Community-Based Precision Agriculture

Sakae Shibusawa

CONTENTS

7.1 INTRODUCTION

The term "precision agriculture" is widely known in Japan. For example, the term scored 120,000 hits on the BIGLOBE website and 1,001,000 hits on Yahoo-Japan in 2014, growing from a few hundred in 2000. The hits cover activities in industry and agriculture, as well as information technology for scientists, engineers, and administrators. In general, they expect that precision agriculture has the potential to offer future solutions to complicated issues in agriculture, such as environment versus productivity and globalization versus localization. National Research Council (1997), SKY-farm (1999), and Vanacht (2001) mention that precision agriculture is a management strategy based on advanced information technology, including describing and modeling soil and plant variability and integrating variable-rate field operations to meet site-specific requirements, all aiming at increasing economic returns as well as reducing energy input and environmental impacts.

In the last decades, multiple concerns such as shortage of food and water, global warming, and energy crises have crept up on people. As for food supply, world food production has increased with food consumption of cereal crops in half a century, although in the last decade, production could not catch up with consumption, as shown in Figure 7.1a. The demand for crops has increased due to increases in population, industrial needs, and meaty, fatty diets accompanying lifestyle changes. A major contribution to increases in the net yield of crops has been the increase in yield per unit area, that is, increased land productivity, while the area of harvest

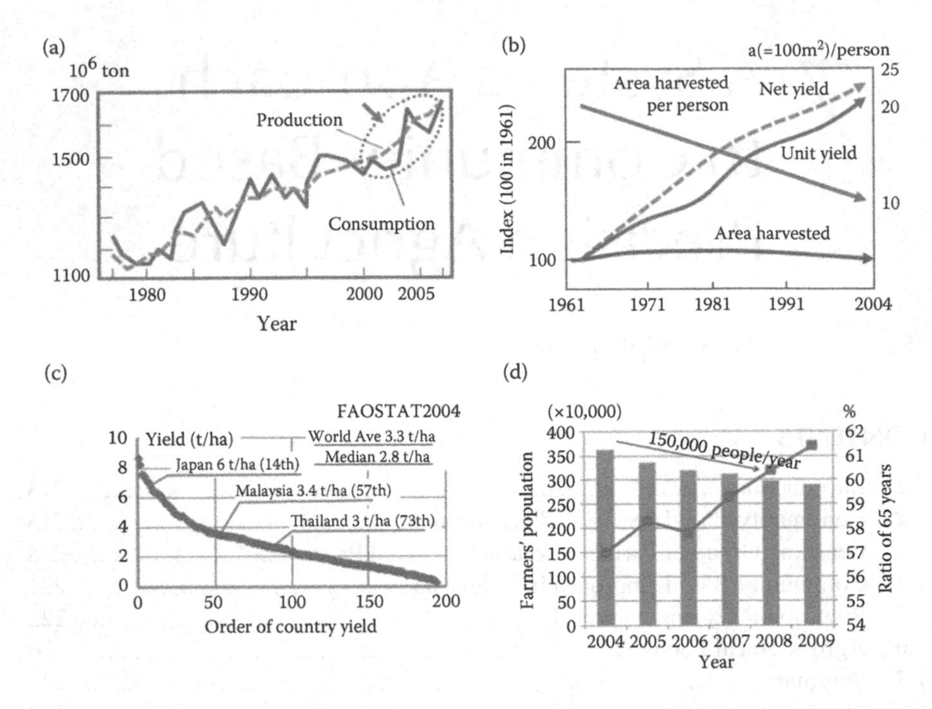

FIGURE 7.1 The position of Japanese agriculture in the world. (a) Worldwide needs and demand of wheat and coarse grains; (b) land productivity; (c) yield per unit area of country; and (d) rapid decrease of Japanese farmers.

has not increased during the last 50 years, as shown in Figure 7.1b. In general, land productivity depends on the crop variety, agricultural materials and facilities, and farm mechanization, as well as socioeconomic factors such as the organization of growers.

Keeping land productivity higher is one of the advantages of Japanese agriculture, as shown in Figure 7.1c, because of its well-organized community of growers but with small-scale farms. In spite of the high land productivity and top-20 net production in the world (FAOSTAT, 2005), the population of Japanese growers has decreased by 150,000 per year during the last decade, resulting in a decrease to one-tenth of 2.5 million by the year 2030 (Figure 7.1d). This is inducing rapid changes in the structure and system of Japanese agriculture, followed by some countries in the world.

Japanese government statistics in 2012 show that the number of growers was 2.5 million, the number of commercial farmers was 1.78 million, and the number of young farmers was 0.17 million (Figure 7.2b). The same statistics show that there is 368 million ha of arable land with an average scale of 2.2 ha, and that 32% of the arable land belongs to 2% of farmers with a farm scale of more than 20 ha. On the other hand, the Japanese population is decreasing dramatically, by 260,000 per year in 2013 and is projected to decrease by a million per year in 2025, which causes big changes in socioeconomic systems. For example, the needs of consumers tend to shift from price and calories to the safety and functions of foods.

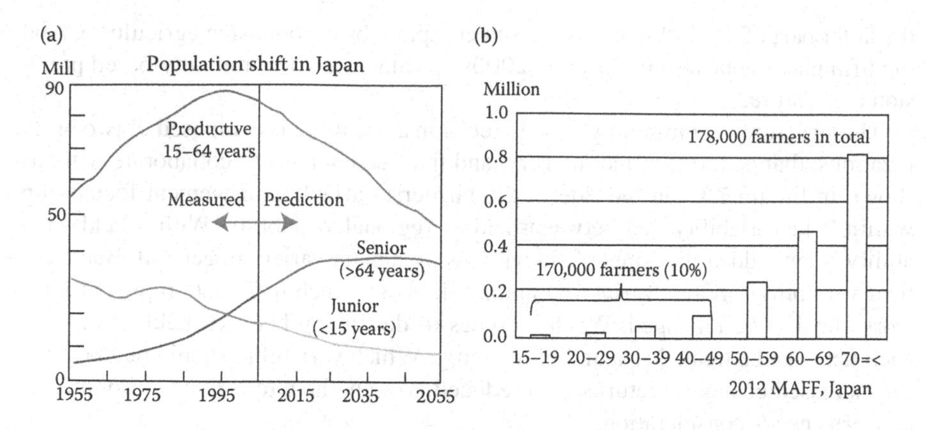

FIGURE 7.2 Rapid aging and depopulation in Japanese society and agriculture. (a) Population decrease in Japan and (b) aging and depopulating of farmers.

Not only land productivity but also consumers' needs have become targets of current farm management in Japan. Transfer of skills and technology has also become big business from generation to generation, from industry to agriculture, and from agriculture to industry. That is why precision agriculture and its players have become the target of investigation. The community-based approaches of this chapter will provide a hint to creating a way of thinking.

7.2 COMMUNITY-BASED PRECISION AGRICULTURE

In this chapter, "community" implies practitioners and/or players of precision agriculture, and precision agriculture implies management practice on the farm. The combination of players and management requires us to rediscover the story of precision agriculture as follows. Community-based precision agriculture is a new regional farming system to gain high profitability and reliability under regional and environmental constraints, promoted by expert farmers and technology platforms, by creating both information-oriented fields and information-added products, with supply chain management from field to table (Shibusawa, 2004). The definition brings us to a home ground where growers, engineers, and business people take action.

During the current quarter of a century, we have experienced five different phases in precision agriculture (Shibusawa, 2004). The first phase was site-specific crop management in the early 1990s. The second phase was mechanization as sensor-based site-specific crop management with variable-rate operation in the mid-1990s. The third phase appeared in the latter part of the 1990s with precision agriculture defined by "a management strategy that uses information technologies to bring data from multiple sources to bear on decisions associated with crop production" (National Research Council, 1997). Furthermore, "a key difference between conventional management and precision agriculture is the application of modern information technologies to provide, process, and analyze multisource data of high spatial and temporal resolution for decision making and operations in the management of crop production" (National Research Council, 1997). The fourth phase appeared in the latter part of the 1990s as cost-driven company-based precision agriculture. And the fifth phase appeared in the early 2000s as value-driven community-based precision agriculture.

The structure of community-based precision agriculture is composed of two organizations, that is, farmers and industry, and five stakeholders to collaborate with, as shown in Figure 7.3. On the side of the farmers, variable management focuses on within-field variability and between-field or regional variability. Within-field variability is embedded in a single field with a single plant variety in general. Between-field variability implies variability among fields in which different crops and farm works tend to be managed. When it comes to describing between-field variability, each field can be treated as a unit of mapping. Which variability should be managed for increased economic returns with reduced cost and how to tackle environmental concerns needs consideration.

There are different stories regarding the practice of management in action when one looks at field variability on different scales. On a single small farm, the farmer can better understand what is going on in each field, which enables variable-rate application for site-specific requirements with the farmers' knowledge and skills. When it comes to covering an area of a few tens of hectares, including lots of small fields, for example, a farm work contractor or a farm company has to manage regional variability due to cropping diversity. They also have to coordinate the farmers with different motivations due to different cropping styles. Here, we have hierarchical variability: within field, between field, and between motivations with different scales and different cropping styles.

Managing hierarchical variability requires two organizations, expert farmers and a technology platform, as shown in Figure 7.3. The groups of expert farmers play

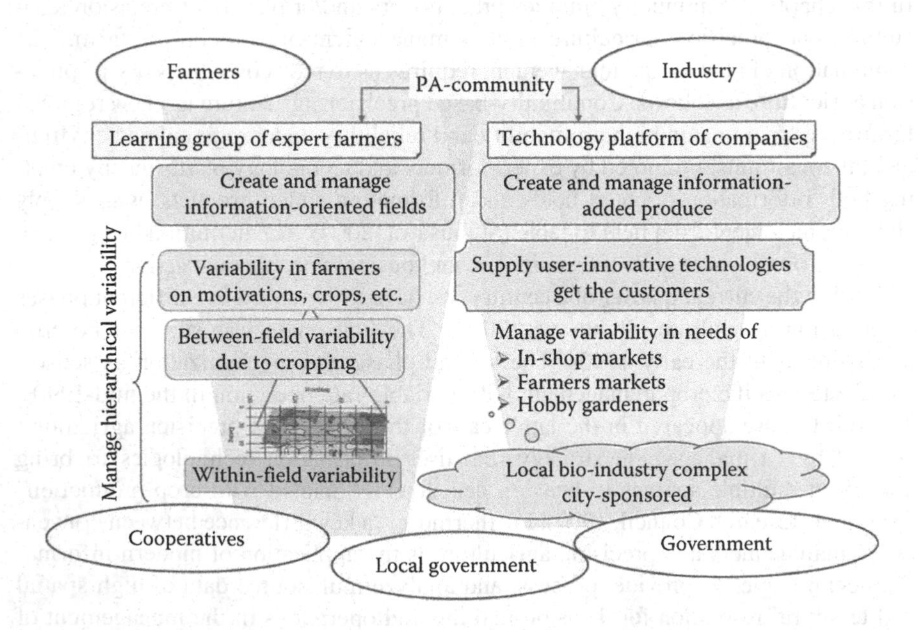

FIGURE 7.3 Structure of community-based precision agriculture.

the role of top management of innovation in the regional farming system, such as the rearrangement of the five factors of the farming system and the development of scenarios for introducing approaches in precision agriculture. The technology platform develops and provides the technologies available with rural constraints as well as marketing channels for high-quality/traceable agro-products.

A combination of the wisdom/experience of the farmers and the technologies of the platform will produce information-oriented fields and information-added products, as shown in Figure 7.4, which can meet compliance as well as farmer's motivation, such as traceability, productivity, and profitability, and environmental concerns.

Rural development by introducing precision agriculture is an attractive proposition in Japan because people face the serious concerns of depopulation, high aging, a downsizing economy, and exhausted infrastructure in rural villages and cities. The information-oriented fields produced by precision agriculture practices are easy to connect with the multifunctions of agriculture so as to manage environmental conservation and design landscape amenity if it merges with a geographical information system (GIS) covering the whole space of a rural area, aiming at gaining the trust of local inhabitants. The information-added products make access to the market with direct communication with consumers easier.

Shibusawa (2004) discussed adoption of precision agriculture in the cases of the United States and Japan in the early 2000s in terms of scale merit and added value. Adoption of precision agriculture in the United States followed a cost-driven scheme of big-farm management with reduced costs, and its profitability threshold was more than 500 ha in farm size (Vanacht, 2001). Cost reduction was 20% for fertilizer and 50% for herbicide, for example, but less or little increase appeared to occur in yield and total sales. Sales were about 1000 US$/ha for crop growers and 30% of that was expenses for fertilizer and chemicals. The cost reduction effect was around 100 US$/ha. On the other hand, they paid about an extra 80 US$/ha/year for the

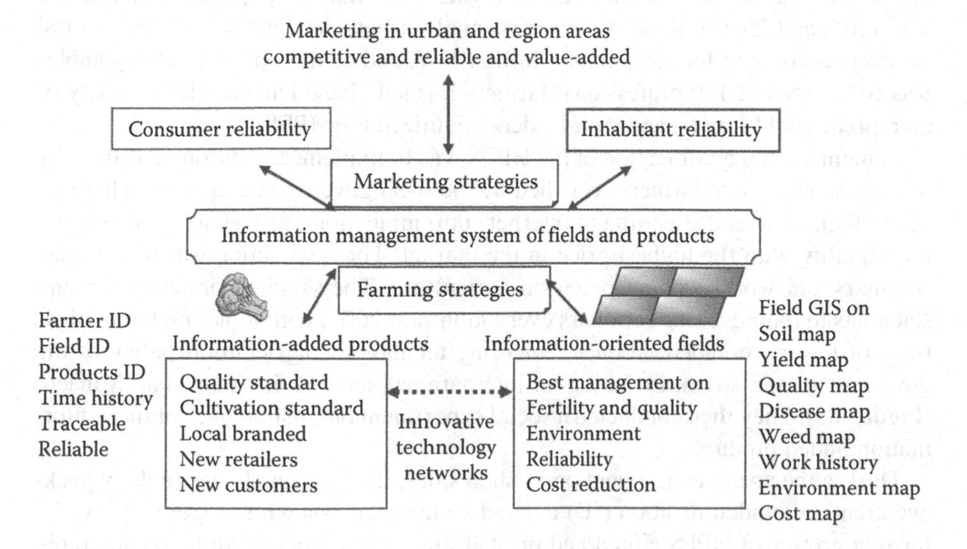

FIGURE 7.4 Strategy of community-based precision agriculture.

precision agriculture service and purchased machines such as variable-rate fertilizing machines costing hundreds of thousands of dollars. Big retailers pushed farmers for lower prices with the pressure of global food markets. The only avenue for commercial farmers was to obtain scale merit for cost reduction. Profitable farm sizes tended to be large in Central America, for example, 200 ha in the 1990s, 500 ha in the 2000s, and 1000 ha in the 2010s (as heard from consultants).

On the other hand, a small farm in Japan had no scale merit. The expenses for machines and labor were relatively high, compared with the cost of fertilizer and chemicals; that is, overequipment with machinery on a small farm was a fatal issue. Evidence-based collaboration was one avenue. Note that sales were about 10,000 US$/ha for rice crop growers, which was about 10 times as high as the sales of U.S. farmers. This motivated farmers to sell their products at high prices. If they could ensure the needs of consumers and supply quality products to the market, they could be competitive in the food supply chain. The distance between growers and consumers might be very close in Japan, compared with the United States.

7.3 LEARNING GROUP OF FARMERS CREATING BRANDED PRODUCE

One learning group was the Honjo Precision Farming Society (HPFS) organized by progressive farmers in April 2002, in collaboration with Waseda University, Tokyo University of Agriculture and Technology, people from the industry, and City Hall. The leader of the farmers recognized that City Hall had promoted zero-emission town planning and was awarded ISO 140001 certification in March 2002, and environment-friendly agriculture was one of the main city projects.

Honjo is a city located 100 km north of Tokyo, having the longest daylight time and rich alluvial soil with rich irrigation water from the Tone river. The population of the city was 80,000, the farmed area was 1300 ha, the population of farmers was 1200, and 25% of these were professional farmers. The net sale of agricultural products accounted for more than 8 billion JPY, and of this, the sale of vegetables was 65%. Around 130 professional farmers formed "New Farmer 21," a society of entrepreneurial farmers, and their leaders organized the HPFS.

A membership qualification of the HPFS was to implement environment-friendly management as "eco-farmers" certified by the local government, creating a homepage of their own, and attending to Internet communications, as well as managing the food quality with the highest price in the market. The next action was to organize seminars and workshops on precision agriculture. They invited professionals and scientists to their evening seminars every month in 2003, with topics on the motivations of buyers, branded produce, emerging technology, agricultural policy of the government, and so on, including an international seminar inviting Marc Vanacht (Figure 7.5). They then conducted a social experiment on in-shop sales of their information-added products.

During the social experiment on in-shop sales, they invented a technology package creating an identification (ID) tag and its usage, as shown in Figure 7.6. At the farm, a grower of HPFS edited and printed his small ID tags with his photograph, and attached it to each package of vegetables in the packing process. At a department

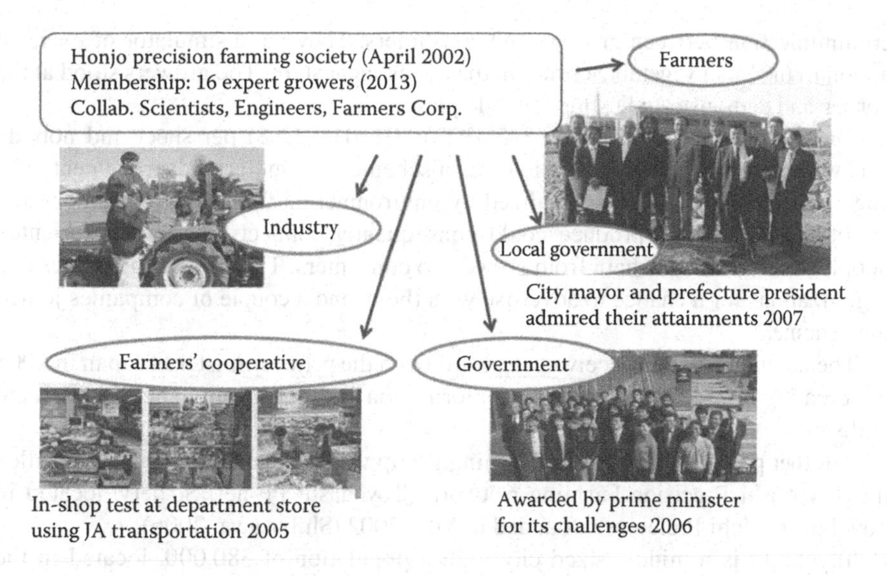

FIGURE 7.5 Activities of the Honjo precision farming society (HPFS).

store and at a wholesaler's, high-quality vegetables with ID tags were put in the fresh vegetable corner at prices 20%–30% higher compared with the normal. The farmers' cooperative to which they belonged transported the vegetables from the farm to in-shop. It was easy for customers to access the respective growers through their websites by mobile phone by clicking the two-dimensional code on the tag. The growers wrote a farm work diary on their homepages every day, which helped direct

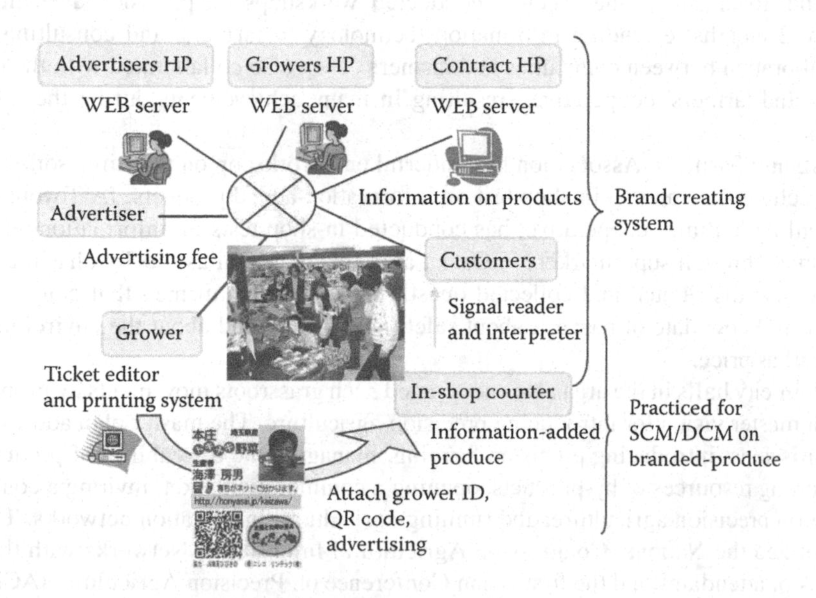

FIGURE 7.6 Scheme of the branded produce of HPFS.

communication between growers and consumers. They put a simulator of retrieval action in the fresh vegetable corner in the department store. The growers stood at the corner and demonstrated using a mobile phone.

The cost of the ID tags was 3–5 JPY (0.03–0.05 US$) per sheet and nobody was willing to pay for it. A solution was a scheme of voluntary advertisement. The vegetables produced were specialized by environmental-friendly management and quality taste, and the produce could consequently connect environment-oriented people across the food chain from growers to consumers. They asked companies and organizations for a chance to advertise with them, and a couple of companies joined the scheme.

The activity of HPFS received an award from the prime minister of Japan in 2005 for creating branded produce using information technology and specific patented skills.

Another precision agriculture learning group was the technology platform called the Toyohashi Precision Farming Network (Toyohashi PF-net) Society, located in Toyohashi, Aichi Prefecture, founded in May 2002 (Shibusawa, 2006).

Toyohashi is a middle-sized city with a population of 380,000, located in the middle of the main island along the coast of the Pacific Ocean, between the big cities of Tokyo and Nagoya. The net sale of agricultural products here was more than 50 billion JPY in 2000, which was the top sale in the cities in terms of agricultural production. The Toyohashi-Atsumi area produced more than 100 billion JPY of agricultural products. The farmed area was about 15,000 ha, the population of farmers was about 10,000, the average farm size was about 1.5 ha, and 25% of the farmers were professional according to the statistics of 2000. People were motivated to maintain the top sale of agriculture in Japan by introducing a new system of precision agriculture.

The Toyohashi PF-net Society conducted workshops on precision agriculture every 2 months, extending information technology to farmers, and consulting on collaboration between companies and farmers. They also collaborated with the city halls and farmers' cooperatives, resulting in many achievements during the last 5 years.

Atsumi Farmers' Association has undertaken a workshop on real-time soil-sensing technologies and an in-shop test on information-added products. JA Toyohashi, a local agricultural cooperative, has conducted in-shop tests on information-added products through supermarkets in Osaka and Tokyo as well as the Toyohashi area. They also distributed and collected questionnaires and confirmed that consumers asked to know date of harvest, about safety and health, and about the environment as well as price.

Four city halls in the area also encouraged such grassroots movements by promoting a master vision for introducing precision agriculture. The master plan addressed six missions: introducing precision farming, managing the traceability of products, enriching resources of by-products, opening a community market, inviting a conference on precision agriculture, and running agricultural information networks. They organized the National Congress on Agricultural Information Networks with thousands of attendants and the first Asian Conference on Precision Agriculture (ACPA) on August 5–6, 2005.

The achievements of the two learning groups have taught us that the participating farmers (1) were familiar with Internet communication; (2) had higher education levels; (3) grew high-quality produce; (4) had good sales and marketing experience; and (5) were greatly outgoing and sociable. The most important factor was that they had the ambition to become good-practice farmers enhancing local communities and industries. The experience above partly followed the aspects mentioned by Blackmore (2002).

Blackmore (2002) identified eight principles in precision agriculture: that (1) precision agriculture is a management process, not a technology; (2) spatial and temporal variability must be measured; (3) the significance of variability in both economic and environmental terms should be assessed; (4) the required outcome for the crop and the farm must be stated; (5) the special requirements of the crop and the country should be considered; (6) ways to manage variability to achieve the stated outcome are to be established; (7) methods to reduce or redistribute the inputs and assess the risk of failure need to be considered; and (8) crops and soil must be treated selectively according to their needs. An attractive aspect is that the development of precision agriculture is characterized by continuous evolution based on independent thinking associated with multidisciplinary collaboration under the crossover of new ideas from other areas.

7.4 COMMUNITY-BASED APPROACH IN INDONESIA

A unique approach that emerged in Indonesia, as a project of education for sustainable development (ESD) based on the concept of precision agriculture, is called the "community learning activity center as the medium for precision agriculture technology implementation with a decision support system to optimize food crop management" sponsored by the Indonesian government (Virgawati et al., 2010). A strong motivation to embark on the project was a shortage of stable food production against demand due to an increasing population and changes of lifestyle in spite of increases in primary food production. Bottlenecks recognized were biophysical factors such as exploitation of land and water resources, economic factors such as shortage of fertilizer because of high cost and low income, social factors such as habits of chemical fertilizer use or stereotypical professional farmers, and technological factors such as less knowledge and poor instrumentation of precision agriculture.

The project team was organized by the Faculties of Agrotechnology, Communication Science, Economics of Development, Agribusiness, and Informatics and Environmental Engineering of the University of Pembangunan Nasional "Veteran" Yogyakarta, Indonesia. The project covered the issues of sustainable economic growth through increased value-added food products, social justice through equal rights and opportunities for access to efficient technologies in food production systems, and preservation of natural resources by maintaining sustainable fertility. Action programs involved mapping the diversity of agricultural land characteristics to build a database; modifying the simulation model of the existing soil–plant–water system; producing precision agriculture technology adapted to local culture; and developing an ESD-based education system. Establishing the Center of Community Learning Activity for Precision Agriculture was a milestone of the project.

FIGURE 7.7 Community-based approach to education for sustainable development (ESD) in Magelang, Java. (a) Salam subdistrict, a flat site with mining and (b) Windusari subdistrict, a mountainous site.

In 2010, the agricultural agency in Magelang district selected six subdistricts for a social experiment, that is, Windusari, Tegalrejo, Secang, Salaman, Muntilan, and Salam. The activities were categorized by preliminary research, community service, and a socialization program.

The preliminary research consisted of mapping soil variability, identifying the farming system, determining the economic aspects, identifying social aspects, developing a crop management model, and recognizing the requirements for structuring a decision support system. The community service provided six districts with 25 students from four faculties for 40 days of activities, including collecting research data and supporting social activities for research and for local communities. The socialization program involved a field trip for professionals, an open lecture for students, and a workshop on precision agriculture.

The author of this chapter was invited to join the field trip and workshop under the scheme (Figure 7.7). The project team had organized learning groups of farmers in collaboration with local government and people from the university using information and communication technology (ICT) tools. Unfortunately, on October 26, 2010, a great eruption of the Merapi volcano struck the Magelang district and the six subdistricts suffered serious damage. The project was halted by the disaster, but the people soon started restoration work.

7.5 PRECISION RESTORING APPROACH

3/11 in 2011 is the day that northeast Japan was hit by a tridisaster: a super earthquake measuring M 9.0, a huge tsunami of more than 10 m high, and the explosion of nuclear power stations. Huge damage was confirmed across the cities and rural communities, including agricultural and industrial sectors. In the last 3 years, the restoration process has changed rapidly. Fukushima prefecture still has issues regarding measuring both radioactive contamination and tsunami damage, while Miyagi and Iwate prefectures are focusing on recovery from tsunami damage.

The Japanese Society of Agricultural Machinery (JSAM, now the Japanese Society of Agricultural Machinery and Food Engineers or JSAM) provided help for recovery from the damage caused by the disasters (Shibusawa, 2012a). They had limited experience of combating such a huge catastrophe of complex disasters. One useful approach was in precision agriculture that was applicable not only to agricultural sectors, but also to the environment and to the field of construction (Shibusawa, 2004; Berry et al., 2005), which led to evidence-based approaches with precision thinking.

On March 12, the first action started with a call for confirmation of the safety of the members of JSAM through e-mails, cellular phone, Internet service, and so on. It took 1 week for the Kanto area and 2 weeks for the Tohoku area to be completed. Unfortunately, information was received that three student members had been killed by the tsunami at Sendai airport.

Information through media and direct calls led to the organization of the working team of JSAM on March 30, 2011. The missions of the team were (1) to validate the facts and information on the disasters since there was confusion and complexity; (2) to investigate the damage in terms of agricultural machinery and farm management; and (3) to propose better solutions to reconstructing community-based agriculture. The working team reconfirmed the potential of Tohoku's agriculture with references. The statistics compiled by the Tohoku regional agricultural administration office in Sendai in 2010 stated that agricultural production was worth 1359 billion JPY, comprising 16% of the total production in Japan, including 496 billion JPY of rice, 383 billion JPY of livestock, and 228 billion JPY of vegetables. The number of growers was 463,000 with a ratio of 16% to all growers in Japan. The ratio of growers above 65 years of age was 30% and it was lower than the national average of 58% in Japan. Local self-sufficiency in food production in the Tohoku region was more than twice the national average. Apples of Aomori prefecture occupied 53% of the entire production of Japan, cherries of Yamagata prefecture accounted for 71%, and the share of peaches of Fukushima prefecture was 20%.

On September 12 and 13, the working team visited paddy fields in Kitakami of Ishinomaki city and fields of protected horticulture in Watari of Natori city in Miyagi prefecture. One site that they visited was the Kitakami riverside around 10 km distant from the coast, as shown in Figure 7.8. The people suffering from the tsunami emphasized the following: (1) The tsunami brought a large amount of rubble on a path of over 10 km distant from the coast and it had still not been removed (Figure 7.8c). (2) They cut and removed the weeds in the paddy fields to prepare for the next cropping season (Figure 7.8a). (3) They were less concerned about salty sludge since the sludge used to be applied in the paddy for soil improvement. (4) They needed recovery of transportation, repair of drain pumps, and recovery of machines and facilities in order to restart farm work. A local dealer continued work on repairing machines flooded with seawater (Figure 7.8d). It was difficult to repair them perfectly because of salt and sludge invading unseen spaces.

FIGURE 7.8 People's combat against tsunami disasters in Ishinomaki, Miyagi. (a) Weed control of overflooded paddy for the next cropping season. (b) Sludge of 10-cm thickness fully covering the paddy field. (c) Dumped rubble produced by the tsunami 7 km from the coastline. (d) On-service local dealer of agricultural machinery.

Based on the results of the survey, the JSAM proposed five recommendations: (1) to develop a strategy for land consolidation and for education of newcomers to be professional farmers; (2) to protect the intellectual properties of farmers; (3) to repair the service network of agricultural mechanization; (4) to simultaneously reconstruct the system for both producers and retailers; and (5) to maintain farm assurance and standard farm management such as GLOBAL G.A.P.

The Ministry of Agriculture, Forestry and Fisheries has launched many national projects for recovery from disasters, such as intensive arable farming (Figure 7.9), a highly automated greenhouse system, highly effective orchard cultivation, and an

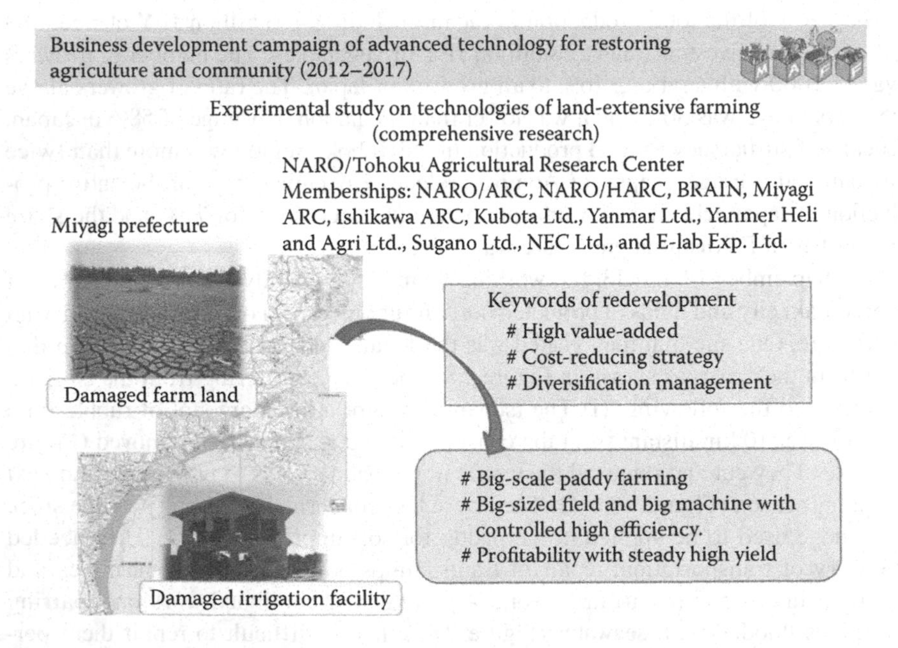

FIGURE 7.9 National project of arable cultivation, Koya Corp. in Miyagi prefecture.

intensive system of aquaculture. A major target was a business development campaign of advanced technology for restoring agriculture accompanied by the local community. Figure 7.9 shows a project of national institutes and private companies for arable farming employing cereal crop rotation using a technology package of precision agriculture, such as a field mapping system and variable-rate technology. Then, the agricultural corporation KOYA joined the project. KOYA Corporation was founded by five local growers in 2003, and 90% of its 100 ha paddy fields suffered damage from the tsunami; in addition, the machinery and facilities were washed away. They restarted cultivation just after the catastrophe and the national project helped them.

Goto et al. (2013) have organized a JST (Japan Science and Technology Agency)-funded 3-year project on precision restoring agriculture in the Fukushima area in 2012, as shown in Figure 7.10. The project team is composed of organizations who had suffered from the tsunami: the agricultural corporation Denpata, the manufacturing company Kanda Ltd., support organizations, ADS Ltd., the National Institute of Advanced Industrial Science and Technology (AIST), and the Tokyo University of Agriculture and Technology (TUAT). These organizations suffered serious damage not only from the East Japan earthquake disaster, but also through rumors related to the collapse of the nuclear power plant. They also had rural issues of farmers' aging, depopulation in the village, the "food desert" phenomenon, and so on. Therefore, they hungered for a future vision as well as sought measures against rumor damage. One idea was an evidence-based farm management scheme.

The goal of the project was to create an information-oriented field to meet the requests of consumers in the market. Within a limited budget, a real-time soil sensor was introduced to monitor the within-field soil condition, and sensor posts were set

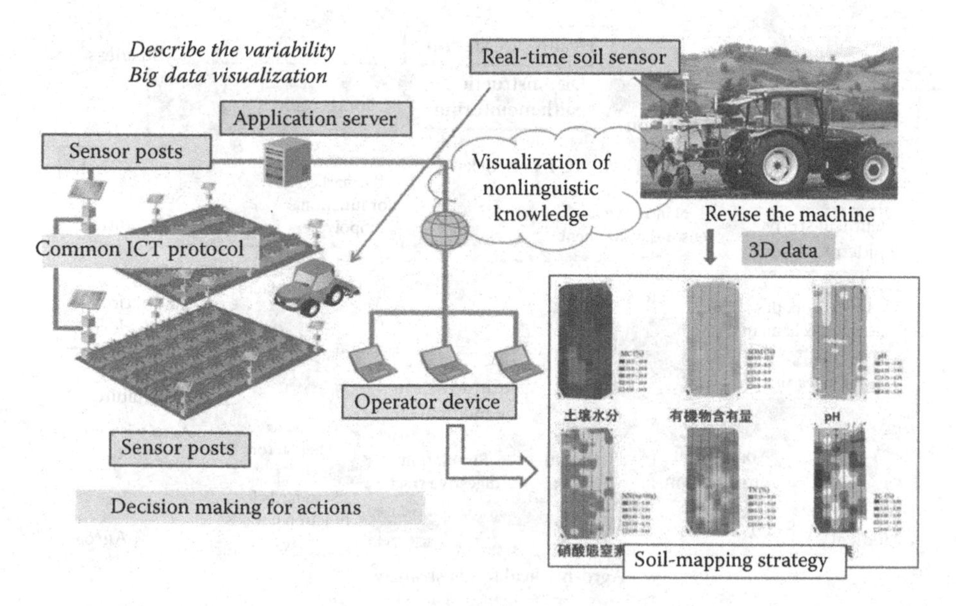

FIGURE 7.10 **(See color insert.)** Precision restoring agriculture in Fukushima toward traceable management against rumor damage.

up in the field to monitor the degree of radiation, wind velocity, wind direction, rainfall, etc. Yanmar Ltd. joined voluntarily to provide a combine harvester with a yield monitor in 2014. The project is still developing with community-based approaches.

7.6 AGRO-MEDICAL FOODS

The strategy for agro-medical foods was formed in 2009 when the concept of community-based precision agriculture encountered the concept of preventive medicine at the meeting of Dr. Sakae Shibusawa and Dr. Toshikazu Yoshikawa, and it then drove many collaborative projects in the fields of medicine, agricultural science, engineering, and industry, though it was not introduced in English (Shibusawa, 2012b). This agro-medical approach promises to expand the fields of precision agriculture, and that is the reason it is introduced here.

Agro-medical foods are defined as agricultural products with a high content of functional materials with evidence of effects on health and wellness produced by precision agriculture, and they are created by an agro-medical initiative, as shown in Figure 7.11. The agro-medical initiative is a research group of medical, agricultural, and engineering scientists, aiming at the cure of lifestyle-related diseases by developing agricultural products with a high content of functional materials.

Figure 7.11 shows a research cycle of agro-medical foods. The agricultural sector supplies fairly controlled products to the medical sector, which requires controlled protocols of production with traceable management. The medical sector confirms the evidence of effectiveness against disease prevention and wellness in medical science.

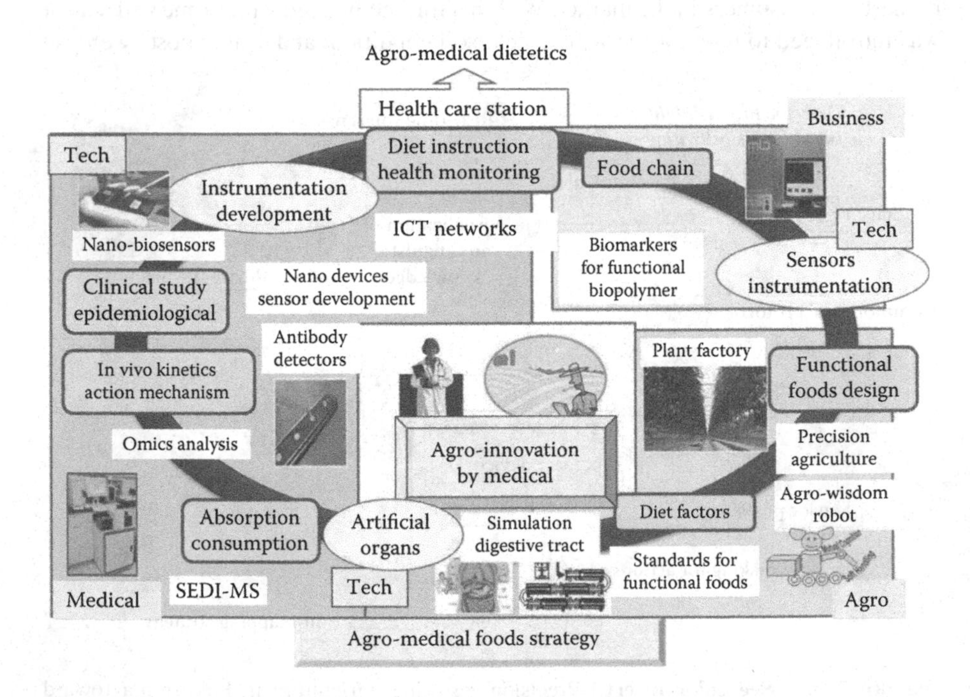

FIGURE 7.11 Concept of producing agro-medical foods (AMF).

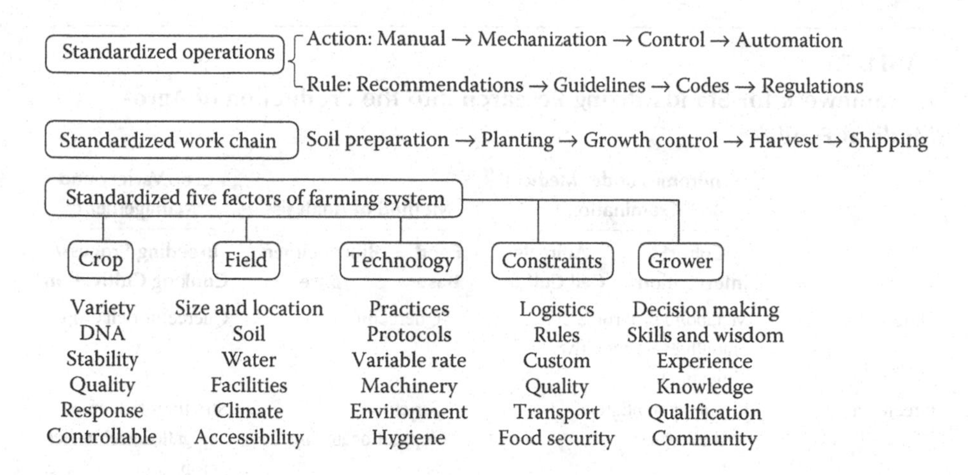

FIGURE 7.12 Needs for a standard in agricultural systems for AMF.

The nutrition and dietetics sector provides personalized diets using agro-medical foods. The business sector commercializes the agro-medical foods and diets. The engineering sector provides biosensing and control technology to manage the system and communicate beyond disciplines.

Figure 7.12 shows a standard scheme of production in the categories of operation, work chain, and farming system. The operation standard involves the specification of mechanization and guidelines. The work chain requires the protocol of process jobs from soil preparation to shipping. The farming system is composed of the five factors of crop, field, technology, constraints, and motivation, and each factor has a substructure of farming elements such as crop variety and tillage machines. At least three production categories need clear description when they are put into practice in the shape of precision agriculture.

Table 7.1 shows a framework or roadmap of how to produce agro-medical foods. There are three control points and nine check items. The first control point is the target syndrome and medical examination with the four check items of cell culture, animals, intervention, and cohort. The second control point is the target material and analysis method with the two check items of food body base and biospecimen base. The third control point is crop variety and management with the three check items of breeding, cultivation, and processing/cooking. The test crops were onion, green tea, orange, soybean, spinach, tomato, and eggplant in 2011. Many more crops and functional materials will be examined in a couple of years.

7.7 SUMMARY

This chapter described the last 15 years' experience of a Japanese model of community-based precision agriculture accompanied by a learning group of farmers and a technology platform of companies. Community-based precision agriculture aims at high profitability and reliability under regional and environmental constraints, promoted by the expert farmers and/or the technology platform, by creating both

TABLE 7.1
A Framework for Standardizing Research into the Production of Agro-Medical Foods

	Syndromes under Medical Examination		Method of Analysis		Crop Variety and Management
Crop	**Cohort Intervention**	**Animals Cell Culture**	**Food Base**	**Biospecimen Base**	**Breeding Process/ Cooking Cultivation**
Onion	Metabolic syndrome dry mouth/eyes, cognitive impatient		Quercetin		Quercetin-rich crop
Green tea	Immunopotentiative, antiallergic		Strictinin, epigallocatechin		Strictinin-rich crop, epigallocatechin-rich crop
Orange	Metabolic syndrome fatty liver/diabetes		Not yet		Beta-cryptoxanthin-rich crop
Soybean	Metabolic syndrome, osteoporosis dry mouth/eyes, macular degeneration		Isoflavone		Isoflavone-rich crop
Apple	Metabolic syndrome, diabetic arteriosclerosis, osteoporosis		Procyanidin		Procyanidin-rich crop
Spinach	Macular degeneration, dry eyes		Lutein		Not yet
Tomato egg plant	Metabolic syndrome, diabetic		Not yet		Osmotin-rich crop

information-oriented fields and information-added products, with aggressive access to food chains. The two participating local learning groups were the technology-driven Precision Farming Network of Toyohashi-Atsumi (PFNET) in Toyohashi and the farmers' learning group Honjo Precision Farming Society (HPFS) in Honjo. The first action of the two groups was market research using information-added produce through in-shop experiments. The scheme of the community-based approach has been applied to a trial of ESD on Java island of Indonesia, to restoration post the catastrophe of the East Japan earthquake and tsunami in 2011, and the production of agro-medical foods in collaboration with professionals in the fields of medicine, agriculture, engineering, dietetics, and business.

REFERENCES

Berry, J.K., J.A. Delgado, F.J. Pierce, and R. Khosla. 2005. Applying spatial analysis for precision conservation across the landscape. *Journal of Soil and Water Conservation*, 60(6):363–370.

Blackmore, S. 2002. Precision farming: A dynamic process. In *Proceedings (on CD-ROM) of the 6th International Conference on Precision Agriculture and Other Precision Resources Management*, July 14–17, Minneapolis, MN, USA, ASA/CSSA/SSSA.

FAOSTAT. 2005. Crops, ProdStat, Production in Statistical Database. Food and Agricultural Organization of the Unite Nations, Statistic Division.

Goto, H., H. Niitsuma, Y. Noguchi, A. Sashima, K. Kurumatani, M. Kodaira, and S. Shibusawa. 2013. Precision restoring agriculture using spatial visualization technique. In *Proceedings (on CD-ROM) of the 5th Asian Conference on Precision Agriculture (ACPA)*, June 25–28, 2013, Jeju, Korea, pp. 120–126.

National Research Council (NRC). 1997. *Precision Agriculture in the 21st Century.* Committee on Assessing Crop Yield: Site-Specific Farming, Information Systems, and Research Opportunities, National Academy Press, Washington, DC, p. 149.

Shibusawa, S. 2004. Paradigm of value-driven and community-based precision farming. *International Journal of Agricultural Resources, Governance and Ecology*, 3(3/4):299–309.

Shibusawa, S. 2006. Community-based precision agriculture with branded-produce for small farms. In *Proceedings (on CD-ROM) of the 8th International Conference on Precision Agriculture, and Other Precision Resources Management*, July 23–26, Minneapolis, MI, USA, ASA/CSSA/SSSA.

Shibusawa, S. 2012a. Precision restoring approach to the East Japan catastrophe—Actions of JSAM. In *Proceedings (on CD-ROM) of the 6th International Symposium on Machinery and Mechatronics for Agriculture and Biosystems Engineering (ISMAB)*, June 18–20, 2012, Jeonju, Korea, pp. 176–181.

Shibusawa, S. 2012b. Agro-medical foods strategy with community-based precision agriculture. *Kyo-sai-sogo-kenkyu*, 62:2–19 (in Japanese).

SKY-farm. 1999. *Opportunities for Precision Farming in Europe*, Updated report 1999, p. 126. Email: Tony@skyfarm.co.uk.

Vanacht, M. 2001. *The Business of Precision Farming*, Private report, p. 91. Email: marcvan8@att.net.

Virgawati, S., S. Sumarsih, W. Choiriyati, D. Nuryadin, E. Murdiyanto, F.R. Kodong, and H. Lukito. 2010. *Community Learning Activity Center as the Media for Precision Farming Technology Implementation with Decision Support System to Optimize the Food Crop Management*, Research report funded by Dp2m Dikti Ministry of National Education, Republic of Indonesia, p. 38.

8 Precision Agriculture in China

Sensing Technology and Application

Hong Sun and Minzan Li

CONTENTS

8.1 INTRODUCTION

The cultivated land is the human survival and development. As is well known, China has 9.6 million square kilometers of land area. According to the *China Statistical Yearbook 2007*, the area of farmland resources in China was about 1.326 billion hectares, in which cultivated land refers to the area of land reclaimed for the regular cultivation of various farm crops, including crop-cover land, fallow, newly reclaimed land, and land lying idle for less than 3 years.

Cultivated land in China is divided into four regions: the eastern, central, western, and northeastern regions. It is also divided into three categories: paddy fields, irrigated land, and dry land (GB/T21010-2007). According to the agriculture census, the distribution of cultivated land is unbalanced in the country, the area of cultivated land in the western region being more than the others, and accounting for 36.9% of arable land; the area in the eastern, central, and northeastern regions being 21.7%, 23.8%, and 17.6%, respectively. For cultivated land categories, the area of dry land accounts for 55.1% of arable land, while the area of paddy field and irrigated land are 26.0% and 18.9%, respectively.

In recent years, economic development has had a profound impact on land-use patterns in China. There is increasing conflict between limited arable land resources and the requirements of agricultural production. Serious land degradation such as soil erosion, depletion, secondary salinization, and pollution is caused by long-term use. Mining activities also damage and take over a lot of cultivated land resources. Precision agriculture (PA) is a farming management method that allows farmers to optimize their resource inputs to achieve high yields (Wang et al., 2003; Wang, 2011; Zhao et al., 2003). Modern technology promotes the development of agricultural mechanization. Thus, information perception, the Internet of Things (IoT), and technology application are introduced in this chapter (Wang, 1999).

8.2 KEY TECHNOLOGIES OF AGRICULTURAL INFORMATION PERCEPTION

In agricultural information perception technology, research in China and abroad all focus on two aspects: agricultural resource investigation and farmland production information perception. Fast acquisition, processing, and understanding of farmland information were the key points of development for PA.

Agricultural resource investigation includes the investigation of cultivated land resources, district plantation, atmosphere, level of agricultural production, etc. 3S (geographic information system [GIS], global navigation satellite system [GNSS], and RS) technology is utilized comprehensively. Some basic information such as cultivated land area, vegetation distribution, and atmosphere dynamics is acquired and analyzed based on remote sensing technology. Distribution characteristics of all kinds of data are calculated based on GIS and global positioning system (GPS). Transition characteristic analysis and visualization expression of agricultural resources in time and space are realized by historical data. At present, much research has been conducted and many application achievements have been made in rating the quality of cultivated land, regional distribution of agricultural plantation, monitoring of crop growth and disease status, and agricultural atmosphere analysis (Kuang and Wang, 2003).

Farmland information perception is based on the research and development of advanced sensors, which concentrates on each link of agricultural production. It mainly focuses on fast acquisition of growth and physiological parameters of soil and plants, and distribution of insect pests and weeds, and can provide decision support for PA.

8.2.1 Rapid Acquisition of Soil Information in Farmland

The detection of soil information could be divided into two types, laboratory measurement and *in situ* measurement. In laboratory measurement, samples should be collected in the field and taken back to the lab to conduct pretreatment such as drying, grinding, and sieving (Peng et al., 1998; Yu et al., 2002; Sha et al., 2003). Then these pretreated soil samples are analyzed by traditional chemical analysis methods or modern atomic absorption spectrometer, and chromatographs. Analysis results are accurate, but time consuming and energy consuming. Therefore, in order to meet the real-time and practical demands in field fertilizer management, the *in situ* measurement technology directly toward the soil is becoming a hot spot of research with many methods of *in situ* measurement attempted. Among those, near-infrared spectroscopy (NIRS) analysis method only simply disposes the original soil to perform the analysis of soil (Bao et al., 2007; Song and He, 2008; Zhu et al., 2008; Zheng et al., 2009), and there is no need to do soil sampling from the field (Sun et al., 2006; Chen et al., 2008; Yuan et al., 2009). The analysis result indicates that there is a high correlation between the NIRS predicted value and the laboratory chemical analysis value. Thus, it is feasible to use NIRS in the determination of contents of soil total nitrogen, soil organic matter, and soil alkaline hydrolysis nitrogen, and it may be used in the rapid analysis of soil in the field. In recent years, Chinese research teams have devoted themselves to the research of soil information acquisition based on spectroscopy, including soil moisture, soil total nitrogen, soil nitrate nitrogen, soil organic matters, etc. They have made breakthroughs by developing highly precise prediction models. Besides theory analysis, they have also focused on the development of soil sensors based on spectral technology and have successfully developed a soil organic matter content sensor and soil nitrogen content rapid detector (Li et al., 2010).

8.2.1.1 Development of Soil Organic Matter Content Detector

A practical portable detector for soil organic matter content was developed based on optical devices (Tang et al., 2007). The working mode is shown as Figure 8.1a. The optical signal at the NIR wavelength was transferred to the crop root zone (a depth of 300 mm). When the incident light reached the target soil, one part of the light was absorbed by the soil, and another part of the light was reflected from the soil as diffuse reflection light. If a certain wavelength is the sensitive waveband to the soil organic matter, the absorbed light is proportional to the content of soil organic matter. In other words, the intensity of the diffuse reflection light is inversely proportional to the content of soil organic matter. As a result, soil organic matter content can be estimated by soil reflectance value.

Based on the working mode above, the developed detector consisted of an optical unit and a circuit unit as shown in Figure 8.1b. The optical unit included light source, incident and reflected optical fiber, and a photoelectricity conversion device. The circuit unit included a light-emitting diode (LED) drive circuit, an amplifier circuit, a filter circuit, an analog-to-digital converter (A/D) circuit, a liquid crystal display (LCD), and a U-disk storage component. When measuring, the probe part was pushed into soil in order to form a confined space. There were incident and reflected optical fibers installed in the probe with the optical fiber opening at the top. The light from the LED was transferred to the top of the probe through the incident fiber. The light then reached the soil around the probe. The reflected light from the soil was transferred to the photoelectrical conversion device through the reflected fiber.

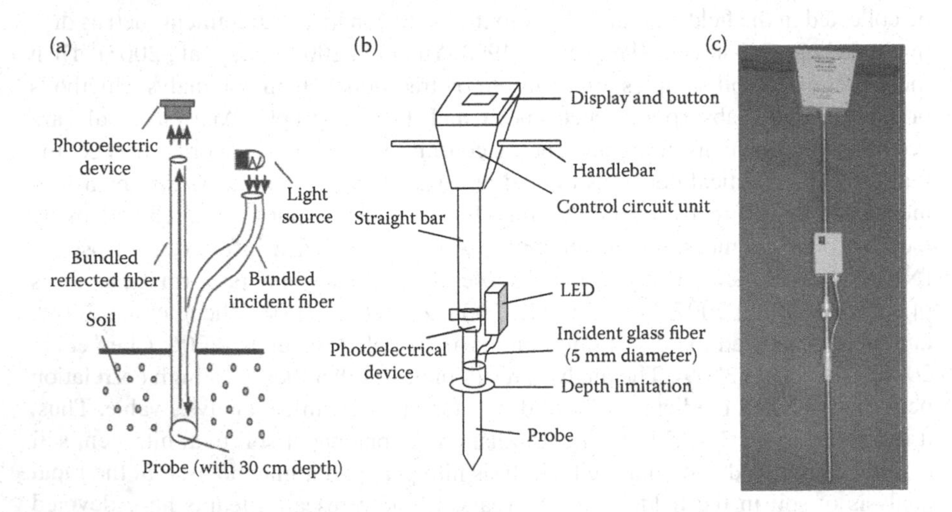

FIGURE 8.1 Overall structure of the soil organic matter detector. (a) Optical system structure, (b) detector sketch, and (c) detector prototype.

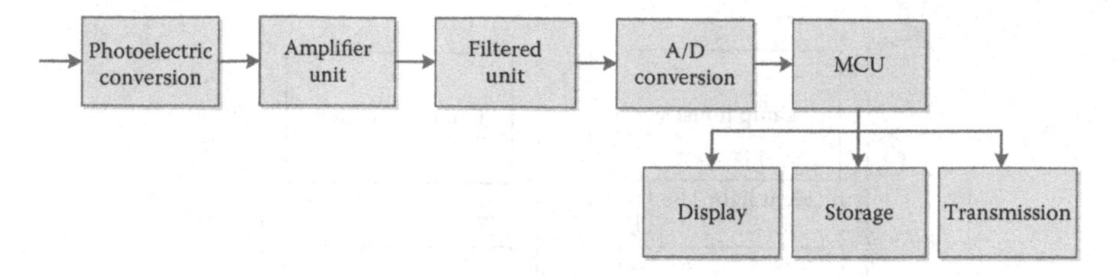

FIGURE 8.2 Block diagram of circuit unit.

Subsequently, the electrical signal was transferred to the circuit unit to be amplified, filtered, A/D converted, displayed, and stored.

To minimize the loss of incident light and reflected light, the Y-type glass fiber consisted of a light-incident port, a light-reflected port, and a port with both irradiation and detection. The diameter of the incident and reflected optical fiber bundle was 5 mm. The incident port was connected with a light source to receive the incident light. The reflected light measuring terminal was connected to the photoelectric sensor through reflective fiber. The incident fiber and the reflected optical fiber were gathered into a bundle at the soil detection port (diameter of 7 mm). The fiber bundle served the purpose of simultaneous transmission of incident light and reflected light. Glass fiber can make a tiny loss in the transmission process, and meet the requirements of Y-type structure because of its soft performance.

The circuit unit included an amplifier circuit, an A/D conversion circuit, an LCD, and a U-disk storage circuit. The block diagram of the circuit unit is shown in Figure 8.2. The optical signal was relatively weak, and the obtained electrical signal was weaker after conversion by the photodetector. Furthermore, there would be the influence of various noises. Therefore, the design of the amplifier played a very important role in the stability and reliability of the whole system.

8.2.1.2 Development of a Portable Soil Total Nitrogen Detector

The operation of the soil total nitrogen detector is similar to the soil organic matter content detector. The soil total nitrogen used seven wavelengths in the NIR region (780–2526 nm) (An et al., 2014). The detector also consisted of an optical unit and a control unit. The optical unit included six near-infrared LEDs in separate housings, a shared LED drive circuit, a shared incidence and reflectance Y-type optical fiber, a probe, and a photoelectric sensor. The control unit included an amplifier circuit, a filter circuit, an analog-to-digital converter (A/D) circuit, an LCD , and a U-disk storage component.

The LEDs were then rotated manually to align them with the Y-type optical fiber. The optical signal at each wavelength was then transferred from the LED to the surface of the target soil. The reflected light from the soil surface was acquired and transferred to the photoelectric sensor, through which the optical signal was converted to an electrical signal. Subsequently, the electrical signal was digitized, and the absorbance at each wavelength was calculated. All six absorbance values were used as input data for the soil TN content estimation model. Finally, the calculated soil TN content was displayed on the LCD and at the same time stored in the U-disk. Figure 8.3 shows the system overall structure design.

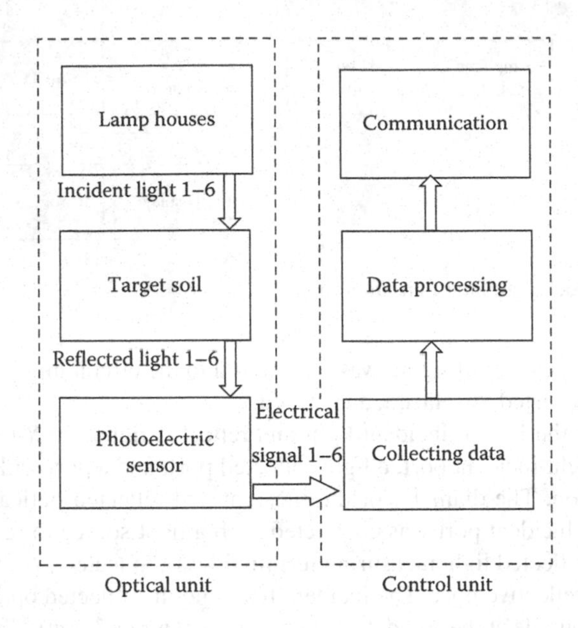

FIGURE 8.3 Overall structure of the portable soil total nitrogen detector.

The designed Y-type fiber is shown in Figure 8.4. In order to reduce the complexity of the circuit, the manual operation mode was adopted to control the seven light sources (1550, 1450, 1300, 1200, 1100, 1050, and 940 nm). The LED traversed the incident optical fiber position in a test cycle. Optical fiber mechanical structure and optical dial distribution are shown in Figure 8.5a and b, respectively.

Because the light source was weak with noise interference, a filtering operation was done before A/D conversion. In this module, an average filtering method was used to perform data filtering, whereas the hardware filtering method was a one-order R–C low-pass filter. The filtering method included two steps: amplitude limiting and mean. All data were taken an average of 10 times when the data were between 20 and 2000 mV. If the data were either below 20 mV or above 2000 mV, the data were deleted as outliers. Then, LCD display, USB storage, and serial communication of the processed data were realized by a single-chip microcomputer.

Figure 8.6 shows the flowchart of the main program. When the detector starts working, the system completes initialization and LCD detection first. The system

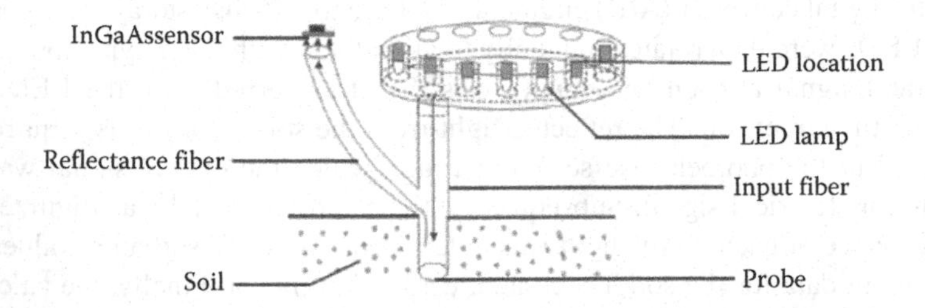

FIGURE 8.4 Overall diagram of optical fiber structure.

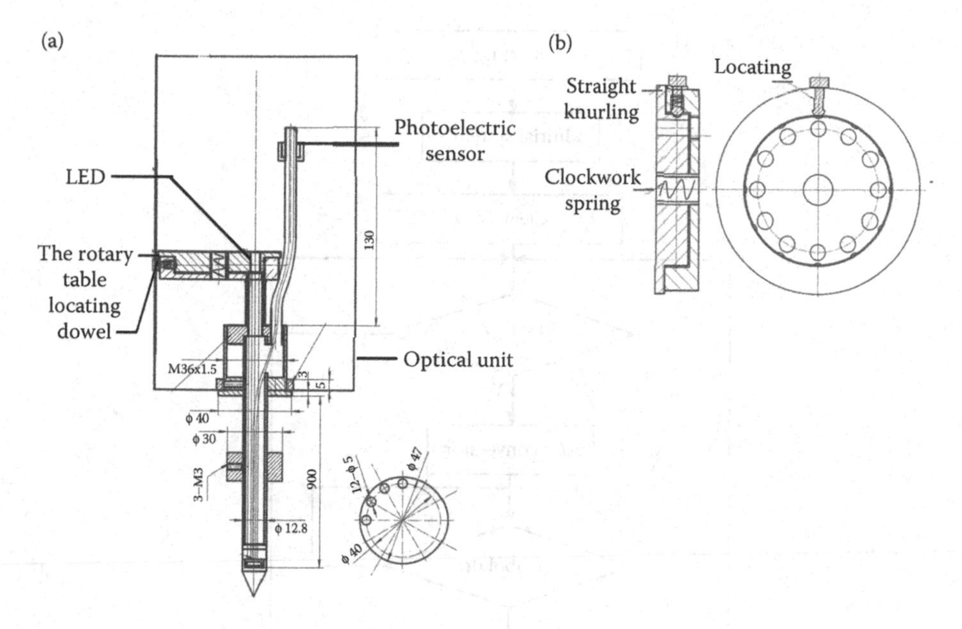

FIGURE 8.5 Structure of optical fiber and distribution of LEDs. (a) Optical fiber structure. (b) Distribution of LEDs.

then enters sleep mode until it receives an interrupt. If an interrupt 0 is present, the detector begins to collect data. After the A/D conversion, the program enters the data processing subroutine. All the processed data are temporarily stored in a variable. When detecting an microcontroller unit (MCU) period, all the collected data are calculated in the MCU. The result can be read in the LCD by using the LCD subroutine. After the result is displayed, the software then judges whether an interrupt 1 is present, in which case the result is stored in the U-disk. The program then returns to the initial interface. Otherwise, the software directly returns to the initial interface to start the next detecting period. The subroutines are shown in Figure 8.7.

8.2.2 Quick Detection of Crop Growth and Physiological Parameters

Acquisition of crop growth status information is very important to the precision management of crops in field. By using advanced detection methods, external and internal crop growth information can be acquired in field (Li et al., 2006). Since the size of field in China is quite small and topdressing is necessary to Chinese farmers, it is important to develop portable and low-cost sensors of crop growth and physiological parameters.

Spectroscopy is an effective method to detect and acquire crop growth information and nutrition status (Xue et al., 2004), which had achieved significant results in both basic theory research and practical application (Liu et al., 2004). Researchers had proposed several spectral indexes to describe crop growth, which were ratio vegetation index (RVI), normalized difference vegetation index (NDVI), agricultural vegetation index (AVI), multitemporal vegetation index (MTVI), normalized

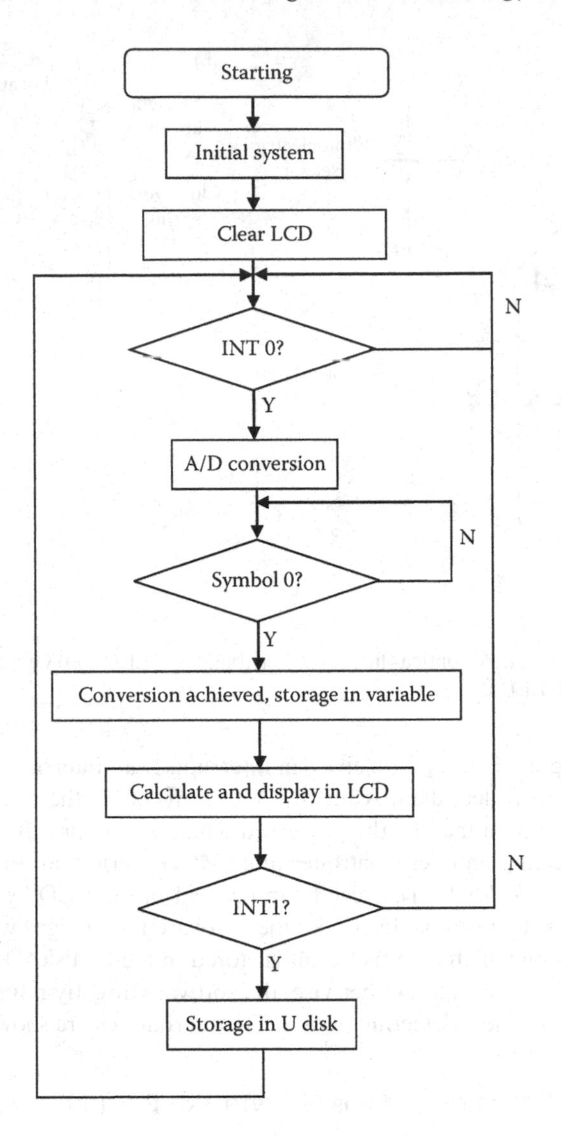

FIGURE 8.6 Flowchart of the main program.

differential green index (NDGI), normalized difference index (NDI), red-edge vegetation index, and differential vegetation index (DVI) (Feng et al., 2009; Yang et al., 2009; Tian et al., 2010). Among all these indexes, NDVI is the most sensitive to green crop, and thus can be used to detect crop growth and predict rainfall in semiarid regions, and can also be used in regional and global vegetation condition research (Chen et al., 2010). It is also the most commonly used vegetation index and plays an important role in vegetation analysis and monitoring in remote sensing.

Crop information detection is an important part of PA, and the development of the NDVI detector has a wide application environment (Yao et al., 2009). There are

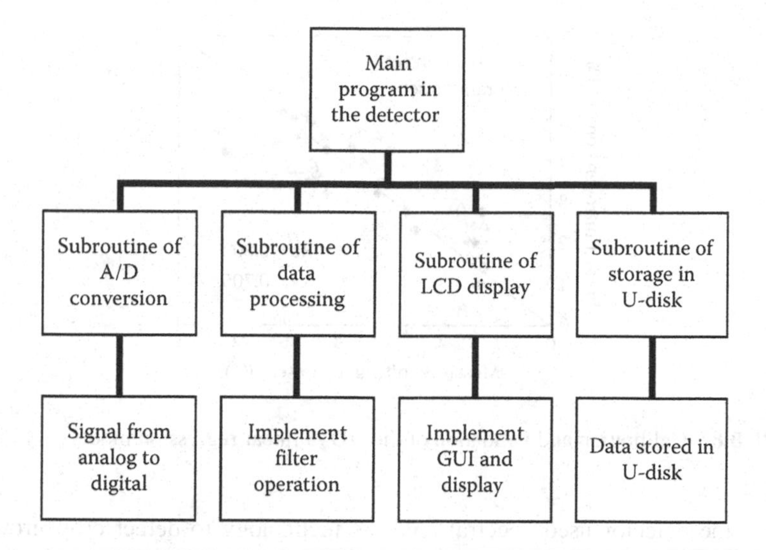

FIGURE 8.7 Subroutines in software.

several marketed instruments such as the GreenSeeker spectral detector, which have been used. Although this equipment can detect the precise NDVI value of vegetation, it comes at a higher price to Chinese farmers. Also, some parameters are not suitable to the agricultural environment in China, thus making the promotion and application of this equipment difficult in China. In order to solve these problems, much research was carried out to develop a low-cost NDVI detector, which could detect and analyze crop growth information, and which would also be suitable to Chinese agriculture.

8.2.2.1 Development of Crop Growth Detector with Optical Fiber

The principle of the crop growth detector with optical fiber was modeled based on NDVI, whose calculation required the reflectance in two different wavelengths, NIR and red. According to the correlation coefficients, two characteristic wavelengths were found in two highly correlated regions.

Choosing one wavelength from 410 to 650 nm and another wavelength from 660 to 850 nm, NDVI values were then calculated based on these two wavelengths. A linear regression analysis was then conducted on NDVI value and leaf nitrogen content. These steps were repeated for a combination of different wavelengths, and the combination with the highest correlation coefficient was chosen as the best combination for the NDVI detector (Zhang et al., 2004).

This detector has been used in greenhouse cucumber, and two characteristic wavelengths are 530 and 765 nm. The correlation results are shown in Figure 8.8.

The correlation coefficient of calibration was 0.808, RMSEC was 0.880, and F test value was 43.730, which meant the model passed the F test. By using data from the validation set to test this model, the complex coefficient of determination was 0.740; RMSEV was 0.836, which proved that the model was solid and reliable.

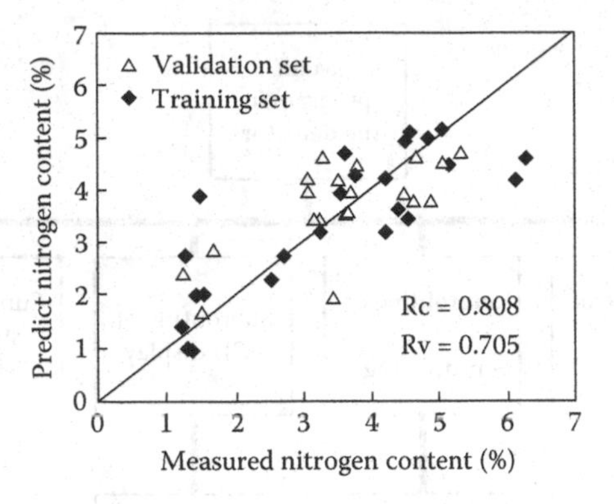

FIGURE 8.8 Calibration and validation of the NDVI linear regression model.

Since the detector used spectral analysis technology to detect crop growth, it needed to conveniently and accurately acquire the leaf reflecting light. As the collecting area for optical fiber was easy to control, optical fiber was chosen for the collecting part to avoid complexity from too many mechanisms. The overall structure of the detector is shown in Figure 8.9. It contains four parts: reflecting light collecting unit, metering unit, signal conditioning unit, and data acquiring unit (Zhang et al., 2006).

The fiber was Y-shaped to separate the collected light for individual filtering and processing. Using this simple mechanism could easily provide incident light at both sensitive wavelengths. The incident light was introduced to the metering unit through the optical fiber, and filtered and converted to an electrical signal in the metering unit. As the incident light under this condition was too weak for photoelectric conversion, a signal conditioning unit was designed after the metering unit, which included amplifier and noise canceler circuits. The signal processing unit was

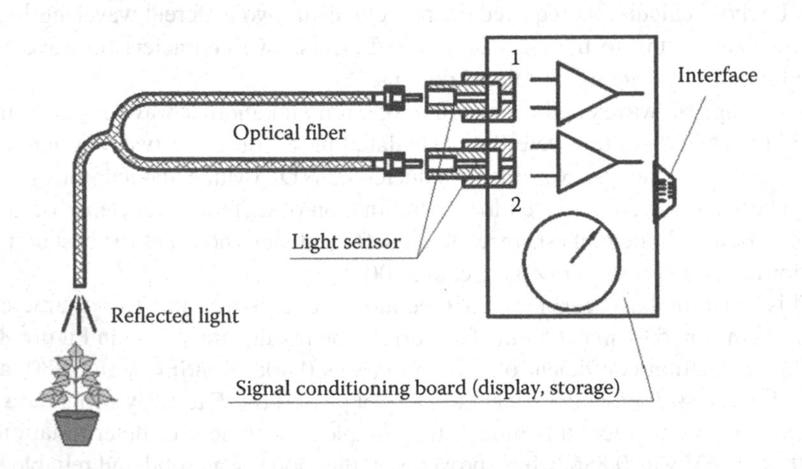

FIGURE 8.9 Overall structure of the crop growth detector with optical fiber.

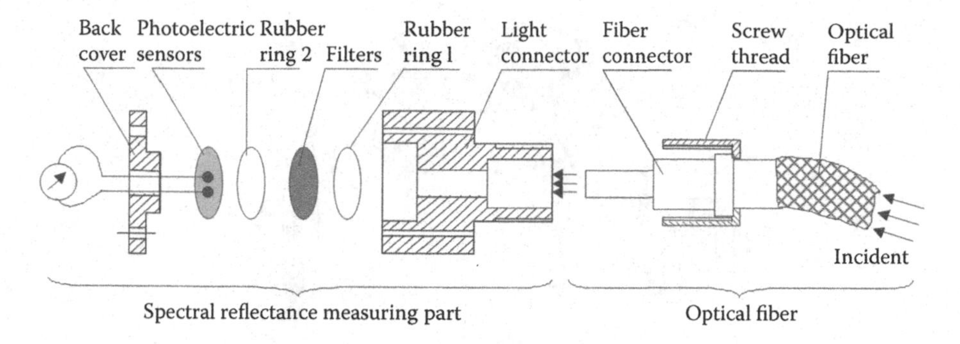

FIGURE 8.10 Spectral reflectance measuring device.

designed to process data afterward, including A/D conversion circuit, data processor, data storage, and result display. In order to store results in other storage devices or in the host computer, the detector had a communication interface (usually serial communication interface).

The structure of the metering unit is shown in Figure 8.10, which includes filter, photoelectric sensor, metering interface, and back cover. The metering interface had a jack on one side to connect with the output end of the optical fiber, and the metering chamber on the other side to place the filter and photoelectric sensor. The back cover was used to seal the metering chamber. The converted electric signal was drawn through the wire passing through the back cover. This unit can measure the reflecting light from the crop canopy and other parts, and change sensitive wavebands when necessary. The seal of the metering chamber can exclude external interference and reduce the distance between the photoelectric sensor and the exit end of optical fiber to increase metering efficiency.

Common photoelectric sensors include photoresistor, photodiode, phototransistor, and photocell. After comparing all four photoelectric sensors, a silicon photocell was chosen as the photoelectric sensor. Using a photocell can provide a large photosensitive area, high-frequency response, and linear photocurrent change.

The center wavelengths of the filter were 530 and 765 nm, and the half band width was 30 nm. As shown in Figure 8.10, the metering unit can collect the reflecting light from the standard plate or the leaf through the optical fiber, and then convert the light to an electrical signal via the photocell. The output signal from the photocell was amplified and converted to a digital signal for data storage and processing to get the final result. The final result could be displayed on the LCD screen and be stored in other storage devices or in the host PC.

8.2.2.2 Development of Hand-Held Crop Growth Detector with ZigBee

In order to easily evaluate crop growth status in a farm, a hand-held crop growth detector based on spectroscopy and ZigBee was developed (Li et al., 2009). As shown in Figure 8.11, the crop growth detector was made up of a sensor and a controller. The sensor and the controller were connected with ZigBee, a kind of wireless sensor network (WSN) technology. Since the distance between the sensor and the controller can vary according to requirement, it was easy for use in an open field. As

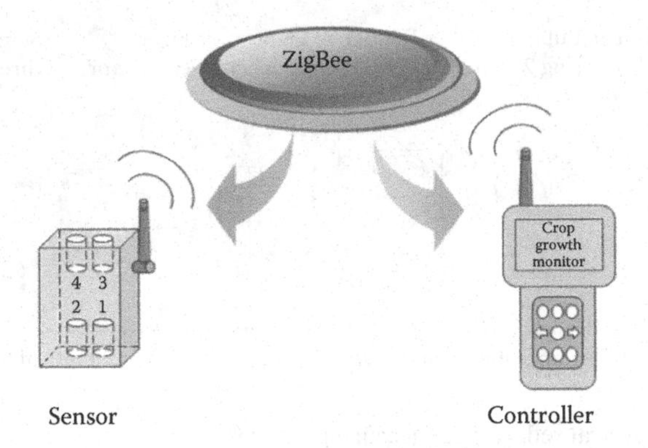

FIGURE 8.11 Structure of the hand-held crop growth detector with ZigBee.

the coordinator of the whole WSN, the controller was used to receive, store, process, and display the data from the sensor. The sensor was designed to collect, amplify, and transmit the optical signals. Because the system used sunlight as a light source, the sunlight intensity should be measured as well besides measuring the crop canopy reflectance spectra.

(1) *Hardware Design:* The block diagram of the sensor hardware is shown in Figure 8.12. The sensor consisted of an optical unit and a circuit unit. There were four optical channels in the optical unit. Channel 1 and channel 2 were used to detect the reflected light of crop canopy, and channel 3 and channel 4 were used to detect the sunlight. Channel 1 and channel 3 were sensitive at the red waveband, and channel 2 and channel 4 were sensitive at the NIR waveband. The circuit unit in the sensor included signal amplification, an A/D converter, and a wireless transceiver. The four optical channels had nearly the same components: an optical window, filter, convex lens, and photodiode. In order to avoid the influence of the changing angle of the incident sunlight, milky diffused glass was used as the optical windows of the two upward channels. The sensor is to be put vertically on the crop canopy while measuring, almost 20 cm, then the light went through the four optical channels and

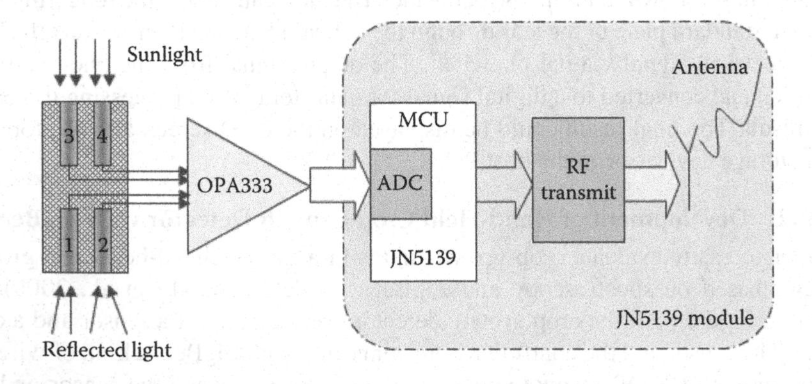

FIGURE 8.12 Block diagram of the hardware of the sensor.

was converted to a current signal by the photodiodes and then amplified to a comparatively large-voltage signal and finally collected and transmitted to the controller by a wireless module JN5139.

The JN5139 wireless module (Jennic Co, UK) was applied as the core element of the detection system. It provided all RF components and various peripherals, and gave users a comprehensive solution with high radio performance. A JN5139 microcontroller as MCU was integrated into that module to implement IEEE802.15.4 or ZigBee compliant systems. This microcontroller included a 4-input 12-bit A/D converter unit and was easy to use.

The block diagram of the controller is shown in Figure 8.13. The wireless transmission modules JN5139 were also used as the MCUs of the controller. An antenna interface and a flash with a volume of 128 KB were integrated into this module, so the controller could implement the functions of receiving and processing data, and displaying and storing the result within the single module. The display was connected to the MCU via two digital IO ports by means of serial communication. The keypad with nine keys provided several common functions such as reset, storage, review, format, and upload data.

(2) *Software Design:* The sensor and controller built up a simplest network (point-to-point network) together. The sensor was the end device of this network. Comparing with the application in the controller, the most distinct characteristic was its sleep mode. Once initialized after being started, the application activated a timer and then entered into sleep mode. It would be wakened by the interrupt, which was caused by the overflowing of the timer, then data would be collected and sent to the controller and then it went into sleep mode again. The sampling frequency was adjustable according to different requirements. One Hz was recommended in this development. The flowchart of the software in the sensor is illustrated in Figure 8.14.

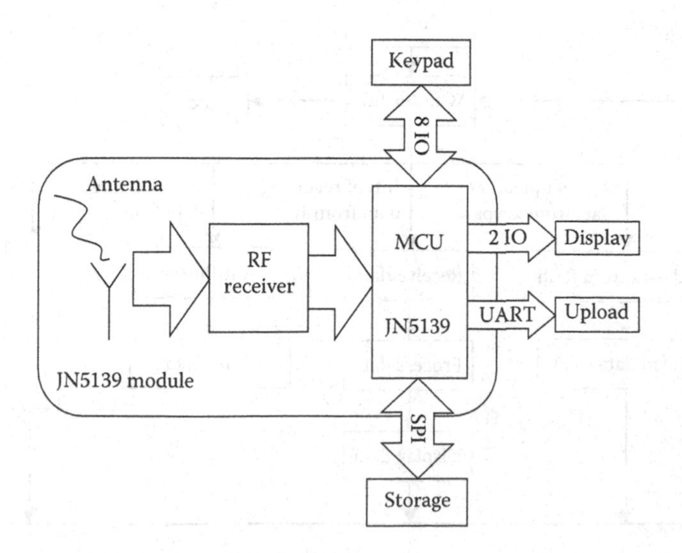

FIGURE 8.13 Block diagram of the controller.

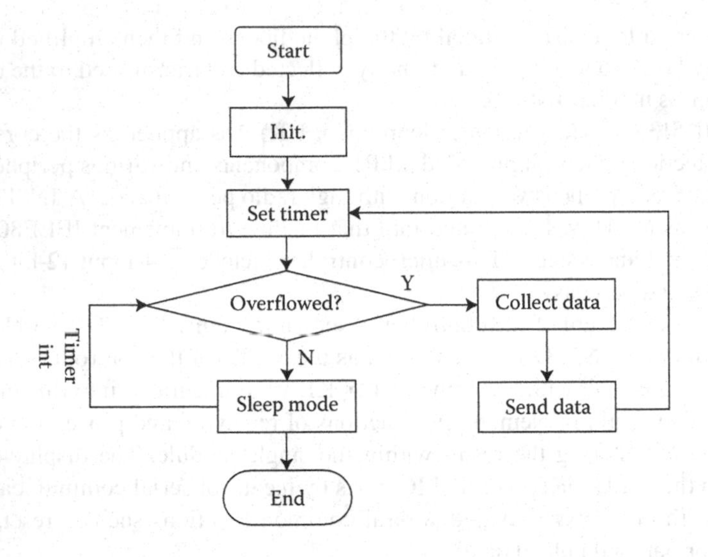

FIGURE 8.14 Flowchart of the software of the sensor.

The controller was the coordinator of this network. The coordinator was the core of the network and kept on working all the time as long as the network was active, and no sleep mode was allowed. The power consumption was about 60 mA, much higher than it was in the sensor. The flowchart of the software in the controller is illustrated in Figure 8.15. The controller would be initialized once powered, and then

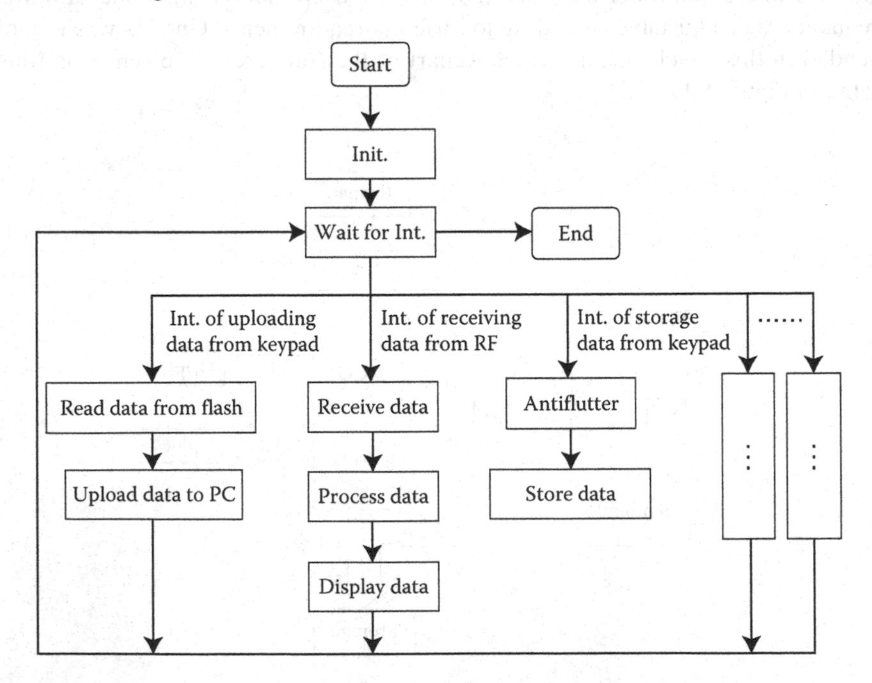

FIGURE 8.15 Flowchart of the software of the controller.

FIGURE 8.16 Field test at the Chinese state farm located in the Heilongjiang Province.

searched and built up an independent ZigBee network. After that, it would enter into an idle mode to waiting for interrupts. Once an interrupt occurred, the system would check which interrupt source it came from; if it was defined, the proper function would be executed, and if it was not defined, the system would return to idle mode. The interrupt from RF reception had top priority and could lead the system to the subfunction of data reception. In this subfunction, the received data were immediately processed and calculated to NDVI value, then displayed on the LCD. Seven other interrupts from the keypad were also defined in this application. Before recognizing which specific interrupt it was, an antiflutter function was strongly requested to eliminate the flutter caused by key pressing. In Figure 8.15, only three typical interrupts were listed.

The detector was applied in a Chinese state farm located in Heilongjiang Province, Northeast China. As shown in Figure 8.16, many field tests were conducted in the paddy fields of Qixing Research Center, Jiansanjiang Sub-bureau of Reclamation, Heilongjiang Province, on June 2008. After calculation, the prediction model of the chlorophyll content of rice was established based on NDVI (550,850), R (650), and R (766).

8.2.2.3 Development of Vehicle-Mounted Four-Waveband Crop Growth Detection System

In order to further extend the function of the crop growth detector, a new four-waveband crop growth detection system was developed to work as a ZigBee WSN with one control unit and one measuring unit. The measuring unit included several sensor nodes, which were used to measure crop canopy (Li et al., 2011). All the units were installed on an onboard mechanical structure so that the detection system could measure crop spectral characteristics on-the-go and in real time.

As shown in Figure 8.17, the system consisted of two parts: the control unit and the measuring unit. The control unit was a CS350 type of personal digital assistant (PDA) with an attached ZigBee wireless communication module (JN5139 module). As the coordinator of the whole wireless network, it was used to establish the wireless network, waiting for the sensor nodes to join in, and receiving, displaying, and

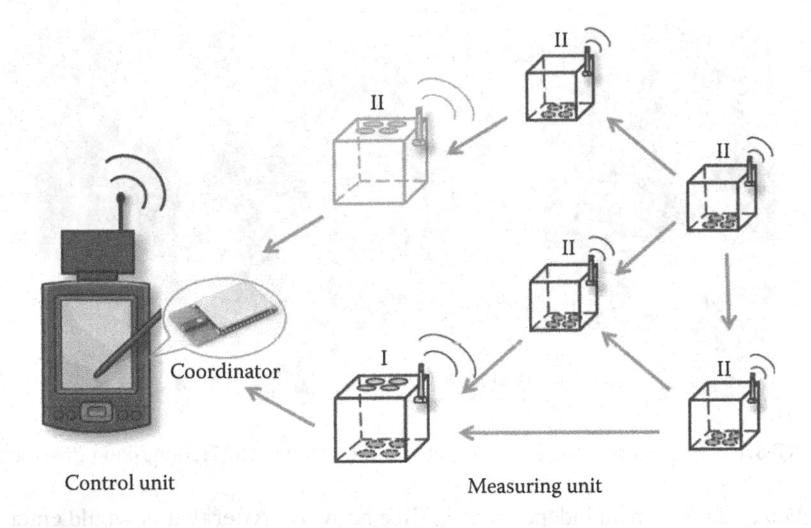

FIGURE 8.17 Structure of the vehicle-mounted crop detection system.

storing all the data from the different sensor nodes. Theoretically, each network can accommodate the maximum number of nodes for 65535.

The measuring unit consisted of several optical sensors, and each optical sensor was used as a sensor node in this WSN. Each sensor node consisted of an optical part and a circuit part. The optical part contained eight optical channels at four wavebands. Since the detection system used sunlight as the light source, besides the reflected light from the crop canopy, sunlight intensity should also be measured as a reference. Therefore, two solutions were put forward:

1. A full function sensor node had to contain eight optical channels, the upward four for the sunlight and the downward four for the reflected light.
2. Select one sensor node to measure the sunlight, as the type I sensor shown in Figure 8.17. Other sensor nodes were used to measure the reflected light, as the type II sensor shown in Figure 8.17.

Under the premise of measurement precision, this kind of design greatly reduced the production cost. As discussed above, the advantages of the four-waveband crop growth detection system mainly reflected in the following:

1. The structure of the optical channel. The four optical channels were designed to integrate with compact structure and light. The filter can be replaced conveniently without opening the sensor node, which enhances the universality of the system.
2. The signal process circuit. In the circuit part, the current signals were amplified and converted to voltage signals. A time-division multiplex chip (ADG704) was applied to share the amplification unit and an OPA333 amplifier, which had the properties of high-precision, low-quiescent current, and low power consumption, was chosen to amplify.

FIGURE 8.18 Field test of the vehicle-mounted crop detection system.

3. Flexibility and portability of the system structure. The sensor and controller can set up the communication network in many ways. The networking mode between the hand-held and vehicle-mounted can be transformed into each other. The transmission distances can be up to hundreds of meters, which realized the real-time, continuous measurements of crops in the field. Furthermore, it increased the flexibility of the detector installation.
4. The independence of the sunlight measuring unit. A sensor node was selected to measure the sunlight, and then the whole network shared the sunlight value. Under the premise of measurement precision, this type of design greatly reduced the cost of the system.
5. Friendly operation platform. Using a personal digital assistant (PAD) as the controller of the system, and developing a visual interface for data acquisition, it was convenient and user-friendly, and easy for further development.

The newly designed system increased the optical channels and realized measured crop spectral characteristics on-the-go and in real time after being installed on an onboard mechanical structure (Zhong et al., 2013). Figure 8.18 shows the field test in Shaanxi Province. The distribution of chlorophyll content of wheat detected by the new system is shown in Figure 8.19.

8.3 APPLICATION OF IoT IN AGRICULTURE

The IoT is defined by the Chinese Academy of Information and Communication Technology (CAICT) as follows: "Internet of Things is an expanded application and a network extension of a communication network and the Internet. It uses sensing technology and intelligent equipment to perceive and recognize the physical world, and communicate through a network to compute, process, and mine data. It can

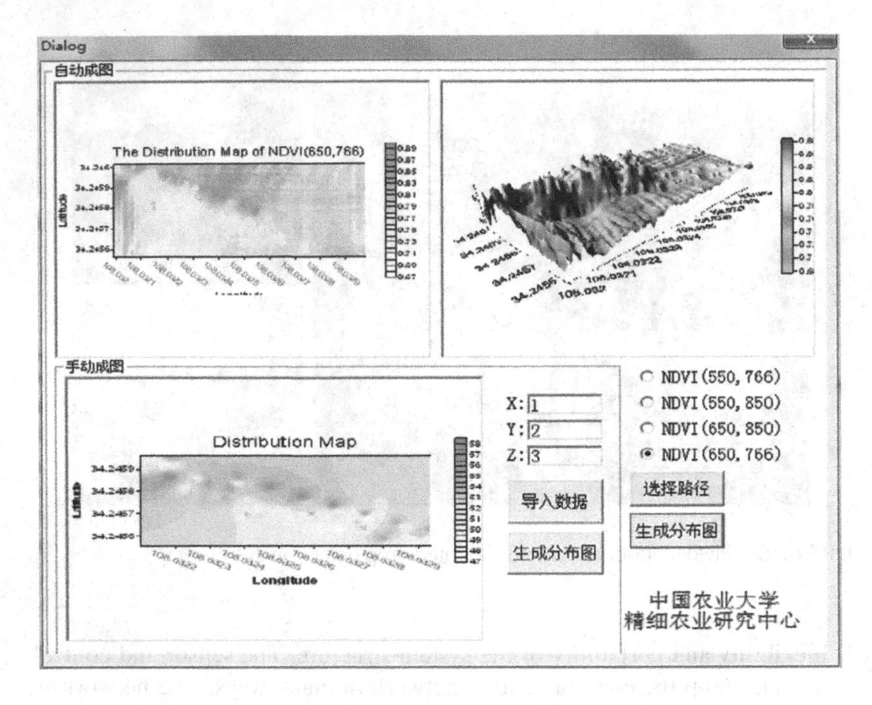

FIGURE 8.19 **(See color insert.)** Distribution of chlorophyll content of wheat.

exchange information and create seamless links between human–things or thing–things, thus to realize real-time control, precise management and scientific decisions of the physical world." Based on this definition, it is concluded that the IoT is an integration of the WSN, microelectromechanical systems (MEMS), and the Internet. Figure 8.20 shows the structure of the IoT. Usually, it includes three layers: perception layer, network layer, and application layer. As the nerve endings of the IoT, the perception layer achieves the function of the acquisition, identification, and control of all necessary information through sensors, radio-frequency identification device (RFID) readers, cameras, GNSS modules, smart meters, mobile phones, IC cards, etc. It is mainly related to sensors, bar codes, RFID, audio and video codec, and GNSS technology. The network layer is the nerve center of the IOT, and is used to transmit information. It uses WSN, Wi-Fi (wireless fidelity), communications networks including the Internet, GPRS (general packet radio service) network, 3G or 4G network, LAN (local area network such as IPV4 and IPV6), radio and television networks, and the next generation of broadcast networks). The application layer is the brain of the IoT, and can realize the data processing and application. The fields used in the application layer include enterprise resource planning, expert system, cloud computing, system integrate, industry application, agricultural application in crop cultivation, husbandry, aquaculture, greenhouses, etc. (Li, 2012).

Currently, Chinese agriculture is in the process of moving from traditional agriculture to modern agriculture, and the development of modern agriculture requires the support of information technologies during the production, sale, management, and service process. With the progress of the IoT, the development of modern

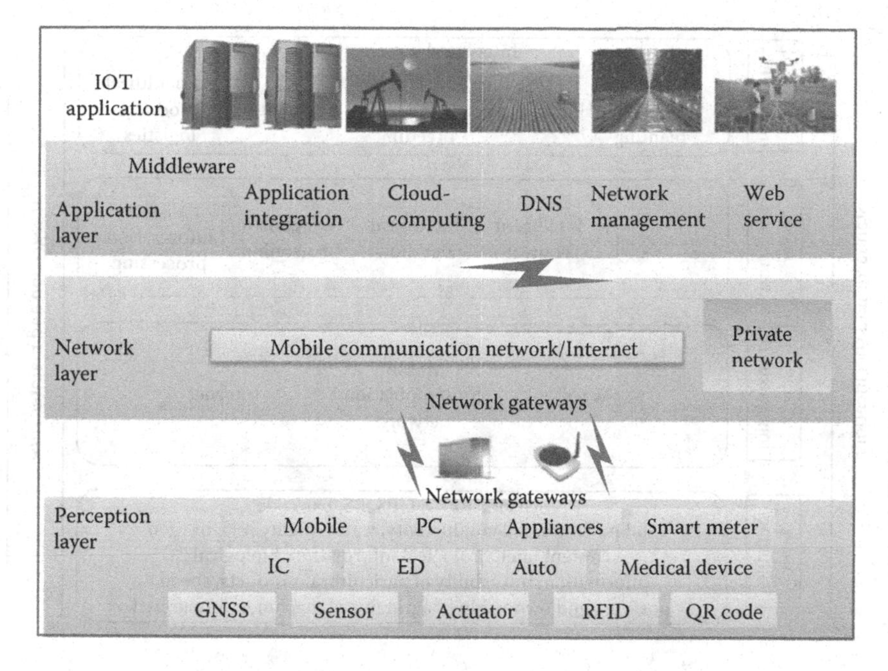

FIGURE 8.20 Structure of IoT.

agriculture has greater opportunity than ever before. Modern agriculture urgently requires the IoT to provide digital design, intelligent control, precise operation, and scientific management to agricultural elements in agricultural industries such as field planting, protected horticulture, livestock breeding, aquaculture, and agricultural logistics. Thus, it is possible to realize "overall perception, reliable transmission, and intelligent processing" for a variety of agriculture elements, and achieve the goal of high yield, high efficiency, ecological sustainability, and safety.

8.3.1 Key Technologies of Agricultural IoT

Figure 8.21 shows the structure of agricultural IoT. The perception layer involves all the factors in field information acquisition with advanced sensing technology. After information acquisition, the network layer connects the sensing equipment to the transmission network, which provides the path for the upload of sensing data. Through a wired or wireless communication network, information and data can interact and share in real time. In the application layer, agricultural information management and intelligent decisions can be made based on the knowledge provided by acquired agricultural information using intelligent computing and processing. Owing to this, transmission and processing of agricultural data are other key technologies besides sensing technology.

WSN and mobile communication are two important technologies of agricultural information transmission. Since WSN is a self-organized wireless communication network system, it can deploy a large number of sensor nodes in the detection area and monitor and collect information about all the subjects in the detection area, and

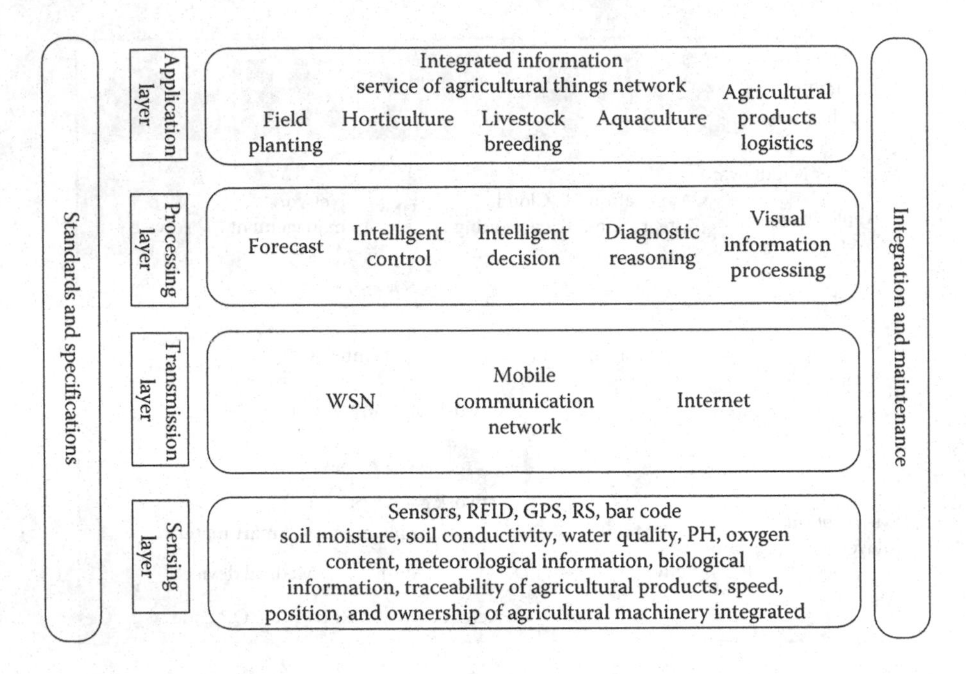

FIGURE 8.21 Structure of agricultural IoT.

then send this information to gateway nodes to conduct detection and tracking on targets within a complex specified range. It is easy to deploy and hard to destruct. Currently, in Agriculture WSN, ZigBee technology is widely used.

Researchers worldwide have been using WSN technology in field information acquisition. By combining ZigBee and GPRS wireless communication technology, NDVI data can be wirelessly transferred to a server thus making it possible to analyze crop growth and support field management. In order to meet the needs of measuring farmland environmental parameters, the monitoring system of soil temperature and moisture in farmland was developed. The system included a field wireless sensing network and a remote data center. Using a JN5121 wireless microprocessor as the core of the sensor nodes, the wireless sensing network was built based on ZigBee protocol. The gateway nodes were developed based on an ARM9 microprocessor embedded Linux system, which could realize data aggregation and remote data forwarding using GRPS. The management system FieldNet was installed in the remote data center, which could monitor the real-time change and analyze spatial variation by using the implemented ESRI GIS ArcEngine Library. The design and development of the system provided an effective tool for the research of spatiotemporal variability and irrigation decisions in PA.

With the improvement of agricultural IT, mobile communication technology has become an important tool in remote transmission of agricultural information. Figure 8.22 shows the structure of a wireless field acquisition system for soil moisture based on GSM technology. It includes a fast positioning system (GNSS device), a GSM module, a terminal computer, and other communication equipment. The

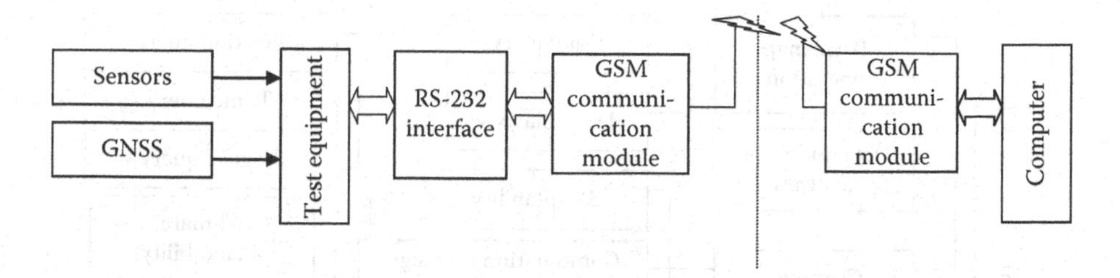

FIGURE 8.22 Structure of wireless field information acquisition system.

GNSS device can detect the position and moisture information of farmland, which will be sent to monitoring computer by text message through the GSM network. It is low-cost and reliable, and can also cover a wide range and transmit through an unlimited distance. It can provide an efficient solution for field information acquisition, transmission, and processing.

8.3.2 Agricultural IoT and Management Decision in Precision Agriculture

Management decisions based on agricultural information play an increasingly important role in agricultural modernization and digitization. Nanjing Agricultural University has developed an agricultural spatial information management and decision supporting system based on WebGIS (Liu et al., 2006). This system provided a great data management platform for PA information. Using a systematic approach and mathematical modeling techniques, a distributed network platform based on B/S structure was built, and a regional agricultural spatial information management and decision support system was designed and developed. Figure 8.23 shows the whole structure of the system. It included several subsystems such as Basic Map Operation, Data Query and Analysis, Cropping System Evaluation, Ecological Zoning, Potential Analysis, Precision Farming Management, Visual Outputs, and System Maintenance. It can perform the position query, topic query, and logical query, and can carry out the evaluation of climate adaptability, soil suitability, and comprehensive conditions. It can also conduct analysis on monoculture production potential and multicrop production potential.

In addition, the decision support system (DSS) of precision fertilization has also made great progress. An information system of soil and fertilizer was developed by the Chinese Academy of Agricultural Sciences (CAAS), which realized functions such as variation prediction of soil and fertilizers, expert system of soil and fertilizers, and output of agricultural maps. Based on this research, the National Engineering Research Center for Information Technology in Agriculture (NERCITA) has proposed a DSS for precision fertilization. The overall structure of the system is shown in Figure 8.24. In order to solve the promotion and extension problems of precision fertilization software in China, this DSS was developed based on component-oriented technology. It had distributed the tasks in precision fertilization into several different service units, which were mapped to corresponding service components. Furthermore, it had also provided a method how to develop a component-oriented

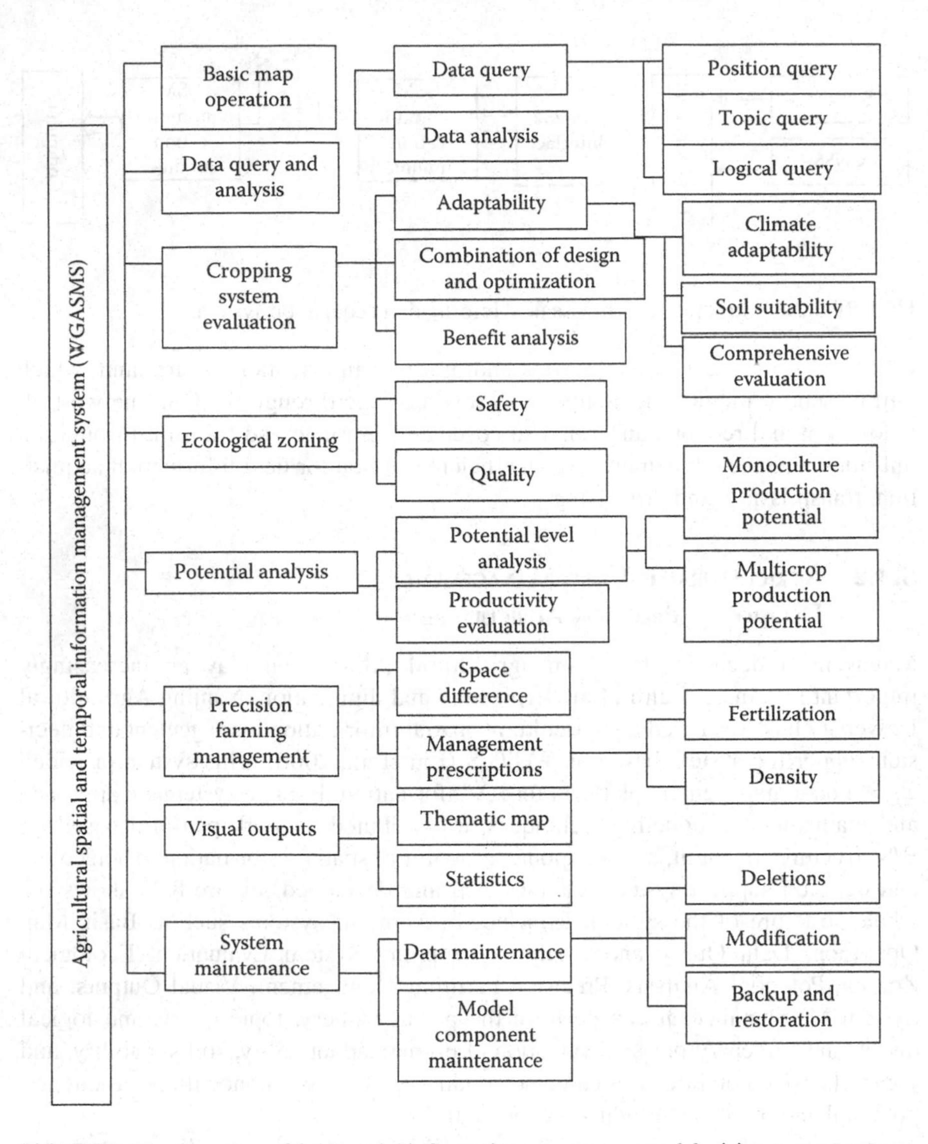

FIGURE 8.23 Structure of field spatial information management and decision support system.

DSS of precision fertilization to meet personalized needs. Experiments showed that the component-oriented DSS of precision fertilization had great advantages in widespread promotion and application.

8.3.3 Agricultural IoT and Field Information Acquisition

8.3.3.1 Soil Moisture Monitoring System in Farmland Based on IoT

In northern China, drought is still a problem in agriculture. In order to prevent spring drought from damaging the growth of winter wheat, it is necessary to use soil moisture information to guide irrigation and prevent drought. There is a serious shortage

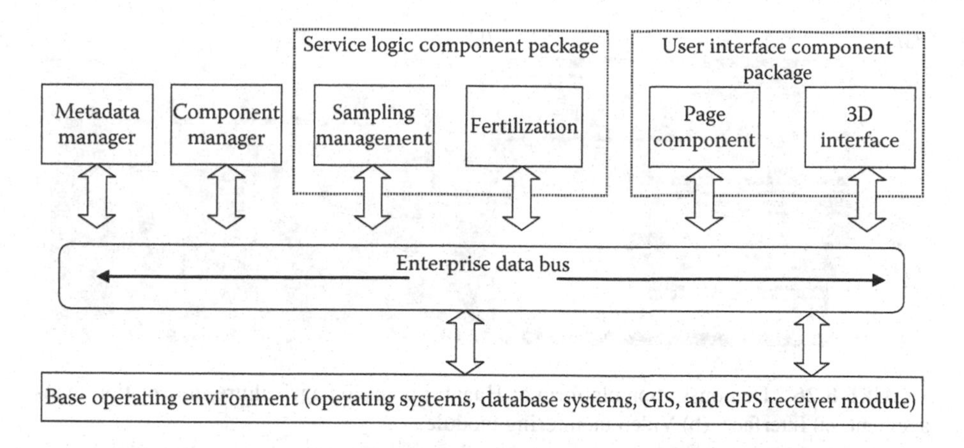

FIGURE 8.24 Overall structure of precision fertilization decision support system.

of water resources in China and other problems such as low utilization rate of water resources. In this case, it is very important to improve water resource efficiency in PA. By adapting advanced communication and sensing technologies, a monitoring system was constructed to realize accurate dynamic monitoring of agriculture water resources, thereby promoting scientific management and rational use of water resources.

A soil moisture monitoring system based on the IoT was constructed in Huaitai County, Shandong Province, China, as shown in Figure 8.25. WSN were constructed

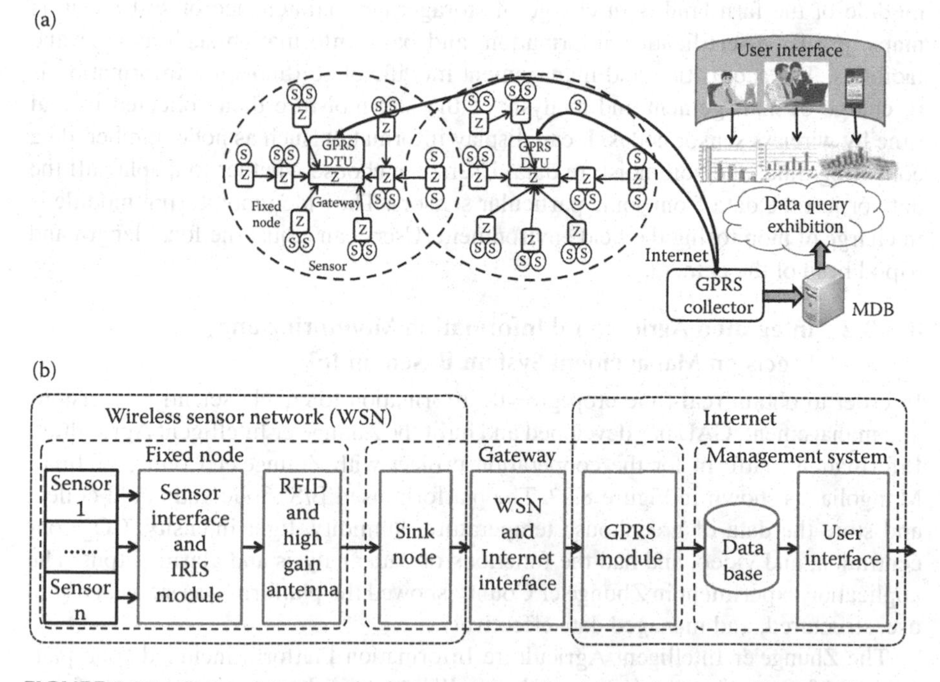

FIGURE 8.25 Soil moisture monitoring system based on IoT. (a) Structure. (b) Block diagram.

(a) (b)

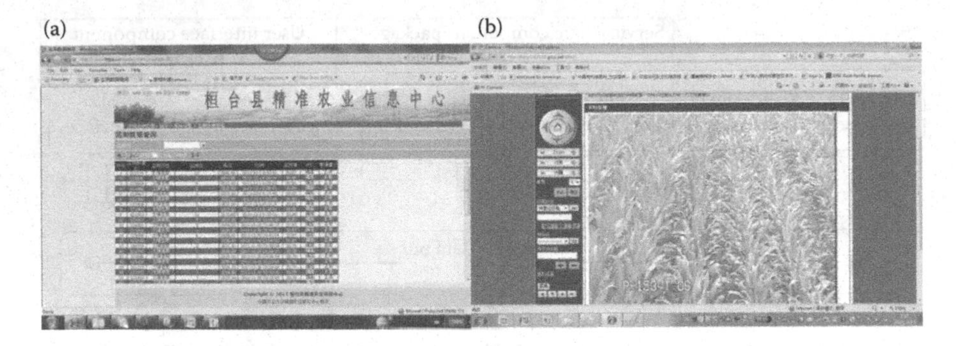

FIGURE 8.26 Demonstration platform of Huantai precision agriculture information management. (a) Interface. (b) Video monitoring module.

to monitor the variation of soil moisture in a farmland using 6–10 monitoring nodes. The data of soil moisture content were integrated to a gateway node, and then uploaded to a web server (database) by using GPRS network or the Internet, dependant on the site's condition. Thus, remote collection and monitoring of soil moisture were performed. The data can be browsed in a webpage (The Demonstration Platform of Huantai Precision Agriculture Information Management, http://www.htpa.cn/).

The Demonstration Platform of Huantai Precision Agriculture Information Management includes basic information management of farmland, soil moisture information acquisition and management, and video monitoring and information publication, which is shown in Figure 8.26. The basic information management module of the farmland is in charge of storage and maintenance of soil nutrition maps, precision fertilization information, and basic information such as area and facilities. The acquisition and management module of soil moisture information is in charge of management and analysis of the soil moisture data collected in real time by wireless sensor nodes. It can display information such as node number, data collecting time, and soil moisture data. Users can choose whether to display all the data or just the data from some particular sensors. The video monitoring module is in charge of monitoring the field environment. Users can adjust the focal length and tripod head of the camera.

8.3.3.2 Integrated Agricultural Information Monitoring and Precision Management System Based on IoT

In order to obtain real-time crop growth information to enable scientific decisions and management, CAU has developed and built the Zhunge'er Intelligent Agriculture Information Platform for the cooperation project with Zhunge'er County of Inner Mongolia, as shown in Figure 8.27. The platform used B/S mode and could collect and store the data of greenhouse temperature, humidity, light intensity, CO_2 concentration, and video, and had the functions of data analysis and alarm output. The application experiment in Zhunge'er County showed the platform was stable, easy to use, structured, and managed data effectively.

The Zhunge'er Intelligent Agriculture Information Platform included "one platform and four systems," which were the intelligent agriculture information platform,

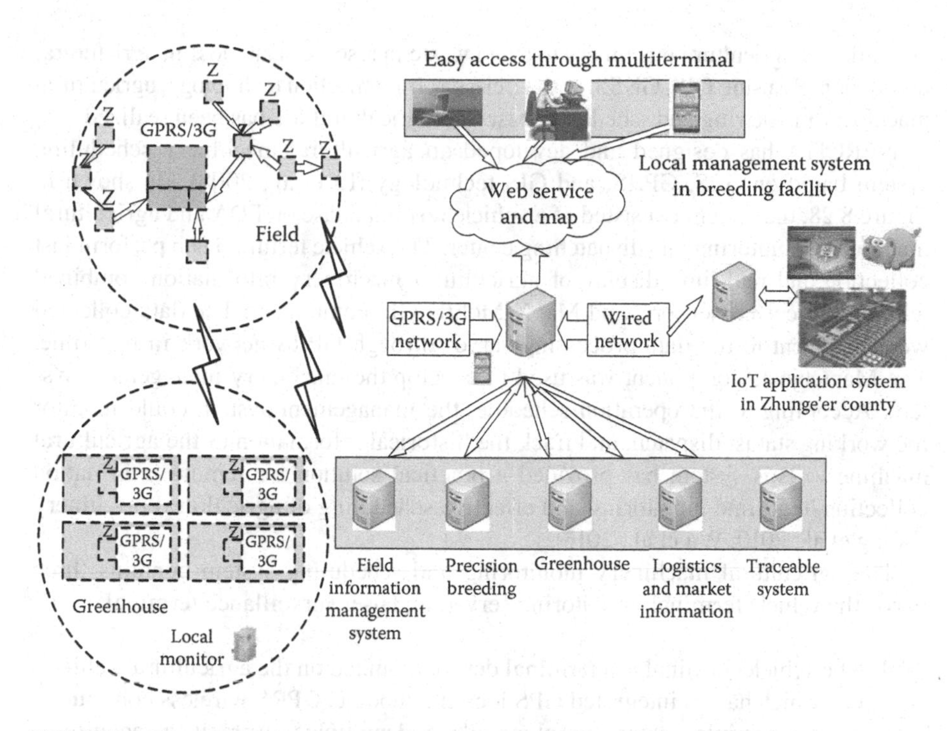

FIGURE 8.27 Zhunge'er integrated agricultural information monitoring and precision management system based on IoT.

and the precision management system for field crop production, precision management system for greenhouse, precision management system for animal production, and traceability system of agricultural products.

In the precision management system for greenhouse, for example, the subsystem of data acquisition and remote transmission consisted of several sensor nodes, gateway nodes, and relay routing nodes. The sensor nodes were connected with temperature, humidity, carbon dioxide, and light, which could be deployed at the center of different greenhouses. Sensing data could be transmitted to the gateway node through the ZigBee wireless network, and were then sent to the local PC through serial communication. The stand-alone monitoring software also runs on this PC to receive data by scanning serial communication and furthermore processing and analyzing these data. The greenhouse administrator could check data in real time on this PC. The video camera was connected to a local PC and server platform through the Internet, which was used to monitor crop growth and pest conditions. The server platform was designed based on B/S mode. Users can access web applications to manage or query monitoring data. Authorized users can watch the greenhouse monitoring video in real time.

8.3.4 Agricultural IoT and Agricultural Machinery Scheduling

With the rapid development of large-scale agricultural production in China, it is important to execute rational allocation and effective scheduling of agricultural machinery resources to ensure the completion of agriculture production on time, and to improve the utilization of agriculture machinery to avoid waste in resources and loss in agricultural production. By using GIS, GNSS, and wireless communication technology, agricultural machinery monitoring and scheduling based on agricultural IoT has been realized.

NERCITA has designed and developed an agricultural machinery scheduling system based on GPS, GPRS, and GIS technology (Li et al., 2008). As shown in Figure 8.28, the system consisted of a vehicle terminal base on PDA and agricultural machinery monitoring and dispatching center. The vehicle terminal can perform fast collection and real-time display of agricultural machinery information combined with GPS receiver, sensors, and MapX Mobile GIS component. The data collected were then sent to the data processing server through GPRS network in real time. The MapObjects component was used to develop the machinery management system. According to the operation schedule, the management system could monitor the working status, dispatch, and track the historical information of the agricultural machinery. This system has provided a practical solution for remote information collection, real-time monitoring and effective scheduling of agricultural machinery (Wang et al., 2010; Wu et al., 2013).

The agricultural machinery monitoring and scheduling system contains three parts: the vehicle terminal, monitoring server, and user surveillance terminal.

1. The vehicle terminal is a terminal device mounted on the agricultural vehicle, which has an integrated GPS locating module, GPRS wireless communication module, center control module, and multiple sensors. It can acquire position data of agricultural machinery by the GPS module, and real-time condition data of agricultural machinery by a series of sensors such as fuel

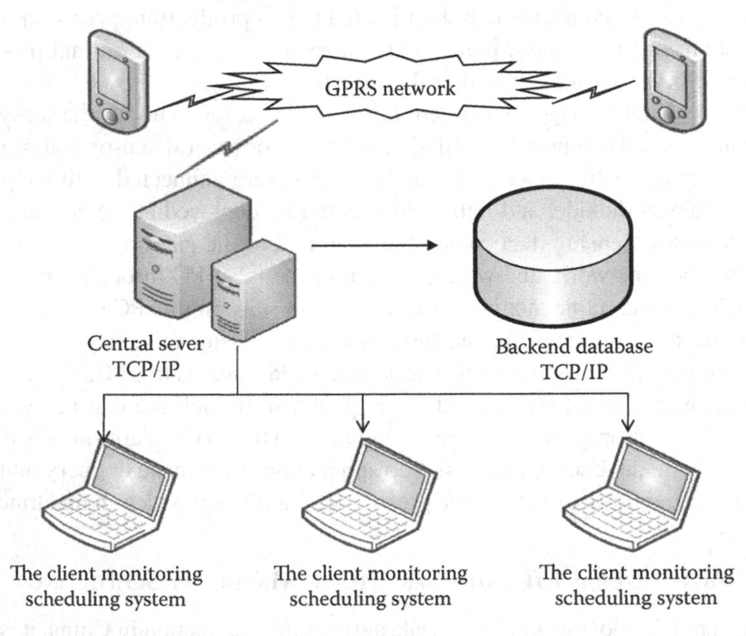

FIGURE 8.28 Agricultural machinery scheduling system based on GPS, GPRS, and GIS technology.

cost sensor, signal sensor, and speed sensor. Finally, it can upload all these data to the monitoring server through the GPRS wireless network.

2. The monitoring server consists of a vehicle terminal server, monitoring terminal server, and database server. The vehicle terminal server is in charge of communicating with the vehicle terminal to receive data from different terminals and to store these data in the database in the scheduling center. It can also send scheduling commands and information to the vehicle terminal. The monitoring terminal server is in charge of interfacing with the client scheduling center, parsing and responding to the request from the clients, and extracting data from the database to clients. The database servers are in charge of storing and managing agricultural machinery data such as position, condition, and operating parameters. It will also regularly back up and dump historical data, which provide data support for the vehicle terminal server and monitoring terminal server.
3. The clients surveillance terminal can provide real-time remote monitoring and processing of agricultural machinery production position and condition, visual display of agricultural machinery position on a digital map, data query and statistical analysis for agricultural machinery operation monitoring data, and the release of agricultural machinery scheduling information to managers. The surveillance terminal can also post scheduling commands by telephone, thus realizing real-time scheduling of agricultural machinery.

In general, an agricultural machinery scheduling management system can provide real-time information of working conditions and positions to agricultural machinery management and agricultural cooperation organizations based on a GSM digital public communication network, GPS, and GIS technology. The agricultural machinery scheduling system can suggest the optimal number and route for agricultural machinery usage by analyzing information such as area and position of production according to tasks given by the manager. Meanwhile, the supportive module can examine the efficiency and fuel cost of historical production, to suggest the optimal operation of agricultural machinery. By processing the data uploaded by the vehicle terminal, the system can accurately obtain information such as real-time position and fuel cost. The current condition of agricultural machinery can be displayed in real time and tracked on the monitor. Statistical analysis of effective mileage of operation and fuel cost can be provided. Also, by providing the historical track of agricultural machinery, remote monitoring of production can be achieved, which can support the scheduling of production, thus increasing the efficiency of agricultural machinery usage.

8.4 SYSTEM INTEGRATION AND APPLICATION OF PRECISION AGRICULTURE

8.4.1 Development and Application of Intelligent Agricultural Equipment

Information and communications technology (ICT) and computer technology have brought about a revolution in traditional agricultural machinery. By introducing sensing and detection technology, automatic control technology, information acquisition and fusion technology, machine vision technology, and field bus technology, agricultural machinery has partly or overall realized automation.

The studies on the automation system of agricultural machinery include sensors, actuating devices or units, data fusion and processing software, field bus, visualization of monitoring interface and control software, and overall system performance and structure. In order to increase performance/price ratio of agricultural machinery, electronic, sensor, power, mechanical, computer, and intelligent control technologies were applied to the design, manufacture, and application of agricultural machines. Consequently, production efficiency was greatly improved. Especially with the development and extension of PA ideas or technologies, it promotes the development and application of intelligent agricultural equipment. Many fruitful results have been achieved in automatic navigation technology, variable operating control technology, vehicle-mounted agriculture operation technology, and agricultural robot technology in agricultural machinery.

8.4.1.1 Application of VRT in Agricultural Machinery

The core of PA is variable agricultural resources management based on spatial and temporal variation. It is in accordance with the crop yield, environmental factors that affect crop growth, and growing requirements. Hence, variable-rate treatment (VRT) technology is the core of PA, while agricultural machines for VRT are important and necessary (Wang et al., 2003).

VRT technology can be divided into two categories: map-based VRT and sensor-based VRT. Thus, the variable targets are taken in two forms: previously generated electronic map and real-time decision data generated during machine movement.

For map-based VRT, a four-step procedure is needed to generate the prescription map. The first step is obtaining the spatial and temporal variation information of a crop yield, soil parameters, etc. The second step includes establishing models on plant growth, environmental conditions, weather, germination rate, growth, and nutrient requirements. The third step involves generating the desired prescription map based on the previous comprehensive analysis by using GIS and DDS. The fourth step is implementing variable inputs in accordance with the prescription map by using corresponding VRT agricultural machinery.

NERCITA has developed a kind of variable fertilization machine matching with the domestic tractor. With the help of the GPS navigation system, it can realize variable fertilization according to the prescription map designed in advance. The structure of the Geneva wheel was used as the fertilizer measuring device of the VRT machinery. By adjusting the rotation speed of the outside Geneva wheel, it can adjust the fertilizer quantity. The structure of the VRT fertilizing machine is shown in Figure 8.29.

After determining the soil fertilizer prescription based on the soil information of a field, the prescription map in the format of a .shp file was input into the AgGPS170 computer by a TF card. A tractor mounted with receiving antennas received GPS signals and differential signals from the radio base station.

After processing the DGPS signals, the system software can determine the geographic location of the machinery. The AgGPS170 computer would put the on-site

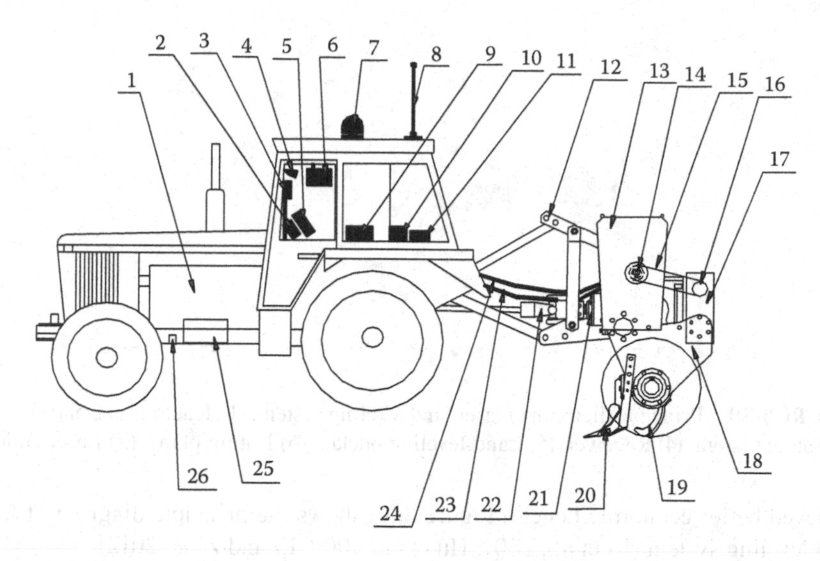

FIGURE 8.29 Variable rotary tillage and fertilizer machinery structure diagram. (1) Tractor. (2) Data exchanger. (3) AgGPS170 computer. (4) Guiding light bar. (5) Fertilization control switch. (6) Fertilization controller. (7) GPS-receiving antenna. (8) Radio-receiving antenna. (9) Junction box. (10) Branch box of power source. (11) GPS moving station. (12) Three-point suspension hitch. (13) Fertilizer can. (14) Transmission shaft sprocket. (15) Transmission chain. (16) Hydraulic motor. (17) Motor supporting structure. (18) Rotary tillage side panel. (19) Rotary blade. (20) Subsoiler. (21) Gearbox. (22) Cardan shaft. (23) Oil return pipe. (24) Oil feed pipe. (25) Storage battery. (26) Speed measuring radar.

fertilizer application rate from the prescription map into the fertilization controller through a data switch exchanger, and then, the fertilizer distributor controlled the hydraulic motor speed to achieve the goal of changing the fertilizer application quantity. The start and stop of fertilizer distribution can also be compulsively controlled by the fertilization control switch in the driving cab, and the running of the tractor in the field is instructed through the guiding light signal.

The AgGPS170 computer will then display the next location coordinates of the tractor, and the fertilizer prescription data on the screen. The fertilizer was discharged from eight rows of distributing wheels installed at the bottom of the fertilizer box, evenly scattered on the surface by the distributing plate, and then the back of the high-speed rotary tillage blade stirred the fertilizer into the soil.

8.4.1.2 Laser-Control Land Leveling System

In a cyclic process of farmland operation, land leveling is an important measure to improve irrigation quality and therefore plays an important role in PA (Jia et al., 1997). It can effectively improve farmland management and seedbed conditions, and realize precision irrigation so as to achieve the purpose of water saving and increased production. As the world's most advanced mode of land leveling, laser control technology has been widely used in Europe, America, and other developed countries in the early 1970s. In the last decade, developing countries such as India, Turkey, Pakistan, China, and others have also successively used laser technology and

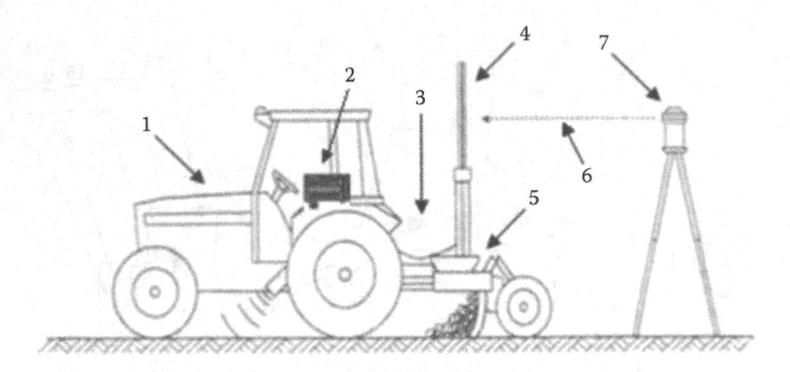

FIGURE 8.30 Principle diagram of laser land leveling system. (1) Tractor. (2) Controller. (3) Hydraulic system. (4) Receiver. (5) Land-leveling bucket. (6) Datum plane. (7) Laser emitter.

achieved better economic benefit. Figure 8.30 shows the principle diagram of laser land leveling system (Li et al., 2007; Hu et al., 2009; Li and Zhao 2012).

Aiming at precision land leveling operation for dry fields in Northern China, China Agricultural University has developed a low-cost laser land leveling system as well as three-dimensional topography measurement system. The receiver of the system adopted double optical filters, the controller utilized a fuzzy control algorithm, and the hydraulic system adopted a gear pump as the power output. Figure 8.31 shows a land leveler in field operation. Equipped with a domestic JP300-type laser emitter, the system can work stably with high accuracy. The receiver has an accuracy of 3 mm, and the controller has good compatibility with both domestic and foreign hydraulic control valves (Lin, 2004; Zhao et al., 2008; Si et al., 2009; Li et al., 2012).

NERCITA has designed and developed a 3D-terrain rapid data acquisition system based on all-terrain vehicle (ATV), which used high-precision RTK-GPS to

FIGURE 8.31 Land leveler in operation.

FIGURE 8.32 Operation of the laser land leveler in paddy field.

automatically measure 3D terrain data. The onboard computers can record real-time 3D terrain data. Furthermore, auxiliary parallel navigation devices can direct the data acquisition vehicle to implement regional coverage measurement, as well as improve the quality and efficiency of data collection. Field measurement tests showed that the 3D terrain automatic data collection system based on ATVs had a good consistency with artificial RTK-GPS measurement, and the maximum average deviation was 3.54 cm, and the largest standard deviation was 2.48 cm (Liu, 2005; Lang et al., 2009; Meng et al., 2009).

South China Agricultural University has developed a laser land leveling system for paddy fields, which has made an important breakthrough and entered the stage of application. The research of the laser control system focused on the level control system of the bucket. Different sensors were used to detect the dip angle of the bucket, among which two ultrasonic sensors were adopted to measure the distance between both ends of the bucket and the surface of reference, and then calculated the dip angle using a triangle relationship. When tested on flat cement ground, the tilt angle measurement error of the level control system was less than 1.0°. Therefore, the paddy field leveling accuracy could be controlled within 3 cm. Figure 8.32 shows the operation of the laser land leveler in a paddy field.

8.4.2 Management Platform of Agricultural Information

The management platform of production-related data is the core of the PA management system. It is responsible for the input, processing, spatial, and temporal variation analysis of all farmland data and the formulation of correct farming and implementation plans. At present, the agricultural information management platform for PA is usually created based on GIS software, such as ArcGIS, MapInfo, and SuperMap, or based on components of GIS for secondary development, such as ArcEngine and MapX. Except for general GIS functions, the platform also supports data interface related to PA, professional models, special analysis functions, etc.

The database is the bas of the farmland information management platform, as shown in Figure 8.33. The data come from field measurement, local investigation, historical data, economic data, etc. According to the functions, it includes the following four parts:

1. Farmland geographic information database
   a. Used as a geographical background of varying resolution satellite image data, aviation image data
   b. Using GPS data of the distribution of farmland infrastructure such as canals, wells, and place for crop drying
   c. Using GPS data of farmland terrain, such as land distribution, and land type (cultivated land, garden land, forest land, grassland, etc.)
   d. Distribution of GPS control units

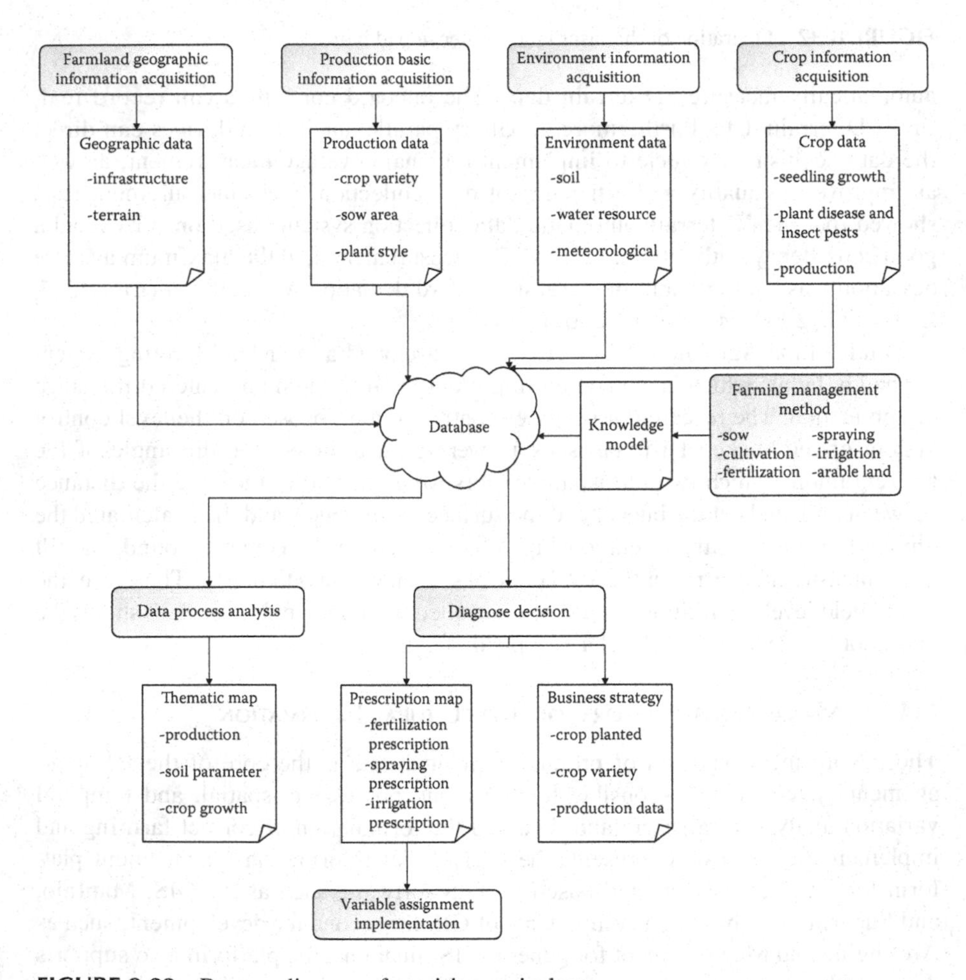

FIGURE 8.33 Process diagram of precision agriculture.

2. Basic database of production
   a. Crop type, crop varieties, ecological adaptability, agronomic shape, resistance, quality, etc.
   b. Sow area, planting method, production level, etc.
   c. Fertilizer input, condition of irrigation, volume of pesticides, etc.
   d. Food prices and market demand, price of seed, fertilizer, pesticide, etc.

3. Environmental database
   a. Soil parameter data: Soil type, soil profile, soil texture, soil bulk density, soil nutrient content, soil organic matter content, soil total nitrogen content, soil total phosphorus content, soil total potassium content, soil alkali solution nitrogen content, soil available phosphorus and available potassium content, soil trace element (boron, manganese, copper, zinc, etc.), soil moisture, soil permeability, field capacity data, etc.
   b. Meteorological data: Daily sunshine duration, average temperature, air relative humidity, wind speed, precipitation, air pressure, etc.
   c. Water resources data: Water quantity, water quality, etc.

4. Crop information collection
   a. Seedling growth data: High-resolution sensors are used in different crop growing periods to comprehensively monitor seedling growth status. Spectrophotometer or multispectral camera can be used to monitor the chlorophyll density, and analyze the relationship with nutrients.
   b. Disease, insect, weeds distribution data: The type, period, distribution, and scale of farmland crop diseases and insect pests can be recorded through analyzing the remote sensing data or portable GPS field inspections.
   c. Production distribution data: When harvesting, a combine harvester with yield monitor can be used to record crop yield distribution in farmland. At the same time, the accumulated historical data can be used for comprehensive analysis.

To establish all the above databases, data collected through various forms, some historic and some real time, are needed. When establishing the platform, data acquisition begins. After gradually accumulating data, databases such as geographic data, basic production data, and some parameters of soil and water resource data are created. These data acquisition is not restricted by the crop growth season, with lower sampling frequency. However, real-time data need to be collected in the crop growth season, such as crop growth, plant disease, insect pests, distribution of production, soil moisture, soil nitrogen content, and meteorological data. It needs higher spatial and temporal resolution and more rapid detection technology support.

Generating a thematic map is the basic function of a GIS system, which is able to show the spatial distribution of attribute data. Therefore, after real-time data such as soil parameters and crop information are collected, the farmland information management platform can analyze the data for the thematic map and understand the status of crop growth and development.

8.4.3 Diagnosis Decision-Making Expert System

Compared with traditional farming, precision farming can help to make better resource management decisions by utilizing various types of information. In order to make a full and accurate diagnosis-based decision in precision crop management, an expert system (ES) in the farmland information management platform is needed for intelligent diagnosis and decision making. ES can integrate the expert's knowledge and the crop growth model into the process of making a production decision scheme. It can make decisions according to the specific circumstances of each sampling point. The decision scheme of the whole area could be calculated through computer interpolation. In this process, the knowledge of an expert can make the decisions more reasonable. Figures 8.34 and 8.35 illustrate the PA practices in wheat and corn production management with a decision-making ES.

A wheat production management ES provides a scientific basis for production targets (Bao and He, 2001; Chen et al., 2008). It can implement the production plan according to special software, and improve the foresight of scientific management. It can also forecast the wheat growth in good time, and adjust and control the population structure with the prediction, ideal plant type, and factors of wheat yield. Therefore, it can increase effective growth and accumulation. The ES can recommend the varieties that are suitable for the region. The following can also be determined by the ES:

1. Quantity of fertilizer, ratio of fertilizer elements, and fertilizing method according to soil fertility and yield target
2. Reasonable planting density according to sowing time and fertility level
3. Irrigation time and water volume according to the soil moisture content, weather, rainfall, and crop growing status
4. Integrated system cultivation and management techniques according to the wheat growing process, population structure, and plant morphology
5. Optimization of the management decision and practical scheme according to the local condition and crop growth information

A corn production management ES can set the proper field target based on local production conditions and the level of productivity. It can predict the growth of corn, adjust and control the group structure, and decide the ideal plant type and factors of the field. It can make decisions for different corn fields according to the requirements of different growth periods. The decisions include farm management technology such as planting farming technology, seed treatment, corn cropping systems before farming, crop varieties making full use of the resources of light and heat, reasonable density, seeding time, planting form, seeding method, seeding rate, seeding inspection, reseeding, replant, timely thinning, fix seeding, hoe weeding, and the use of growth regulators. According to the production targets, reasonable fertilization techniques, water saving irrigation technology, and disease pest prevention and control technology are adopted. The reasonable harvest time and irrigation postharvest can be determined according to the corn growth indexes.

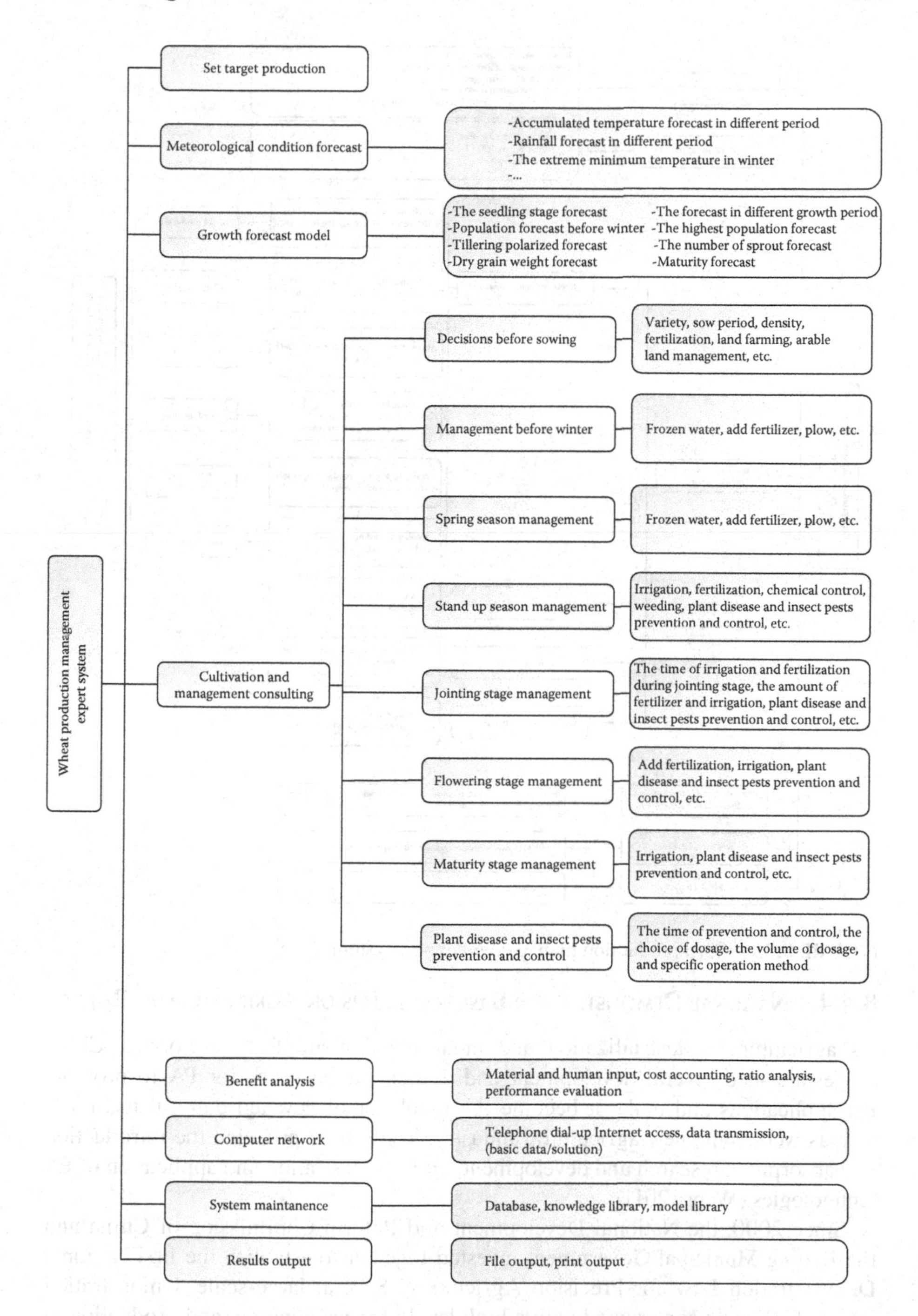

FIGURE 8.34 Wheat production management expert system.

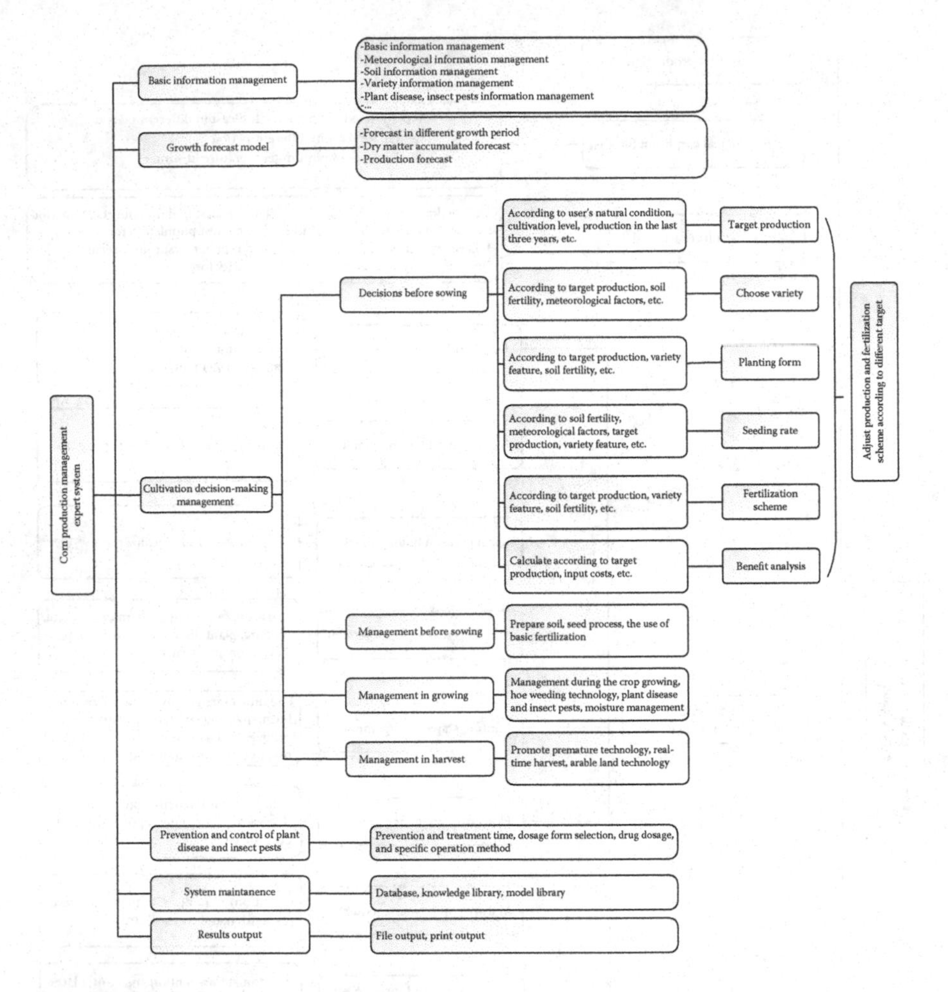

FIGURE 8.35 Corn production management expert system.

8.4.4 National Demonstration Base for Precision Agriculture in Beijing

As agricultural industrialization and modernization are fast developing, China has established several development and demonstration bases for PA to promote PA applications and make it become the incubator of new agricultural technologies as well as the new agricultural industry. These have propeled the introduction and absorption, research and development, and demonstration and application of PA technologies (Wang, 2011).

Since 2000, the National Development and Reform Commission of China and the Beijing Municipal Government invested together to establish the first National Demonstration Base for Precision Agriculture. Several large-scale demonstration bases of PA were constructed with a high level of mechanization and production in the northeast of China, Inner Mongolia, Xinjiang, etc. Here, we take two examples, one in Beijing and one in Heilongjiang, to introduce the application of practices related to PA as well as to provide a research overview.

FIGURE 8.36 National Demonstration Base for Precision Agriculture (Beijing).

The National Demonstration Base for Precision Agriculture is located in Xiao Tangshan Modern Agriculture Technology Demonstration Park in Beijing (Figure 8.36). It has introduced and applied a large number of advanced domestic and foreign PA machines, devices, and instruments, which included the combine harvester with automatic yield monitor, the large-scale lateral move sprinkling machine, the DGPS positioning and navigation system, the VRT control system, etc. The base has the functions of scientific experiments, data analysis, system integration, and exhibition of achievements. Currently, the base has achieved fruitful results in terms of PA resource management GIS, field information collection systems, airborne remote sensing platform applications, intelligent production and measurement systems, and VRT machinery applications.

8.4.4.1 Precision Agriculture Resource Management GIS

In PA practice, the establishment of the agricultural resources management platform based on GIS is one of the crucial steps. Through efficient management of field information and timely spatiotemporal difference analysis of all kinds of farmland data, it can provide accurate information for the generation of prescription map and production management decisions. The various farmland databases in the demonstration base have been developed as shown in Figure 8.37.

Except the basic GIS functions, the system designed and developed the display functions including layer operations, and remote sensing images superposition according to the specific practice of PA. Especially for non-GIS professional users, the functions of map labeling, land measurement, hot links, data query, data analysis, and management were developed to realize the visual marking of the software interface, the record of field special features in the form of images, the accurate measurement of fields, assisting complete data conversion, analysis, and management. For example, a set of yield data from the combine harvester and the soil data

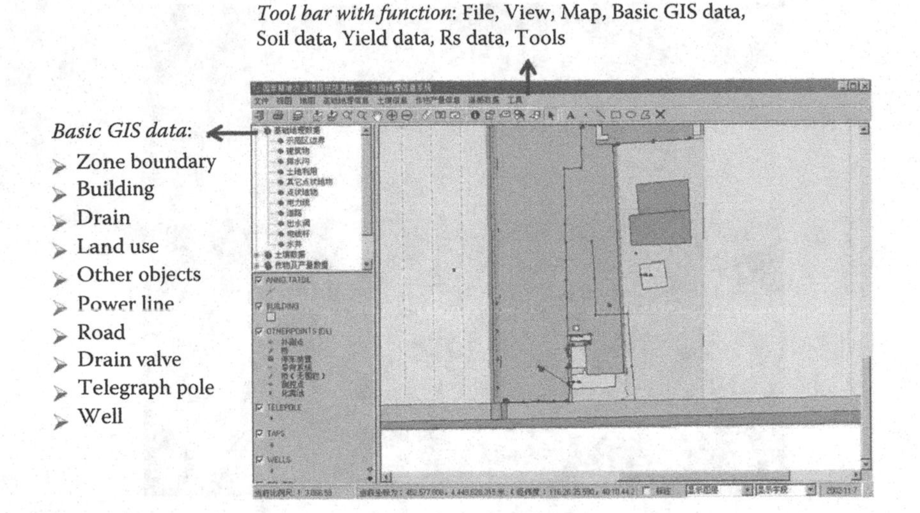

FIGURE 8.37 Precision agriculture resource management GIS.

of a field can be imported and then used to generate a map. Through the functions of the software, the yield data were analyzed and processed to meet the needs of the following analysis. According to the soil nutrient data collected, the software can achieve the calculation of fertilization scale effect to determine the best fertilization scale unit that should be used in production management.

8.4.4.2 Field Information Collection System

Field information collection system based on a PC or PDA can collect farmland data through the coordination between GPS positioning equipment and different sensors, as shown in Figure 8.38. The system has the functions of communication and data processing of GPS systems from different manufacturers, the basic GIS functions, collecting and recording spatial distribution and attribute information of farmland objects, soil grid sampling, and navigation. It can also acquire the position of farmland and various factors that affected crop growth environment, such as soil nutrients, crop diseases and insect pests, and water content of weeds. It provides the basis for PA management decisions.

8.4.4.3 Airborne Remote Sensing Platform Applications

China has made great progress on aviation remote sensing and analysis system of crop information based on airborne hyperspectral imaging device PHI and operative modular imaging spectrometer (OMIS). It included a remote sensing platform and related application methods and laid the foundation of airborne remote sensing applications in PA. The system has been successfully applied in research fields such as assessment of winter wheat growing, wheat yield analysis, and crop growth analysis.

In terms of winter wheat growing assessment, several statistical models of remote sensing were established by using red edge position and Red Valley location. Through analysis of false color composites, crop growth and nutrient distribution

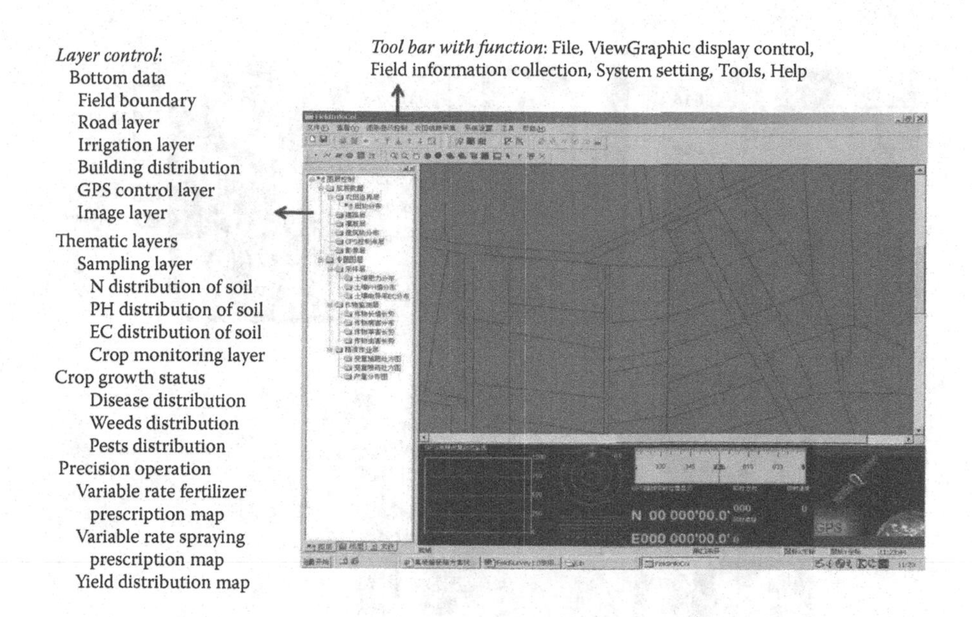

FIGURE 8.38 Field information collection system for precision agriculture.

maps on chlorophyll, total nitrogen, soluble sugar, leaf water, and other biochemical parameters were obtained as shown in Figure 8.39.

8.4.4.4 Intelligent Production and Measurement Systems

The National Demonstration Base for Precision Agriculture introduced a 2366 Combine Harvester with an AFS (advanced farming systems) yield monitoring system. It was used to harvest wheat and corn in summer and autumn, respectively.

After harvesting, yield maps were obtained according to test data, and the causes of yield variations were analyzed to provide a reference for future decisions. Different types of yield maps are shown in Figure 8.40. The yield was divided into six grades. It can be observed that the boundary section of the plots have lower yields, mainly due to the serious soil compaction near the block boundary. In addition, in some plots with lower yields, a lot of weeds were found. The analysis results of wheat and corn production show that errors were introduced in the data and could mainly be attributed in cutting amplitude and width setting error, filling time error, and delay time error. Thus, the processing of error correction was necessary.

The National Demonstration Base for Precision Agriculture introduced a U.S. Mid-Tech variable fertilization control system. The system mainly consisted of a console, a hydraulic control mechanism, and a fertilizing executing mechanism, and Figure 8.41 shows the structure diagram of the variable fertilizer spreader. To cope with variable fertilizing operations, the system also needs a field computer, a GPS, assisted navigation, radar guns, and other equipment.

In VRT fertilization, it is first necessary to import a fertilizing prescription map to the field computer. The computer also received real-time GPS data as well as radar speed data and real-time fertilizer amount in the prescription map. The fertilizing control instructions in the current fertilizing location were transmitted to the

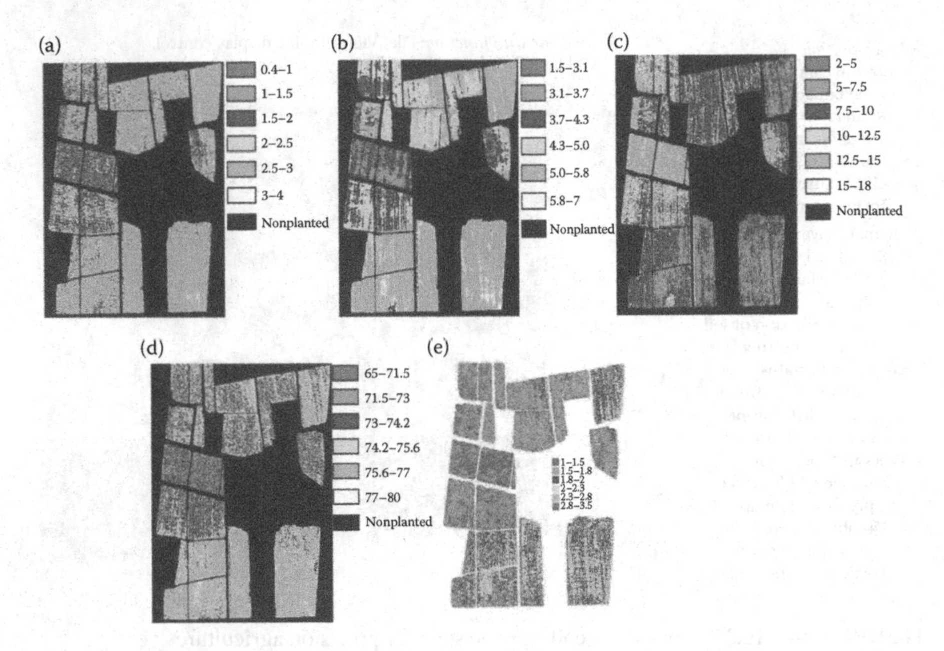

FIGURE 8.39 **(See color insert.)** Maps of chlorophyll, total nitrogen, soluble sugar, and water of leaf biochemical parameter. (a) Chlorophyll content (mg/g), (b) nitrogen content (%), (c) soluble sugar content (%), (d) water content (%), and (e) LAI.

central controller of fertilizing control system, and the central controller converted the digital signal into an analog signal, and then the valve opening was adjusted by an electrohydraulic proportional control mechanism. At the same time, the actual fertilizer amount was fed back to the field computer by the control systems, which would later be used for data processing and analysis.

The data were monitored and stored using a Trimble Ag170 field computer. The fertilizing operation was controlled and navigated by using a Trimble Ag132 DGPS receiver and a navigation light bar. The data with submeter positioning accuracy can meet the needs of the VRT fertilizing operations. The system showed good static and dynamic performances with better work precision. Comparing the prescription map with the practical distribution map of the fertilizer amount, it was indicated that the system could conduct the VRT fertilizing operation in accordance with the fertilizing prescription map.

8.4.5 Practice of Precision Agriculture in Heilongjiang Province

Heilongjiang Province, as the major grain base of China, has already established several demonstration farms of PA across the province. Heilongjiang Agricultural Reclamation Bureau (HARB) is an administrative institution of state farms located in Heilongjiang Province. HARB started the practice of PA in 2002 and introduced a great number of advanced PA machines from John Deere and Case IH, and DGPS systems from Trimble. As an example, the Precision Agriculture Center of Hongxing

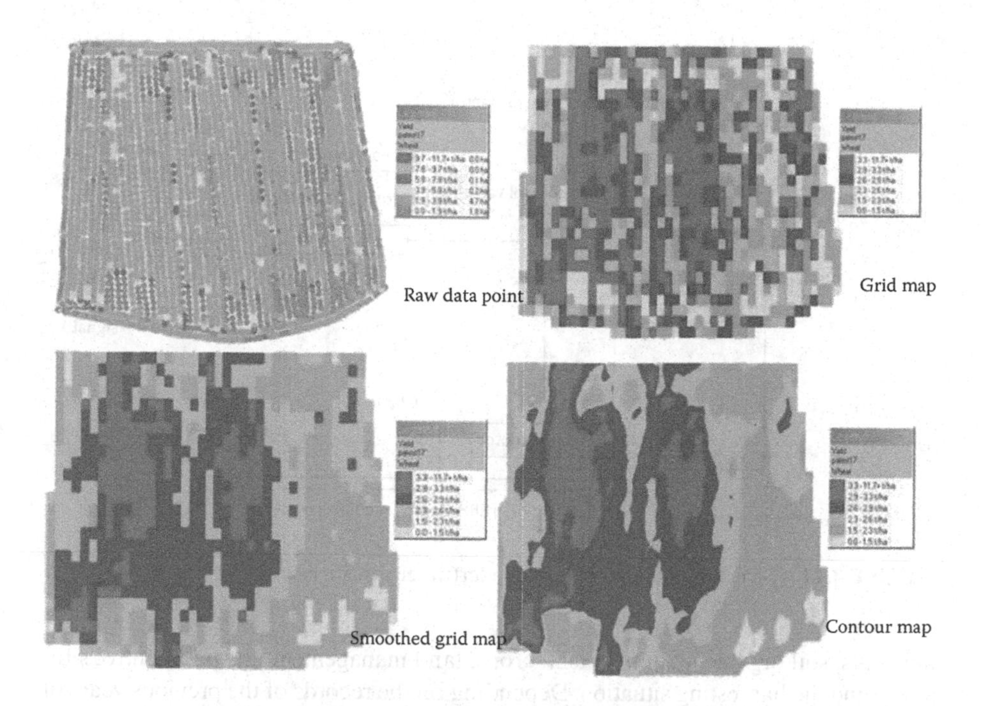

FIGURE 8.40 (**See color insert.**) Four different yield maps of winter wheat from the same field.

Farm, one of the 113 state farms in HARB, has integrated and utilized PA technologies such as farmland information computers, wireless sensor networks, navigation and management of agricultural machines, VRT agricultural machinery, and other PA equipment. The intellectualization, informatization, and mechanization of agricultural production were all improved. The practice of PA also brought economic benefit for farmers, and played a demonstrable and leading role in agricultural production.

8.4.5.1 Remote Sensing Image and Data System (RS)

Hongxing Farm has uploaded remote sensing images of 2.5 m resolution to the digital information network from September 2007. At this resolution, farmers can find buildings, roads, reservoirs, and other infrastructure on the farm and measure the distance between any two points and the area of any region on the map. Since October 2008, Hongxing Farm has updated the remote sensing images to a 1.0 m resolution. Farmers can clearly find trees and buildings on both sides of the road on the farm from updated remote sensing images.

8.4.5.2 Geographic Information System

Hongxing Farm has built a GIS for its field based on RS data. Different management zones were divided into different colors based on individual farmers. The farmers can use the system to measure the distance between any two points and the area of any region. Clicking the query button, users can visually see brief information of each zone (zone name, area) and the details of the land, such as soil pH, soil

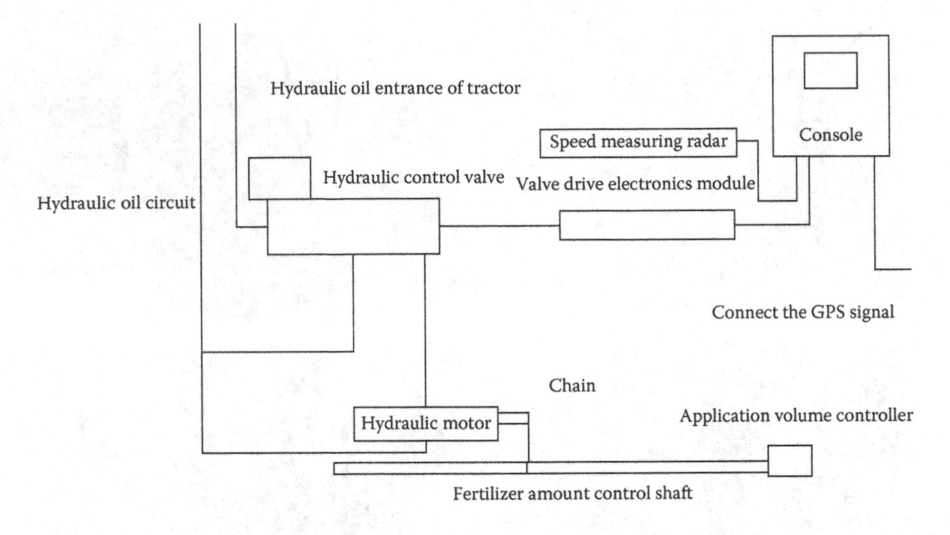

FIGURE 8.41 Structure diagram of variable fertilizer spreader.

nutrients, soil organic matter content, crops, land management, the pest control situation, and the harvesting situation. Depending on the records of the previous years of each zone, it is convenient and rational to design or plan rotation and fertilization for the next year. This GIS system established the foundation of precision crop management in Hongxing Farm.

8.4.5.3 Dynamic Tracking System of Agricultural Machines

The dynamic tracking system of agricultural machines of Hongxing Farm was also built on the basis of remote sensing and GIS. With this system, farmers can see the operation situation of agricultural machines within a farm or across farms. Some machines were equipped with GPS navigation devices. Those machines can upload parameters such as location and speed to the website of the dynamic tracking system of agricultural machines by mobile signals. The staff in the control center can log on to the system at any time to query the parameters of the agricultural machines such as latitude and longitude, operating time, speed, direction, status of network, and other data. Within the effective range, the controller can talk with the driver through the intercom, and observe the workspace, profit and loss of the machine and other information. In addition, the dynamic tracking system can also store the records of a year and provide services for historical data queries and program management in the future. Based on the technologies mentioned above, the Hongxing Farm PA systems integration platform was developed, including public map service subsystem, digital farm management subsystem, and intelligent decision subsystem. The public map service subsystem provides basic map information, basic map tools (such as zoom, display, and measurement), query tools, and other general-purpose modules. The digital farm management subsystem provides management tools for managers, including production information management, soil information management, and crop pest information management. The production information management subsystem includes zone

- *Basic layer*:
  - Street map
  - Image map
  - Farm field
- *Business layer*:
  - Temporary layer
  - Sampling point
  - Plant diseases and insect pests

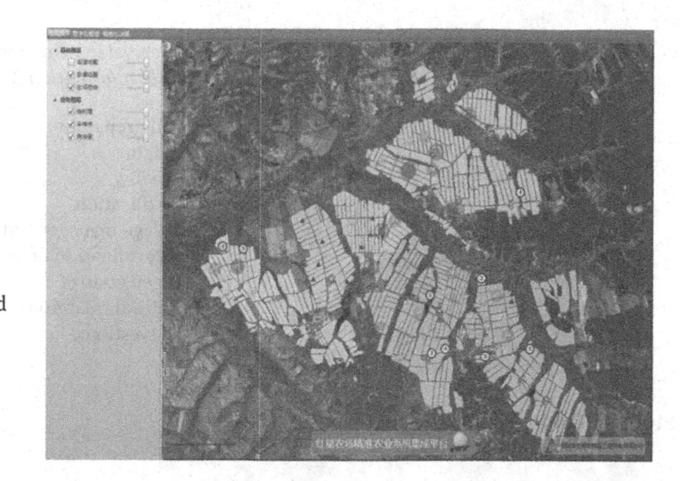

FIGURE 8.42 Main interface of the dynamic tracking system for agricultural machines.

archives, information maintenance, and production plan information maintenance. The information management subsystem includes soil samples maintenance, administrative region maintenance, and index maintenance. Crop pest management includes pest, diseases, and pathogens control. The Hongxing Farm intelligent decision subsystem provides analytical tools for the decision maker, including production archives statistical analysis, soil sensing and fertilization, and pest diagnostics. The main interface of the software is shown as Figure 8.42, which was divided into three modules: public map module, digital management module, and intelligent decision-making module.

The Map Search toolset was used for querying production archives of farm plots, attributes, and sampling point attributes. The query tool for production archives was used to query the production archives information of farm plots over the past several years. Clicking the farm plots button, production archives information can be sorted and shown by year, as shown in Figure 8.43. Clicking the more information button, the production archives information of the selected year can be displayed in detail, as shown in Figure 8.44.

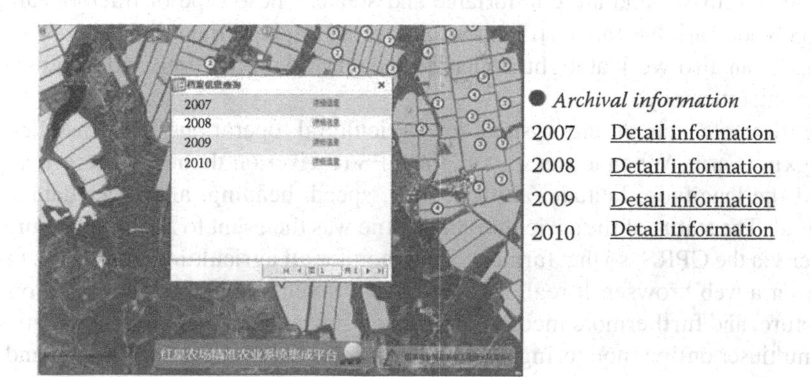

FIGURE 8.43 Production archive query.

Archival information
Field information: Field number, area, year, crop, variety et al.
Soil preparation
Ridging
Sowing
Fertilization
Tillage management
Extermination of disease and insect pests
Weed control
Natural calamities
Harvesting

FIGURE 8.44 Production archive details.

The attribute query tools for plot and sampling point were used to query the attribute information of a plot or a sampling point, respectively, including basic information, natural conditions, production conditions, crop information, soil information, and nutrient information.

8.4.5.4 Straight Navigation of Agricultural Machines and Remote Scheduling of Field Operations

By using a good GPS base station infrastructure construction, a straight navigation system of agricultural machines and a remote scheduling system of agricultural operations were developed to improve work quality and operationing efficiency.

The straight navigation system of agricultural machines was developed for PA. It consisted of navigation software and a light target. Navigation software can receive positioning signals from the GPS receiver. After setting the navigation path, it can conduct straight and automatic navigation. By using a DGPS device, it can navigate the farm machine precisely straight without repeat or miss, and calculate the operation area. The system was installed on CASE 450 tractor (450 hp) and John Deere 9520 tractor (450 hp). Since tractors have electrohydraulic control, they are easy to operate and drive, and are comfortable and stable. These type of tractors can pull a large wide variable-rate fertilization seeder and improve the reliability of the unit work. It can also work at night, which extends the operating time and improves the operating efficiency.

A rRemote scheduling system for agricultural operations was installed for Hongxing Farm. When it was started, the GPS receiver on the mobile terminal provided the longitude, latitude, altitude, time, speed, heading, and other data in 1 s interval. The status of the agricultural machine was then sent to a remote monitoring server via the GPRS, so that farmers could monitor all agricultural operations in real time on a web browser. It realized network GPS vehicle monitoring based on B/S structure, and furthermore incorporated user rights management to make it possible for multiuser online monitoring. The system supported massive spatial data and had a variety of statistical functions for users to compare with historical data. It also had an alarm function to make it intelligent for vehicle monitoring.

8.5 SUMMARY AND DISCUSSION

China is at a critical stage for the practice of agricultural ICT. Thus, it is necessary to attach importance to international scientific progress and experience in the research of PA, digest and absorb advanced and relatively mature foreign technology, and focus attention on the research of application and innovation technologies suitable for the situation in China. With the development of scientific technology, a large number of emerging information collection and processing measures, such as the IoT and cloud computing, have provided a new platform for the development of PA. PA can guarantee the sustainable development of agriculture in technology and with the efforts of agricultural scientists and the attention and support of the Chinese government, the practice of PA in China will make great progress.

ACKNOWLEDGMENTS

We would like to express our deepest appreciation to colleagues and the faculty members of the Research Center for Precision Agriculture at China Agricultural University, particularly, Professor Wang Maohua, Professor Liu Gang, Dr. Zhang Man, Dr. Zheng Lihua, Dr. Zhang Miao, Dr. Li Li, and Dr. Yang Wei for their support and contribution to Chapter 8. We would also like to thank our students, Zhang Yao, Pei Xiaoshuai, Zhang Meng, Mi Xiuchen, Wu Lixuan, Song Yuanyuan, Zhao Yi, and Wen Yao for collating the text and drawing figures.

REFERENCES

An, X.F., M.Z. Li, L.H. Zheng, Y.M. Liu, and H. Sun. 2014. A portable soil nitrogen detector based on NIRS. *Precision Agriculture*, 15:3–16.

Bao, Y.D. and Y. He. 2001. Study on multimedia decision support system of agricultural machinery. *Journal of Zhejiang University (Agriculture. & Life Sciences)*, 27(2):187–190.

Bao, Y.D., Y. He, H. Fang, and A.G. Pereira. 2007. Spectral characterization and N content prediction of soil with different particle size and moisture content. *Spectroscopy and Spectral Analysis*, 27(11):62–65.

Chen, P.F., L.Y. Liu, J.H. Wang, T. Shen, A.X. Lu, and C.J. Zhao. 2008. Real-time analysis of soil N and P with near infrared diffuse reflectance spectroscopy. *Spectroscopy and Spectral Analysis*, 28(12):295–298.

Chen, P.F., N. Tremblay, J.H. Wang, P. Vigneault, W.J. Huang, and B.G. Li. 2010. New index for crop canopy fresh biomass estimation. *Spectroscopy and Spectral Analysis*, 30(2):512–517.

Chen, T.E., C.J. Zhao, L.P. Chen, and H. Chen. 2008. Research on component-oriented decision-making support platform of soil testing and formulated fertilization. *Application Research of Computers*, 25(9):2748–2750.

Feng, W., Y. Zhu, X. Yao, Y.C. Tian, T.C. Guo, and W.X. Cao. 2009. Monitoring nitrogen accumulation in wheat leaf with red edge characteristics parameter. *Transaction of the CSAE*, 25(11):194–201.

Hu, L., X.W. Luo, Z.X. Zhao, Q. Li, and W.T. Chen. 2009. Evaluation of leveling performance for laser-controlled leveling machine in paddy field based on ultrasonic sensors. *Transactions of the CSAM*, 40(S1):73–76 + 81.

Jia, G.R., R.L. Tang, and Y.F. Dai. 1997. Study on laser plane system for levelling machine. *Transactions of the CSAE*, 13(S1):308–312.

Kuang, J.S. and M.H. Wang. 2003. Application of GIS, GPS and RS for field surveying, mapping and data updating. *Transaction of the CSAE*, 19(3):220–223.

Lang, X.Z., G. Liu, and X.F. Xie. 2009. Tractor-mounted field 3-D topography surveying system. *Transactions of the CSAM*, 40(S1):69–72.

Li, D.L. 2012. *Introduction to Agricultural Internet of Things*. Beijing, China: Science Press.

Li, H., G.Q. Yao, and L.P. Chen. 2008. Farm machinery monitoring and scheduling system based on GPS, GPRS and GIS. *Transactions of the CSAE*, 24(S2):119–122.

Li, M.Z., L. Pan, L.H. Zheng, and X.F. An. 2010. Development of a portable SOM detector based on NIR diffuse reflection. *Spectroscopy and Spectral Analysis*, 30(4):1146–1150.

Li, Q., S.X. Huang, S.M. Ruan, and B.W. Li. 2012. A paddy field laser flat shovel horizontal position control algorithm. *Modern Agricultural Equipment*, 8:48–50.

Li, Q., X.W. Luo, M.H. Wang, Z.X. Zhao, Y.J. Xu, Y.G. Ou, G. Liu, J.H. Lin, and Y.S. Si. 2007 Design of a laser land leveler for paddy field. *Transactions of the CSAE*, 23(4):88–93.

Li, X.H., M.Z. Li, and C. Di. 2009. Non-destructive crop canopy analyzer based on spectral principle. *Transaction of the CSAM*, 40(supplement):252–255.

Li, X.H., F. Zhang, M.Z. Li, R.J. Zhao, and S.Q. Li. 2011. Design of a four-waveband crop canopy analyzer. *Transactions of the CSAM*. 42(11):169–173.

Li, Y.J. and Z.X. Zhao. 2012. Design of attitude measurement system for flat shovel of laser-controlled land leveler for paddy field. *Journal of Agricultural Mechanization Research*, 34(2):69–75.

Li, Y.X., Y. Zhu, Y.C. Tian, X. Yao, X.D. Qin, and W.X. Cao. 2006. Quantitative relationship between leaf nitrogen concentration and canopy reflectance spectra. *Acta Agronomica Sinica*, 32(3):358–362.

Lin, J.H. 2004. *Research & Development on Receiver and Controller for Laser Controlled Land Leveling System*. Beijing: China Agricultural University.

Liu, L.Y., J.H. Wang, W.J. Huang, C.J. Zhang, B. Zhang, and Q.X. Tong. 2004. Improving winter wheat yield prediction by novel spectral index. *Transaction of the CSAE*, 20(1):172–175.

Liu, X.J., Y. Zhu, X. Yao, Y.C. Tian, and W.X. Yao. 2006. WebGIS-based for agricultural spatial information management and aided decision-making. *Transactions of the CSAE*, 22(5):125–129.

Liu, Z.C. 2005. *Research and Development on 3-D Intelligent Topography Measurement System for Land Leveling*. Beijing: China Agricultural University.

Meng, Z.J., W.Q. Fu, and H. Liu. 2009. Design and implementation of 3D topographic surveying system in vehicle for field precision leveling. *Transactions of the CSAE*, 25(S2):255–259.

Peng, Y.K., J.X. Zhang, X.S. He, and E.S. Lu. 1998. Analysis of soil moisture, organic matter and total nitrogen content in loess in China with near infrared spectroscopy. *Acta Pedologica Sinica*, 35(4):553–559.

Sha, J.M., P.C. Chen, and S.L. Chen. 2003. Characteristics analysis of soil spectrum response resulted from organic material. *Research of Soil and Water Conservation*, 10(2):21–24.

Si, Y.S., G. Liu, Z. Yang, X.F Xie, and M.H. Wang. 2009. Development and experiment of laser land leveling system. *Journal of Jiangsu University (Natural Science Edition)*, 30(5):441–445.

Song, H.Y. and Y. He. 2008. Determination of the phosphorus, kalium contents and pH values in soils using near-infrared spectroscopy. *Journal of Shanxi Agricultural University (Natural Science Edition)*, 28(3):275–278.

Sun, J.Y., M.Z. Li, L.H. Zheng, Y.G. Hu, and X.J. Zhang. 2006. Real-time analysis of soil moisture, soil organic matter, and soil total nitrogen with NIR spectra. *Spectroscopy and Spectral Analysis*, 26(5):426–429.

Tang, N., M.Z. Li, J.Y. Sun, L.H. Zheng, and L. Pan. 2007. Development of soil organic matter fast determination instrument based on spectroscopy. *Spectroscopy and Spectral Analysis*, 27(10):2139–2142.

Tian, Y.C., J. Yang, X. Yao, Y. Zhu, and W.X. Cao. 2010. A newly developed blue nitrogen index for estimating canopy leaf nitrogen concentration of rice. *Chinese Journal of Applied Ecology*, 21(4):966–972.

Wang, M.H. 1999. Development of precision agriculture and innovation of engineering technologies. *Transaction of the CSAE*, 15(1):1–8.

Wang, M.H. 2011. *Precision Agriculture*. Beijing, China: China Agricultural University Press.

Wang, X., C.J. Zhao, Q.J. Meng, L.P. Chen, Y.C. Pan, and X.Z. Xue. 2003. Design and experiment of variable rate fertilizer applicator. *Transactions of the CSAE*, 20(5):114–117.

Wang, Z., L.P. Chen, and Y.S. Liu. 2010. Design and implementation of agricultural machinery monitoring and scheduling system. *Computer Engineering*, 4(11):232–237.

Wu, C.C., Y.P. Cai, M.J. Luo, H.H. Su, and L.J. Ding. 2013. Time-windows based temporal and spatial scheduling model for agricultural machinery resources. *Transactions of the CSAM*, 44(5):237–241.

Xue, L.H., W.X. Cao, W.H. Luo, and X. Zhang. 2004. Correlation between leaf nitrogen status and canopy spectral characteristics in wheat. *Acta Phytoecologica Sinica*, 28(2):172–177.

Yang, J., Y.C. Tian, X. Ya, W.X. Cao, Y.S. Zhang, and Y. Zhu. 2009. Hyperspectral estimation model for chlorophyll concentrations in top leaves of rice. *Acta Ecologica Sinica*, 29(12):6561–6571.

Yao, J.S., H.Q. Yang, and Y. He. 2009. Nondestructive detection of rape leaf chlorophyll level based on vis/NIR spectroscopy. *Journal of Zhejiang University (Agriculture & Life Sciences)*, 35(4):433–438.

Yu, F.J., S.G. Min, X.T. Ju, and F.S. Zhang. 2002. Determination the content of nitrogen and organic substance in dry soil by using near infrared diffusion reflectance spectroscopy. *Chinese Journal of Analysis Laboratory*, 21(3):49–51.

Yuan, S.L., T.Y. Ma, T. Song, Y. He, and Y.D. Bao. 2009. Real-time analysis of soil total N and P with near infrared reflectance spectroscopy. *Transactions of the CSAM*, 40(S1):150–153.

Zhang, X.J., M.Z. Li, D. Cui, P. Zhao, J.Y. Sun, and N. Tang. 2006. New method and instrument to diagnose crop growth status in greenhouse based on spectroscopy. *Spectroscopy and Spectral Analysis*, 26(5):887–890.

Zhang, X.J., M.Z. Li, Y.E. Zhang, P. Zhao, and J.P. Zhang. 2004. Estimating nitrogen content of cucumber leaf based on solar irradiance spectral reflectance in greenhouse. *Transaction of the CSAE*, 20(6):11–14.

Zhao, C.J., X.Z. Xue, X. Wang, L.P. Chen, Y.C. Pan, and Z.J. Meng. 2003. Advance and prospects of precision agriculture technology system. *Transaction of the CSAE*, 12(4):7–12.

Zhao, Z.X., X.W. Luo, Q. Li, B. Chen, X. Tian, L. Hu, and Y.J. Li. 2008. Leveling control system of laser-controlled land leveler for paddy field based on MEMS inertial sensor fusion. *Transactions of the CSAE*, 24(6):119–124.

Zheng, L.H., M.Z. Li, L. Pan, J.Y. Sun, and N. Tang. 2009. Application of wavelet packet analysis in estimating soil parameters based on NIR spectra. *Spectroscopy and Spectral Analysis*, 29(6):1549–1552.

Zhong, Z.J., M.Z. Li, H. Sun, L.X. Wu, and Q. Wu. 2013. Development and application of a smart apparatus for detecting crop nutrition. *Transactions of the CSAM*. 44(S2):215–219.

Zhu, D.S., D. Wu, H.Y. Song, and Y. He. 2008. Determination of organic matter contents and pH values of soil using near infrared spectroscopy. *Transaction of the CSAE*, 24(6):196–199.

9 Good Agricultural Practices, Quality, Traceability, and Precision Agriculture

Josse De Baerdemaeker and Wouter Saeys

CONTENTS

9.1 INTRODUCTION

Agricultural production is part of a long chain of activities that starts from seeding (or even earlier) and stretches all the way to the consumer. It should meet consumer expectations in terms of quality, safety, and also value or price. Many intermediate steps are involved and these often involve handling, storage, and transportation across national borders or continents. Information should be transferred across this chain. The automation that will be a major part of future agricultural and biological production systems also faces some challenges posed by system characteristics that will have to be dealt with. When we look at the processes in agricultural production systems, we can say that they are complex in nature. Indeed, as we gain a better understanding of biological processes, we also find that they have a great complexity and that in many cases this complexity remains difficult to formulate in exact terms. Complexity means that the system comprises numerous parts or processes that interact and yield outcomes that are not easily predicted. These processes and interactions occur in and across different spatial and temporal scales.

Crop growth is the result of photosynthetic activity and transport processes in the cells, in the leaves, and in the different organs of single plants. There is a close interaction with the physical environment around these single plants, such as solar radiation, temperature, humidity, soil texture, and its nutrient or water content. Of course, there is also the influence of neighboring plants within a field. There is also an interaction with many associated biota such as insects, pests, and microorganisms on the plant or near the plants and fields as well as soil microorganisms and invertebrates. All these biotic and abiotic effects can be variable in a field and also vary with time. At the farm level, there are complex interactions within the enterprise where many different activities occur, and with communities and economic operators. All these interactions affect the decision making at this level.

Society has high expectations from food production. However, at the same time, food production is increasingly subject to international agreements on trade. This makes competition between producers or regions of production an important factor in decision making. Nevertheless, this competition should not impair food safety to consumers or long-term food security for society. Transparency of the entire food chain for ensuring safety is a must. There are also needs for technology development because of the need to reduce land degradation or to optimize water use. For example, as a result of the (bio)technological revolution, genetically modified crops or crops for green chemicals need different planting, tending, harvesting, and handling equipment. There is also a growing concern for maintaining biodiversity to preserve the abundant genetic resources as well as to have a basis for more efficient crop production or pest management. A broad range of definitions dealing with transparency exist in the literature. However, that transparency is only reached if everybody with stakes and interests in food production and consumption understands the relevant aspects of products, processes and process environments, and other factors that allow them to make informed decisions (Schiefer and Deiters, 2013).

Since agricultural products are stored and shipped over long distances and time periods, there can be a considerable change in quality. So, one would like to know how quality will evolve after harvest. This may affect the timing of the harvest, the required storage conditions for maintaining a certain level of quality, or the available time between harvest and consumption.

9.2 FOOD SAFETY AND GOOD AGRICULTURAL PRACTICES SCHEMES

Consumers show increasing concerns about food safety and about the properties of the food they eat. Indeed, food scandals and incidents in the food supply chain have raised public concern over agricultural practices and the handling and processing of food. Reports of food poisoning incidents and deaths due to contamination of fresh, minimally processed, and processed fruit and vegetables and the occurrence of other emerging food pathogens have reduced consumer confidence in the safety of food systems (Opara and Mazaud, 2001). As a result, there have been major developments in the world related to food safety and traceability. Some of the initiatives have come from governments to protect the health of their citizens, while others are private initiatives by growers and retailers in order to meet the expectations of their customers with respect to food safety and environmental sustainability. Everyone in the food chain assumes that these expectations can be satisfied if production is done in line with good agricultural practices (GAP). To ensure this, the qualified authorities or food safety departments at manufacturers or retailers demand that the origin and destination of animal feed, materials, and food in all stages of production and distribution are known and available as information.

All stakeholders in the food production chain now consider food safety to be an important issue and producers of food are increasingly subjected to greater scrutiny of their production practices. It is also then recognized that there is an increasing need for greater quality assurance, transparency, and traceability in the food supply chain (Opara and Mazaud, 2001). It has also been shown that traceability, in the absence of quality verification, is of limited value to individual consumers. Bundling traceability with quality assurances has the potential to deliver more value (Hobbs et al., 2005).

High-quality food, integrity, and associated services and information should be guaranteed. Consumers call for food that can be fully trusted. They ask for safety guarantees and information with integrity to confirm their trust. In this context, integrity of information is defined as follows: "the information provided is in conformance with the reality it depicts." Information that is accurate, relevant, precise, timely, and complete for a particular purpose can be termed to be "fit for purpose" (Trites, 2013). It also implies that tampering of the data is not possible. The call for integrity of information is voiced in particular by retailers who state transparency requirements to be met by their suppliers. Part of that transparency is concerned with realizing tracking and tracing systems as primary objectives to enable efficient recalls on the chain level when necessary, on proactive monitoring quality along chain processes, with an objective early warning in case of a possible emerging problem, and/or aiming at optimizing the remainder of processes along the supply chain downstream (Beulens et al., 2005). In recent years, there have been major developments related to food safety regulations and international trade. Van Plaggenhoef et al. (2003) reviewed the legislation and standards and classified the regulations as follows.

9.2.1 International Institutions That Deal with Food Safety

These institutions are (Van Plaggenhoef et al., 2003)

- Codex Alimentarius. The Codex Alimentarius Commission (CAC) was created in 1963 by the Food and Agriculture Organization (FAO) and the World Health Organization (WHO) to develop food standards, guidelines, and related texts such as codes of practice under the Joint FAO/WHO Food Standards Programme.

- Sanitary and Phytosanitary (SPS) Agreement of the World Trade Organization (WTO). The SPS agreement relates to protection of human, animal, and plant health and life. The basic aim of the SPS agreement is to maintain the sovereign right of any government to provide the level of health protection it deems appropriate, but to ensure that these sovereign rights are not misused for protectionist purposes and do not result in unnecessary barriers to international trade.
- Legislation from the European Community. The European Food Law (European Commission, 2012; Regulation EC No 178/2002) establishes the general principles upon which international trade in food shall be based. Food producers have the primary responsibility for the safety of food and the member states have to develop codes of good practice at the national level.

9.2.2 Internationally Acknowledged Food Safety Systems

A number of food safety systems with a worldwide application range are listed by SAI Global (http://www.saiglobal.com/assurance/food-safety/), a company specialized in certification of quality management systems:

- **FSSC/FS 22000** (Food Safety System Certification standard) is a certification scheme for food manufacturers.
- **ISO 22000** takes a whole chain approach to food safety, providing a standard that goes all the way from the farm to the fork, including packaging and ingredient suppliers, caterers, storage and distribution facilities, and chemical and machinery manufacturers, and can be applied to primary producers such as farms.
- **BRC** is one of the choices for retailers worldwide looking for confidence from food suppliers.
- **SQF** is one of the world's leading food safety and quality management systems to assure that a supplier's food safety and quality management system complies with international and domestic food safety regulations.
- **HACCP** (hazard analysis and critical control points) is a risk management system that identifies, evaluates, and controls hazards related to food safety throughout the food supply chain.
- **IFS** (International Food Standard) is a quality and food safety standard for retailer (and wholesaler) branded food products, which is intended to assess suppliers' food safety and quality systems, with a uniform approach that harmonizes the elements of each.
- **GFSI**: Under the umbrella of the Global Food Safety Initiative (GFSI), seven major retailers have come to a common acceptance of GFSI-benchmarked food safety schemes.
- **GlobalGAP** (GlobalGAP, 2012) was introduced by FoodPLUS GmbH, but is now managed in the form of a retailer–producer alliance to raise standards in primary agricultural production. Certification to the standard ensures a level playing field in terms of food safety and quality, and proves that growers are prepared to constantly improve systems to raise standards.

The GAP standards can be considered as the basis for food safety. They are in general based on the following concepts:

- Food safety: The standard is based on food safety criteria, derived from the application of generic HACCP principles.
- Reducing the inappropriate use of chemicals in general, and especially the use of chemical plant protection products (PPPs), or reducing the level of residues found on food crops.
- Environmental protection: The standard consists of environmental protection GAP, which are designed to minimize negative impacts of agricultural production on the environment.
- Occupational health, safety, and welfare: The standard establishes a global level of occupational health and safety criteria on farms, as well as awareness and responsibility regarding socially related issues.
- Animal welfare (where applicable): The standard establishes a global level of animal welfare criteria on farms.

As an example, the GlobalGAP (GlobalGAP, 2012) scheme covers the whole agricultural production process of the certified product, from before the plant is in the ground (seed and nursery control points) to nonprocessed end products (produce handling control points). In response to the challenges posed by fast-changing crop protection product legislation, the GlobalGAP organization developed guidance notes to help farmers and growers to become fully aware of the maximum residue levels (MRLs) in operation in the markets where the product will be sold. A general regulations document explains the structure of certification to the GlobalGAP standard and the procedures that should be followed in order to obtain and maintain certification. The requirements for GAP certification are bundled in a document with control points and compliance criteria (Figure 9.1). Several other GAP schemes also have similar requirements although the emphasis may be different depending on the country where it was initiated or applied.

9.3 PRECISION AGRICULTURE, GAP, AND "LICENSE TO OPERATE"

Precision agriculture (PA) technologies share the underlying ideas of GAP and may become important tools for complying with regulations and for documentation of the production conditions as a proof of compliance.

PA can be seen as a summary of GAP that rely on (De Baerdemaeker, 2013)

- Correct information (soil, previous crops and treatment, etc.)
- Correct observation
- Correct analysis
- Correct genotype
- Correct dose
- Correct chemical/biological compound
- Correct place
- Correct time
- Correct (climatic) conditions
- Correct equipment

Nº	Control Point	Compliance Criteria
CB 8.3.6	Justification for application?	The name of the pest(s), disease(s) and/or weed(s) treated is documented in all plant protection product application records. If common names are used then they must correspond to the names stated on the label. No N/A.
CB 8.3.7	Technical authorization for application?	The technically responsible person making the decision of the use and the doses of the plant protection product(s) being applied has been identified in the records. No N/A.
CB 8.3.8	Product quantity applied?	All plant protection product application records specify the amount of product to be applied in weight or volume or the total quantity of water (or other carrier medium) and dosage in g/l or internationally recognized measures for the plant protection product. No N/A.
CB 8.3.9	Application machinery used?	The application machinery type (e.g. knapsack, high volume, U.L.V., via the irrigation system, dusting, fogger, aerial, or another method), for all the plant protection products applied (if there are various units, these are identified individually), are detailed in all plant protection product application records. No N/A.
CB 8.3.10	Pre-harvest interval?	The pre-harvest interval has been recorded for all plant protection product applications where a pre-harvest interval is stated on the product label or if not on label, as stated by official source.

FIGURE 9.1 Example of control points in the fruit and vegetables checklist of GlobalGAP. (Adapted from http://www.globalgap.org/uk_en/for-producers/crops/FV/.)

It is clear that when such principles are adhered to, the requirements of GlobalGAP can be met. For example, a GAP scheme requires that fertilizer application doses are based on soil analysis and should be applied at a rate that can be taken up by the crop. The application equipment should be in good condition such that the operator can be sure about the dose. Pesticides should only be applied in the framework of a pest management scheme, for example, integrated pest management (IPM), that is based on the observation or risk of a pest or disease, while beneficial organisms must be preserved as much as possible. Application should not be done within the required preharvest time interval as stated by the government conditions for use. Of course, only approved pesticides or herbicides can be used. While applying the chemical protection, sufficient distance from water sources must be maintained to avoid contamination of the surface waters. To ensure that all these principles have been respected, a record has to be kept of all the steps and treatments carried out during production.

A report on the environmental impacts of products (EIPRO) (Tukker et al., 2006) has identified those products with the greatest environmental impact. The results are based on a life cycle analysis of the products consumed in the European Union. They found that three areas of consumption have the greatest environmental impact in Europe: housing, food and drink, and private transport. There is no clear ranking, as products in the three areas identified are of approximately equal importance. Together, they are responsible for 70%–80% of the environmental impact of consumption, and account for some 60% of consumption expenditure. Life cycle analyses help to understand the environmental impacts of individual products on carbon, water, eutrophication, etc. across all the stages of the value chain: from the production of agricultural inputs, farming, processing, transport, and storage on the production side, to shopping, cleaning, cooking, home storage, and recycling behavior on the consumer side (European Food SCP Round Table Working Group 2 on "Environmental Information Tools," European Food SCP Round Table Report, 2011; European Food Sustainable Consumption and Production (SCP) Round Table). The principles of PA can become a major tool for communicating along the food chain, including to consumers, farming activities that have an environmental impact. The technology makes it possible to do so in a scientifically reliable and consistent way, understandable and not misleading.

Changes in society and consumer attitudes are such that agricultural practices will be increasingly questioned in the future. This will go further than "say what you do" and "do what you say," but will also imply that communities will give a "license to operate" only when stringent production requirements are met and documented. It is not only that global consumers require GAP when buying products, but that local consumer action groups will only allow production when certain conditions are met and documented.

9.4 MEETING THE TRACEABILITY REQUIREMENT

Precision farming and the use of global positioning systems (GPS) on agricultural machinery provide location and time information for all treatments. This is of course very important for automation such as navigation during the different treatments or the collection of data on crop status, diseases, and yields.

9.4.1 Site History and Site Management

Planting a suitable crop (and variety) at the correct place implies that the farm manager or the decision support tool is aware of the soil condition and of what crops were grown in the previous seasons and what treatments were given. In a number of cases, residues from fertilizers, herbicides, or pesticides from treatments in a previous season may still be high because of environmental conditions that were less favorable for their degradation or breakdown. It is then important that the farmer or decision support algorithm can retrieve the data (dose, time, and location) about these earlier treatments to make informed decisions. The risk of chemical leaching in the soil may vary by location and soil type and can be taken into consideration for crop production decisions. In other cases, a sequence of crop rotations should be respected to avoid the effect or the spreading of soil-borne diseases. This means that there is also a need for a traceability system that is linked to a field and not just to a crop that is grown and commercialized.

9.4.2 Fertilizer Application

GAP implies that the correct dose of fertilizer is applied at the correct moment and in the correct way. Automation and control in fertilizer application can be of great value toward satisfying this GAP requirement. Accurate measurements of soil macronutrients (i.e., nitrogen, phosphorus, and potassium) are needed for efficient agricultural production, including site-specific crop management (SSCM), where fertilizer nutrient application rates are adjusted spatially based on local requirements (Kim et al., 2009). Optical diffuse reflectance sensing has been reported to show potential for rapid, nondestructive quantification of soil properties, including nutrient levels (Roy et al., 2005; Maleki et al., 2008; Chacon et al., 2014). Kim et al. (2009) also discuss electrochemical sensing based on ion-selective electrodes or ion-selective field effect transistors that have been recognized as useful in real-time analysis because of their simplicity, portability, rapid response, and ability to directly measure the analyte with a wide range of sensitivity. They also give examples of optical and electrochemical sensors applied in soil analyses, while advantages and obstacles for their adoption are discussed.

9.4.3 Crop Protection and Integrated Pest Management

9.4.3.1 Weed Control

Core technologies (guidance, detection and identification, precision in-row weed control, and mapping) are required to meet the GAP criteria for weed control (Christensen et al., 2009). Detection and identification of weeds under the wide range of conditions common to agricultural fields remains the greatest challenge. Various methods have been developed for weed detection (Vrindts et al., 2002; Slaughter et al., 2008). They are all in some stage between research and commercial application. Most are based on spectral characteristics and/or image-based shape recognition to discriminate between weeds and the crop. In case population dynamics models are sufficiently developed, they can help to decide not to treat if the weeds pose no direct threat to crop production or quality. An overview of the modeling approaches to field weed dynamics is given by Holst et al. (2007). These models may become more accurate after each observation in time. The subsequent treatment can be a mechanical or thermal action or herbicide application. Precise herbicide treatment using microdosing nozzles on the most sensitive parts of the plant further reduces the chemical use (Young and Giles, 2013). When the detection and application systems are equipped with a GPS receiver, place and time of weed populations and the applied treatments could be automatically registered in the GAP database as well as in the field database (in a field passport).

9.4.3.2 Pest and Disease Management

GAP reduce the incidence and intensity of pests and diseases, and also the use of chemical control methods. This also implies that observation and monitoring practices are established and that nonchemical approaches must be considered. Where possible, biological control and the use of natural predators should be favored. Specific chemical control should only be considered when the economic value of the crop would be affected if this is not done.

The European Community Directive 128/2009 on the Sustainable Use of Pesticides establishes a strategy for the use of PPPs in the European Community to reduce risks to human health and the environment. Integrated Pest Management (IPM) is a key component of this strategy, which will become mandatory in 2014. IPM is based on dynamic processes and requires decision making at strategic, tactical, and operational levels. Rossi et al. (2012) state that, relative to decision makers in conventional agricultural systems, decision makers in IPM systems require more knowledge and must deal with greater complexity. Different tools have been developed for supporting decision making in plant disease control and include warning services, on-site devices, and decision support systems (DSSs). These decision support tools operate at different spatial and temporal scales, are provided to private sources, focus on different communication modes, and can support multiple options for delivering information to farmers (Rossi et al., 2012).

There are indications that automatic observation of diseases may be possible at an early stage, but at this moment, a good visual and instrumental strategy must be used for scanning the crop for disease initiation and if possible combined with population dynamics models to make a treatment decision. Sankaran et al. (2010) reviewed advanced techniques for detecting plant diseases. Some of the challenges in these techniques are (i) the effect of background data in the resulting profile or data, (ii) optimization of the technique for a specific plant/tree and disease, and (iii) automation of the technique for continuous automated monitoring of plant diseases under real-world field conditions. The review suggests that these methods of disease detection show a good potential with an ability to detect plant diseases accurately. Spectroscopic and imaging technology could be integrated with an autonomous agricultural vehicle for reliable and real-time plant disease detection to achieve superior plant disease control and management. Some examples are the detection of diseases in wheat using spectroscopic methods, which could potentially be developed further into airborne hyperspectral detection systems (Bravo et al., 2003; Mewes et al., 2011).

The spatiotemporal challenge for disease detection is also discussed by Mahlein et al. (2012) in the case of sugar beet. They reported that sugar beet diseases differed in their temporal and spatial development as well as in their effects on plant tissue associated with reflectance characteristics. High spatial resolution is crucial in particular for the detection of leaf diseases with discrete, roundish symptoms. The spatial resolution of the hyperspectral camera used in their study provided information even on subareas of disease symptoms. Nevertheless, the tiny uredinia of *Uromyces betae* and limited spatial resolution of the sensor resulted in a high number of mixed pixels. Depending on the shape of the symptoms, pixel size should be smaller than the object of interest by a factor of 2–5. This rule from remote sensing still restricts the (early) sensing of plant diseases to proximal sensing technologies. Specific effects of diseases, disease stage, and the impact of disease severity on spectral characteristics of plants are complex. The development of patterns in time and space, recorded by hyperspectral imaging, may help to identify disease or stress influencing crops at the tissue level and on the canopy level.

Since diseases are stressors of plants, this usually also affects the production and emission of volatile chemical compounds. If these could be detected in the field at an early stage and with sufficient spatial resolution, they could be the basis for decision making. Sensing systems of insects or animals or even plants are also a source of inspiration for novel developments, because of their uniqueness in type or sensitivity or also in the amount of information that is acquired and processed. For example, insects are able to perceive volatiles released by damaged plants in order to find food sources or mating partners. In order to use the highly developed olfactory sense of insects for analytical purposes, the biological nose of insects has to be combined with some electronic instrument via a bioelectronic interface to yield a bioelectronic nose (Schütz et al., 2000). Such a bioelectronic sensor system is very sensitive to detect volatiles released from damaged plant parts or at the onset of fungal infection. For *Phytophtora* detection in potatoes, this could lead to interesting applications.

The same is the case for pest control where traps are frequently used, but the readout of the traps is still time consuming and requires a lot of field travel because the traps must be spread out over a large area. However, there are also indications that it may be possible to identify insects and their population density through optical detection of the wing beat characteristics (van Roy et al., 2014).

9.4.3.3 Application Equipment

It is clear that any chemical treatment must be registered and correct application can only be done if the equipment is in good working condition. In the future, application systems may be made such that the use of a specific chemical compound is only possible according to the license as specified on the label: the site or crop, pest stage or crop stage, application rate depending on the pest or soil type, the timing of application according to season, application method and type of equipment, and number of applications allowed per season. In addition, one has to respect a preharvest interval in order not to exceed the MRLs, which can be country specific. At the time of pesticide application, all information about the crop would already be up to date in the farm database. The label information for a specific compound is also available or could be scanned before the active ingredient is loaded in the sprayer. In that case, an alarm could be given if an erroneous treatment is planned, or maybe the equipment might be locked into a safe mode. Of course, such a system must be made reliable and foolproof to be effective. Measures should also be taken to avoid some chemicals contaminating neighboring crops by monitoring wind speeds and estimating the spray drift (Nuyttens et al., 2011). The development and use of such technology should be part of a management and decision system. Dose level and disease threats are one aspect of the decision; the other ones are the harvest plan and decision. Moreover, both aspects must be fine-tuned and can be spatially and temporally dependent.

Another possibility would be to use smartphones in the field to take pictures of perceived diseases or pests and send these together with the GPS coordinates of the location in the field where the picture was taken to the cloud. After some computations in the cloud, the system could provide information on the kind of disease and the potential or desirable treatment. This treatment advice would then be based on the crop information (type of crop, planting date, and expected harvest date) that is stored in the cloud. The advice can also include the required dose depending on the biomass density (Uschkerat, 2013) or even the microclimate variations in the field. The risk of spray drift and required distances to waterways can be calculated based on information on local weather conditions. Next, a scan of the barcode on thc package of the pesticide will tell the operator if the treatment is allowed. Afterward, the applied dose and dose variation, together with the relevant information, would be recorded in the database of the field and the crop as part of the traceability system. It is expected that such a traceability system could improve disease and pesticide management as well as reinforce the confidence of the consumers in the safety of agricultural crops. Most elements of such a system have already been demonstrated in Japan (Nanseki, 2007).

9.4.4 Microbial Safety

Microbial contamination can occur during the field stage and at harvest and postharvest. Worker hygiene is very important here, and systems could be contemplated to enforce hygiene of workers and repeated cleaning of harvesting and transport equipment.

The early detection and removal of an infected item, if possible even before it reaches the main parts of the harvesting machine or grading line, can help to avoid problems. This implies that design engineering must now also have a strong emphasis on design for food safety. For example, modular design with suitable cleaning procedures and the use of noncontact sensing tools are one way for reducing risks. Eventually, additional microbial sensing technology should be installed to warn the user in case of a problem item. This may alter the future concepts of harvesting, handling, sorting, and packing equipment. All detections and subsequent removal and cleaning actions should be registered as part of the traceability system.

The core of this enhancement would allow farmers to include either climate forecasts or the latest measurable site-specific field condition data into the resource management decision-making process by best utilizing the historic yield data in similar conditions to adjust the input(s) responsively to the situation. Mid- to late-vegetative growth stage variable-rate nitrogen side-dress application is a good example of responsive control. Either based on the data obtained from in-season canopy reflectance sensing or from late spring soil nitrate tests, N-deficient crop plants will respond to additional nitrogen fertilizer being side-dress applied. It could potentially achieve higher yield efficiency with a smaller amount of total nitrogen fertilizer being applied if the amount of side-dressed fertilizer could be correctly determined.

9.5 CROP CONDITION SENSING

For all treatments such as fertilizer use, irrigation, or harvest scheduling, it is very important that the crop condition is known and also that the crop response to a treatment is observed, such that this can be taken into account for subsequent actions. It is of interest that the acquired data can yield information on physiological processes through the use of underlying physiological models rather than just statistical correlation models. In this way, control actions can be based on a better understanding of the physical and physiological processes. Optical measurement methods are considered to be the most appropriate for observing crop conditions and will be briefly discussed here.

Photonics offers many opportunities because photons are ultrafast, extremely focusable, and function contactless. This opens a number of possibilities for agricultural diagnostics. Photonics can be the basis for measurement systems to observe plant responses at different spatial and temporal scales. Indeed, growers can also visually recognize when a problem arises or when there is a large variation in crop condition in the field. Human observation is mostly limited to a qualitative interpretation. In the search for a more quantitative approach, numerous articles and reviews have been published on optical properties of crops and image analysis in relation to fertilizer use, crop stress, disease or weed detection, and product quality. In most of these cases, correlations have been established between a spectrum or an image and the particular crop characteristic that one wants to evaluate. These are then mainly empirical studies that have resulted in some practical implementations (Sims and Gamon, 2002; Reyniers et al., 2004, 2006; Lenk et al., 2007; Saeys et al., 2009; Gorbe and Calatayud, 2012; Tremblay et al., 2012).

There is a growing desire to link the measurable optical characteristics to physical and physiological processes in the crop, to increase the understanding of what is happening and then to better pinpoint potential actions. One approach is the use of biophysics-based mathematical models that link physiological processes to observed radiation and then apply model inversion. This model inversion may not always yield sufficient sensitivity to the different physiological components that can affect radiative transfer. Models at different spatial scales are used and, sometimes, they are integrated, which increases the computational complexity.

From the beginning of optical remote sensing, radiative transfer models, based on biophysical theory, have helped in the understanding of light interception by plant canopies and the interpretation of vegetation reflectance in terms of biophysical characteristics. The canopy radiative transfer models attempt to describe absorption and scattering, the two main physical processes involved, and are useful in designing vegetation indices, performing sensitivity analyses, and developing inversion procedures to accurately retrieve vegetation properties from remotely sensed data (Jacquemoud et al., 2009).

The processes and mathematical formulae that are used to simulate sensing signals depend on the scale of the system. In general, most models either simulate leaf-scale signals or canopy-scale signals (Atherton, 2012). At the canopy level, the model inversions typically require inputs of many canopy parameters that cannot be readily estimated from remote sensing data.

Four principal crop characteristics determine the reflection, absorption, and transmission of electromagnetic waves (Tucker and Garratt, 1977):

1. Internal structure or the histological arrangement of tissues and cells is responsible in part for the diffusion or internal scattering of incident irradiance. Spectral absorbance, reflectance, and transmittance are thereby greatly determined by the mean optical path length of incident radiation.
2. The pigment composition, concentration(s), and distribution(s) control the absorption of UV, visible, and IR radiation. Light absorption in food matrices is molecule-specific and theoretically described by Beer's law.
3. The concentration and distribution of leaf water determines the absorption of radiation in the NIR and IR region of the spectrum.
4. The surface roughness characteristics and the refractive index of the cuticular wax of the upper epidermis determine the spectral reflectance from this surface.

It should be noted that a number of crop characteristics or processes of interest are linked and highly correlated, and that inverse modeling based on optical measurements does not easily allow separate estimation of these characteristics.

Plant growth or vegetation development involves several processes that each occur on a different spatial as well as temporal scale (De Baerdemaeker, 2013). Examples of these different scales are disease symptoms on a leaf, growth, or the vegetative biomass in a field or a larger area. At those different scales, information is required for correct identification or classification of quality characteristics, of diseases, or of plants or crops. In many cases, this identification can rely on optical information taken at a high spatial resolution, but it can just as well happen that this identification is only possible through the use of high temporal frequency information. In other cases, rapid scans with low spatial resolution may indicate that uneven changes occur in the canopy or field and the cause of these may then be investigated by high spatial resolution inspection of locations of interest. Also, in case one wants to use information for statistical process control in order to detect abnormal deviations, it is required to have high temporal frequency information. There is usually a trade-off to be made between fine (or coarse) spatial resolution and low (or high) temporal frequency information since it may be impossible to have a high spatial and temporal resolution. It is a challenge to combine the data obtained at different temporal and spatial scales such that useful information is obtained (Robin et al., 2005).

Variation of crop characteristics over time can be due to normal development or also due to emerging stress conditions. Again, there may be different scales at which these changes occur. Patterns in spectra or hyperspectral image changes can be observed using time-lapse acquisition. Obtaining the information from subtle changes may require advanced image processing. For example, Wu et al. (2013) described a method to reveal temporal variation that are difficult or impossible to see with the naked eye in videos and display them in an indicative manner. The method, which they call Eulerian video magnification, takes a standard video sequence as input, and applies spatial decomposition, followed by temporal filtering to the frames. The resulting signal is then amplified to reveal hidden information.

9.6 CHAIN OF TRACEABILITY

After harvest, the GPS coordinates of the harvest location may be added to the shipping documents such that the origin of the product (the region, the farmer, the field, and the location in the field) can be traced and the consumer can be assured about the origin claims. It is also possible in mixed final products to state where the different components of such a mixture originated from. For retailers or stores that claim to sell locally produced food and for their clients, it offers the possibility to trace the product and verify the claims as long as the system has been made foolproof.

A crop goes through a number of operations, transactions, or shipments in the chain from the field to the customer. This is even more complicated when feed and animal production are part of the chain. At each step, there should be a possibility to trace the crop either upstream or downstream. As the chain can be relatively long, it has been suggested to implement this as a distributed system where only one step in either direction at every stage is traced instead of centralizing all data. This requires a good communication network between potential sites where the traceability data are stored, as well as access control. Cloud computing may be a way to proceed here. A benefit of accessibility of data can be that in the longer term, field variability related to weather and soil conditions can be extracted from such a database allowing farmers or their advisors to optimize production strategies. It is also a way to increase the expert knowledge or models for predicting what the outcome of a treatment this year can be, given that similar production conditions may have occurred in the past. In this way, the historical traceability information is not only valuable for consumers, but also for producers or other operators in the chain.

9.7 VARIABILITY MODELING AND TRACEABILITY

The advantages of having a nondestructive sensor reach far beyond the fact that it is just nondestructive. Indeed, they offer the possibility to monitor individual products during the experimental period, which in turn allows for modeling the change of quality attributes or other characteristics.

An approach based on mechanistic models further improves the interpretation of postharvest behavior (Tijskens et al., 2001). By definition, such a model will be based on a simplification of the food product and, therefore, will never be "true" as the only true model is the product itself. The aim of modeling food quality attributes is, however, not to develop true models but to develop valid models. That is, models that are consistent with the current knowledge level and that contain no known or detectable flaws of logic (Tijskens et al., 2001). Also, models should be detailed enough for the intended purpose, but at the same time simple enough to give robust manageable models. The basic strategy to develop a suitable model is to apply a systematic process of problem decomposition, dissecting the problem into its basic building blocks and then reassembling them leaving out the unnecessary detail. What is essential and what is redundant depends largely on the intended application of the model. In the end, the models are to be used to provide an appreciation of the quality of the logistic handling chain and to translate this into the impact the logistic conditions have on product quality attributes (Hertog et al., 2014).

The major challenge is to develop predictive models that assess the uncertainty of the predicted result. Given a simulation model, this problem reduces the propagation of errors from the simulation input to the simulated result. With an increasing number of random factors, it becomes practically impossible to establish the correct model response. Generally, some reduction is required by identifying the most important (combinations of) input parameters that capture most of the variability.

With the availability of nondestructive techniques, the quality of individual product items can be monitored over time, fully characterizing biological variance within a given batch. To properly analyze such data, biological variance has to be explicitly included in the (statistical) data analysis. De Ketelaere et al. (2006) proposed a novel statistical approach ("mixed models") to model such repeated quality measures and demonstrated its potential for a practical example in which the firmness change of different tomato cultivars was considered. Both types of data analysis allow quantifying different sources of variance such as variance within a tomato cultivar and within a tomato and how those sources of variance change during storage. These approaches open the door to an improved measurement, understanding, and prediction of postharvest batch behavior. As such, these approaches enable postharvest management to optimize logistics, taking into account the full range of product variation that will be encountered.

If biological variance is included in (statistical) models describing postharvest quality change, propagation of the initial biological variance at harvest throughout the entire postharvest chain can be predicted when all relevant aspects affecting postharvest fruit behavior are taken into account (Hertog et al., 2014).

Shelf life prediction is an important issue for fruit handling. As mentioned before, not only the average quality trajectory a batch follows has to be estimated, but also how much the quality is dispersed around the batch average, since we are generally interested in an estimation of the time at which, for example, 5% of the fruits reach a preset lower bound for their quality.

The implementation and validation of such a stochastic quality change model was tested in a traceability system for tomato (Hertog et al., 2008). Experimental results showed the potential benefits of integrating quality change models with traceability systems to satisfy consumer expectations. As the temperature logging radio-frequency identification (RFID) labels are too expensive to put on individual boxes, the alternative to use a single RFID label per pallet seems to be feasible given the limited effect of temperature differences within the palletized fruit. The model-based traceability systems to monitor product quality throughout the chain can then assist in identifying poor temperature control or temperature abuse at a given point in the logistic chain as the cause of unacceptable quality at the receiving point. Furthermore, such monitoring and modeling can help in identifying locations in a field or orchard where there is a large deviation in quality or shelf life. It also can help in differentiating the harvest time within a field or between fields.

These approaches enable postharvest management to optimize logistics, taking into account the full range of product variation that will be encountered.

9.8 MODEL-BASED STATISTICAL PROCESS CONTROL

Nowadays, agricultural production performance is usually assessed and monitored by comparing mean values of a recent measurement period (e.g., week or month) with past performances or predetermined performance standards. This is usually done without the interference of statistical analysis. However, excessive biological variation interferes with the evaluation of performance. High variability makes the performance outcome unpredictable and difficult to interpret. Therefore, understanding variability is the diagnostic key for improving process performance (Reneau and Lukas, 2006). Two concepts that are especially interesting for performing process optimization through monitoring are engineering process control (EPC) and statistical process control (SPC). EPC is the set of activities that focus on the mathematical modeling of (production) systems (del Castillo, 2002), and SPC is a collection of tools that aim at discerning between normal and abnormal process variation (Montgomery, 2005). An SPC tool that is widely used for the detection of abnormal variability is the quality control chart. The use of control charts in agricultural production, and especially in livestock production, is gaining considerable interest (de Vries and Conlin, 2003; Reneau and Lukas, 2006). The signal of the control chart can be used for early detection of problems. This synergistic concept has only recently been applied to agricultural production. Since the data of many agricultural production processes evolve over time (nonstationary) and subsequent measurements are correlated (dependent), they cannot be monitored as such with the control charts. To overcome these limitations, the concept of synergistic control was proposed (De Ketelaere et al., 2011).

In a synergistic procedure for early problem detection, the concepts of EPC and SPC are combined. For example, by using the EPC adjusted data, by means of a recursively estimated trend and ARMA model, as the input to the cusum control chart (SPC), it was shown to be possible to detect registrations that result from an out-of-control situation as a result of an emerging problem or disease. The potential of this concept was already demonstrated for monitoring laying hens (Mertens et al., 2011) and dairy cows (Huybrechts et al., 2014). This procedure can form the basis for the development of an intelligent management support tool for agricultural production systems such as dairy production, pig production, and crop production. The synergistic concept is in most cases applied for processes changing with time, but it can also be applied for assessing spatial variability or the sensitivity of, for example, varieties or treatments to spatially variable soil conditions. However, it should be noted that this approach can only be successful if reliable sensor data are available.

9.9 SUMMARY AND CONCLUSIONS

In PA and automation, many measurements are carried out at different spatial scales (from single plants to entire fields) and at different times during crop production. Precision farming and the use of GPS on agricultural machinery can provide location and time information of all treatments. It started with yield sensors, but at this time, tools are available for on-the-go measurement of the type and dose of treatments, for identification of crop condition, and possible infection with pests or diseases. Wireless communication can be used to transfer field data to record keeping software. Thanks to these technological developments, the control points and compliance criteria of certification systems for GAP, such as GlobalGAP or other GAP schemes, can to a large extent be automatically addressed using PA technology for automatic record keeping. PA technology can be made smart such that the requirements for environmentally friendly and sustainable production are implemented in real time in crop treatment and fertilizer equipment. This also includes the identification and registration of operations or treatments on the crop in the growing stage. At the time of harvest, the technology can help in the identification and, if possible, the measurement of the quality parameters depending on where in the field the crop was grown. Different batches can be made with labels linking to all the information. As such, PA technology can evolve to being great instruments for food safety and quality assurance.

Novel crop sensing techniques during growth or after harvest give information on crop stress, quality, diseases, pests, or weeds. Now, information is available about variability of crop or product characteristics. The repeated nondestructive measurements allow for modeling of the process evolution or the evolution of quality over time (or maybe also in space), thereby separating inherent biological variability from variations caused by external process conditions. These models and observations form the basis for SPC and informed decision making for interventions.

The frequently asked question about the economic benefits of PA is also raised about the economic effects of food safety and safety risks along the chain. In this respect, Valeeva et al. (2004) state that acceptable levels of food safety hazards need further elaboration to clarify the process of food safety improvement for producers. They also note that it is furthermore important to gain more insight into cost-effective ways of food safety improvement throughout the entire chain and that valuation of producers' benefits along the chain and their distribution are urgently needed. Perhaps, the combined economic benefits of PA and GAP for food safety and consumer confidence are underestimated at this moment.

REFERENCES

Atherton, J.M. 2012. *Multiscale Remote Sensing of Plant Physiology and Carbon Uptake*. PhD thesis, The University of Edinburgh, June.

Beulens, A.J.M., D.-F. Broens, P. Folstar, and G.J. Hofstede. 2005. Food safety and transparency in food chains and networks—Relationships and challenges. *Food Control*, 16:481–486.

Bravo, C., D. Moshou, J. West, A. McCartney, and H. Ramon. 2003. Early disease detection in wheat fields using spectral reflectance. *Biosystems Engineering*, 84(2):137–145.

Chacon Iznaga, A., M. Rodriguez Orozco, E. Aguila Alcantara, M. Carral Pairol, Y.E. Diaz Sicilia, J. De Baerdemaeker, and W. Saeys. 2014. Vis/NIR spectroscopic measurement of selected soil fertility parameters of Cuban agricultural Cambisols. *Biosystems Engineering*, 125:105–121.

Christensen, S., H.T. Søgaard, P. Kudsk, M. Nørremark, I. Lund, E.S. Nadimi, and R. Jørgensen. 2009. Site-specific weed control technologies. *Weed Research*, 49:233–241. doi: 10.1111/j.1365-3180.2009.00696.x.

De Baerdemaeker, J. 2013. Multi-scale photonics for precision agriculture. In *Invited Lecture at the 1st International Conference on Sensing Technologies for Biomaterial, Food and Agriculture 2013 (SeTBio)*, April 23–25, 2013, Pacifico Yokohama, Japan.

De Ketelaere, B., K. Mertens, F. Mathijs, D.S. Diaz, and J. De Baerdemaeker. 2011. Nonstationarity in statistical process control—Issues, cases, ideas. *Applied Stochastic Models in Business and Industry*, 27:367–376.

De Ketelaere, B., J. Stulens, J. Lammertyn, N. Cuong, and J. De Baerdemaeker. 2006. A methodological approach for the identification and quantification of sources of biological variance in postharvest research. *Postharvest Biology and Technology*, 39(1):1–9.

De Vries, A. and B.J. Conlin. 2003. Design and performance of statistical process control charts applied to estrous detection efficiency. *Journal of Dairy Science*, 86:1970–1984. doi: 10.3168/jds. S0022-0302(03)73785-0.

Del Castillo, E. 2002. *Statistical Process Adjustment for Quality Control (Wiley Series in Probability and Statistics)*. New York, USA: John Wiley & Sons, Inc.

European Commission. 2012. Regulation (EC) No 178/2002 of the European Parliament and of the Council of 28 January 2002 laying down the general principles and requirements of food law, establishing the European Food Safety Authority and laying down procedures in matters of food. *Official Journal* L031, 01/02/2002 P. 0001–0024. http://www.europa.eu.int/eur-lex/en/search/search_lif.html.

European Food SCP Round Table Working Group 2 on "Environmental Information Tools". 2011. *Report: Communicating Environmental Performance along the Food Chain*, December 2011. http://www.food-scp.eu/files/ReportEnvComm_8Dec2011.pdf.

European Food Sustainable Consumption and Production (SCP) Round Table. www.food-scp.eu/files/Guiding_Principles.pdf.

GlobalGAP. 2012. *Integrated Farm Assurance*. http://www.globalgap.org/uk_en/what-we-do/.

Gorbe, E. and A. Calatayud. 2012. Applications of chlorophyll fluorescence imaging technique in horticultural research: A review. *Scientia Horticulturae*, 138:24–35.

Hertog, M.L.A.T.M., I. Uysal, U. McCarthy, B.M. Verlinden, and B.M. Nicolaï. 2014. Shelf life modelling for first-expired-first-out warehouse management. *Philosophical Transactions of the Royal Society A*, 372:20130306. http://dx.doi.org/10.1098/rsta.2013.0306.

Hertog, M.L.A.T.M., R.F. Yudhakusuma, P. Snoekx, J. De Baerdemaeker, and B.M. Nicolaï. 2008. Smart traceability systems to satisfy consumer expectations. *Acta Horticulturae*, 768:407–415 (presented during International Horticultural Congress—IHC2006, Seoul 2006).

Hobbs, J.E., D. Bailey, D.L. Dickinson, and M. Haghiri. 2005. Traceability in the Canadian red meat sector: Do consumers care? *Canadian Journal of Agricultural Economics/Revue canadienne d'agroeconomie*, 53:47–65. doi:10.1111/j.1744-7976.2005.00412.x.

Holst, N., I. Rasmussen, and L. Bastiaans. 2007. Field weed population dynamics: A review of model approaches and applications. *Weed Research*, 47:1–14.

Huybrechts, T., K. Mertens, J. De Baerdemaeker, B. De Ketelaere, and W. Saeys. 2014. Early warnings from automatic milk yield monitoring with online synergistic control. *Journal of Dairy Science*, 97:3371–3381.

Jacquemoud, S., W. Verhoef, F. Baret, C. Bacour, P.J. Zarco-Tejada, G.P. Asner, C. François, and S.L. Ustin. 2009. PROSPECT + SAIL models: A review of use for vegetation characterization. *Remote Sensing of Environment*, 113(1):S56–S66.

Kim, H.-J., K.A. Sudduth, and J.W. Hummel. 2009. Soil macronutrient sensing for precision agriculture. *Journal of Environmental Monitoring*, 11:1810–1824. doi: 10.1039/B906634A.

Lenk, S., L. Chaerle, E.E. Pfündel, G. Langsdorf, D. Hagenbeek, H.K. Lichtenthaler, D. Van Der Straeten, and C. Buschmann. 2007. Multispectral fluorescence and reflectance imaging at the leaf level and its possible applications. *Journal of Experimental Botany*, 58(4):807–814.

Mahlein, A.K., U. Steiner, C. Hillnhütter, H.W. Dehne, and E.C. Oerke. 2012. Hyperspectral imaging for small-scale analysis of symptoms caused by different sugar beet diseases. *Plant Method*, 8(1):3. doi: 10.1186/1746-4811-8-3.

Maleki, M., A. Mouazen, B. De Ketelaere, H. Ramon, and J. De Baerdemaeker. 2008. On-the-go variable-rate phosphorus fertilisation based on a visible and near-infrared soil sensor. *Biosystems Engineering*, 99(1):35–46.

Mertens, K., E. Decuypere, J. De Baerdemaeker, and B. De Ketelaere. 2011. Statistical control charts as a support tool for the management of livestock production. *Journal of Agricultural Science*, 149:369–384. doi: 10.1017/S0021859610001164.

Mewes, T., J. Franke, and G. Menz. 2011. Spectral requirements on airborne hyperspectral remote sensing data for wheat disease detection. *Precision Agriculture*, 12:795–812. doi: 10.1007/s11119-011-9222-9.

Montgomery, C. 2005. *Introduction to Statistical Quality Control*, 5th Edition. Hoboken, NJ, USA: John Wiley & Sons, Inc.

Nanseki, T. 2007. *A Navigation System for Appropriate Pesticide Use and Food Safety*. http://www.fftc.agnet.org/htmlarea_file/library/20110704173849/bc54010.pdf, accessed on January 29, 2015.

Nuyttens, D., M. De Schampheleire, K. Baetens, E. Brusselman, D. Dekeyser, and P. Verboven. 2011. Drift from field crop sprayers using an integrated approach: Results of a five-year study. *Transactions of the ASABE*, 54(2):403–408.

Opara, L.U. and F. Mazaud. 2001. Food traceability from field to plate. *Outlook on Agriculture*, 30:239–247.

Reneau, J. and J. Lukas. 2006. Using statistical process control methods to improve herd performance. *Veterinary Clinics of North America: Food Animal Practice*, 22:171–193.

Reyniers, M., E. Vrindts, and J. De Baerdemaeker. 2004. Optical measurement of crop cover for yield prediction of wheat. *Biosystems Engineering*, 89(4):383–394.

Reyniers, M., E. Vrindts, and J. De Baerdemaeker. 2006. Comparison of an aerial-based system and an on the ground continuous measuring device to predict yield of winter wheat. *European Journal of Agronomy*, 24(2):87–94.

Robin, A., S. Mascle-Le Hégarat, and L. Moisan. 2005. A multiscale multitemporal land cover classification method using a Bayesian approach. *Proc. SPIE 5982, Image and Signal Processing for Remote Sensing XI, 598204*, Bruges, Belgium: SPIE. (October 18, 2005); doi: 10.1117/12.627604.

Rossi, V., T. Caffi, and F. Salinari. 2012. Helping farmers face the increasing complexity of decision-making for crop protection. *Phytopathologia Mediterranea*, 51(3):457–479.

Roy, S.K., S. Shibusawa, and T. Okayama. 2005. Site-specific soil properties prediction using hyperspectral signatures of topsoil coverage and underground image by real-time soil spectrophotometer. In Stafford, J.V. (ed.), *Precision Agriculture* '05. Wageningen, The Netherlands: Academic Publishers, ISBN 9076998698.

Saeys, W., B. Lenaerts, G. Craessaerts, and J. De Baerdemaeker. 2009. Estimation of the crop density of small grains using LiDAR sensors. *Biosystems Engineering* 102(1):22–30.

SAI Global. http://www.saiglobal.com/assurance/, accessed on January 27, 2015.

Sankaran, S., A. Mishra, R. Ehsani, and C. Davis. 2010. A review of advanced techniques for detecting plant diseases. *Computers and Electronics in Agriculture*, 72:1–13. doi: 10.1016/j.compag.2010.02.007.

Schiefer, G. and J. Deiters (eds.). 2013. *Transparency in the Food Chain*. Germany: Universität Bonn-ILB, ISBN 978-3-941766-17-4.

Schütz, S., M.J. Schöning, P. Schroth, Ü. Malkoc, B. Weißbecker, P. Kordos, H. Lüth, and H.E. Hummel. 2000. An insect-based BioFET as a bioelectronic nose. *Sensors and Actuators B: Chemical*, 65(1–3):291–295.

Sims, D.A. and J.A. Gamon. 2002. Relationships between leaf pigment content and spectral reflectance across a wide range of species, leaf structures and developmental stages. *Remote Sensing of Environment*, 81:337–354.

Slaughter, D.C., D.K. Giles, and D. Downey. 2008. Autonomous robotic weed control systems: A review. *Computers and Electronics in Agriculture*, 61:63–78.

Tijskens, L.M.M., M.L.A.T.M. Hertog, and B.M. Nicolaï (eds.). 2001. *Food Process Modelling*. Cambridge, UK: Woodhead Publishing Limited, p. 496.

Tremblay, N., Z. Wang, and Z.G. Cerovic. 2012. Sensing crop nitrogen status with fluorescence indicators. A review. *Agronomy for Sustainable Development*, 32:451–464.

Trites, G. 2013. *Information Integrity*. AICPA® Assurance Services Executive Committee's Trust Information Integrity Task Force. January 2013. http://www.aicpa.org/InterestAreas/FRC/AssuranceAdvisoryServices/DownloadableDocuments/ASEC-Information-Integrity-White-paper.pdf, accessed on January 31, 2015.

Tucker, C.J. and M.W. Garratt. 1977. Leaf optical system modelled as a stochastic process. *Applied Optics*, 16(3):635–642.

Tukker, A., G. Huppes, J. Guinée, R. Heijungs, A. de Koning, L. van Oers, S. Suh, T. Geerken, M. Van Holderbeke, B. Jansen, and P. Nielsen. 2006. Environmental Impacts of Products (EIPRO). Analysis of the Life Cycle Environmental Impacts Related to the Total Final Consumption of the EU-25. JRC/IPTS 2006, Institute for Prospective Technological Studies, Sevilla. http://ec.europa.eu/environment/ipp/pdf/eipro_report.pdf.

Uschkerat, U. 2013. *Crop Sense: Radar Technique for the Determination of Ear Biomass*. http://www.fhr.fraunhofer.de/en/businessunits/Energy-and-Environment/Crop-Sense-radar-technique-for-the-determination-of-ear-biomass.html.

Valeeva, N.I., M.P.M. Meuwissen, and R.B.M. Huirne. 2004. Economics of food safety in chains: A review of general principles. *NJAS-Wageningen Journal of Life Sciences*, 51(4):369–390.

Van Plaggenhoef, W., M. Batterink, and J. Trienekens. 2003. International food safety: Overview of legislation and standards. http://www.globalfoodnetwork.org

van Roy, J., J. De Baerdemaeker, W. Saeys, and B. De Ketelaere. 2014. Optical identification of bumblebee species: Effect of morphology on wingbeat frequency. *Computers and Electronics in Agriculture*, 109:94–100.

Vrindts, E., J. De Baerdemaeker, and H. Ramon. 2002. Weed detection using canopy reflection. *Precision Agriculture*, 3:63–80.

Wu, H.-Y., M. Rubinstein, E. Shih, J. Guttag, F. Durand, and W.T. Freeman. 2012. Eulerian video magnification for revealing subtle changes in the world, *ACM Transactions on Graphics (Proc. SIGGRAPH 2012)*, 31(4).

Young, S.L. and D.K. Giles. 2013. Targeted and Microdose Chemical Applications. West Central Research and Extension Center, North Platte. Paper 81. http://digitalcommons.unl.edu/westcentresext/81.

10 State of the Art and Future Requirements

Hermann Auernhammer and Markus Demmel

CONTENTS

10.1 INTRODUCTION

Precision agriculture (PA) means more than site-specific farming, it also deals with more than just variability. In discussions worldwide, "precision agriculture" and "precision farming" are often used interchangeably.

Agriculture is one sector in the entire land use scenario, while PA is specifically associated with precision forestry and precision fishery. "Precision (crop) farming" and "precision livestock farming" can be thought of as categories within PA. In most countries, viticulture and horticulture are seen as parts of agriculture; another way to look at these categories is as farms where operations are carried out only outdoors, on the one hand, and those where operations take place both outdoors and indoors, on the other. Of these, outdoor farming is dominant as it covers a wide range of precision farming activities (Figure 10.1).

Whichever classification is used, precision farming must be seen from the farm-level perspective. Activities of interest are farm management itself, crop management, machinery management, and labor management. In all of these areas, PA measures can be seen and may contribute to sustainability and traceability (Figure 10.2).

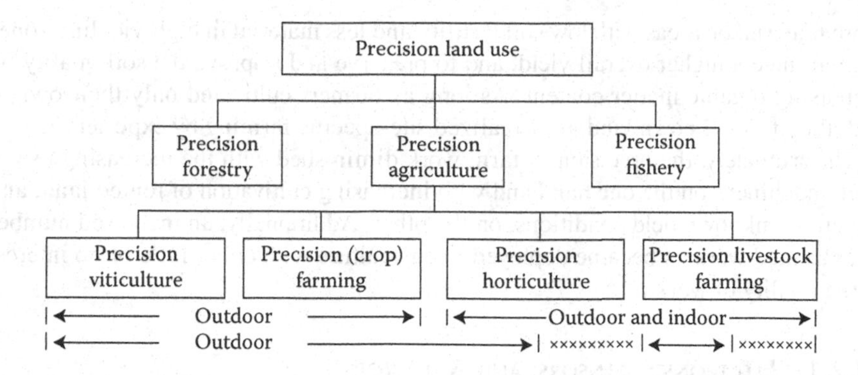

FIGURE 10.1 Precision agriculture in precision land use broken down to outdoor and indoor systems.

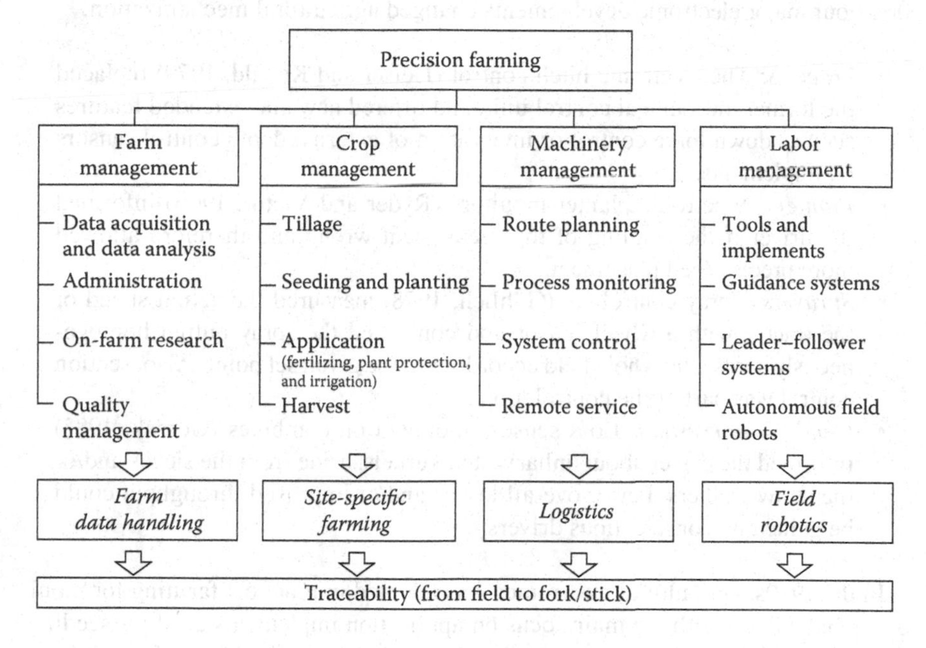

FIGURE 10.2 Precision farming sections and items. (From Auernhammer, H. 1999. *Zeitschrift für Agrarinformatik*, 3:58–67. With permission.)

10.2 BASIC TECHNOLOGIES

Farmers have for millennia strived to farm more precisely, first with simple hand tools such as the sickle for exact cutting of crops, and later with mechanical implements, such as the plow, with its ability to cut cleanly and to turn the soil on a large scale. Also, very early, the knowledge of given field conditions such as soil type, water availability, and topography, together with the experience gained out of previous field work and the previous harvest, were integrated into measures for the new vegetation cycle. Manure handling is a good example, where farmers distributed more material on areas with low soil fertility and less material in high-yielding zones to guarantee a higher overall yield, and to preserve and improve the soil quality by increasing organic matter content. As long as farmers cultivated only their own—and, therefore—known land, they realized site-specific farming by experience!

Unfortunately, the precision of farm work diminished with the increasing use of farm machinery on the one hand and with increasing cultivation of rented land, and therefore unknown field conditions, on the other. Additionally, an increased number of untrained workers became deployed in agriculture with either little or no interest in the quality of work.

10.2.1 Electronics, Sensors, and Actuators

In the early 1970s, the first electronic solutions entered agricultural technology and mechanization. Agriculture began using calculators with on/off and revolution sensors. Four major electronic developments changed agricultural mechanization:

- *Tractors:* The electronic hitch control (Heiser and Kobald, 1979) replaced the former mechanical control unit, and offered new and extended features such as down force control or integration of external depth control sensors on implements.
- *Planters:* Electronic planter monitors (Ryder and Victor, 1966) informed the driver if the singling of the seeds went wrong and therefore allowed more precise seed placement.
- *Sprayers:* Spray controllers (Göhlich, 1978) measured the actual speed of the tractor with a wheel sensor and controlled the spray output homogeneously across the whole field according to the given set point. Also, section control was part of the control loop.
- *Combine harvesters:* Loss sensors mounted on combines (Gorsek, 1983) informed the driver about unharvested kernels, aside from the sieves and/or the straw walkers. Fewer overall losses and/or improved throughput could be gained by conscientious drivers.

In the 1980s, agricultural electronics was adopted in practical farming for monitoring and control with the main focus on application implements used for seeding, planting, fertilizing, and spraying. During that period, two other major developments took place on farms:

- *On-farm data processing:* Based on personal computers (PCs), field records went from written records to digital files with enhanced capabilities in analyses and predictions.
- *Implement control:* Specified implement controllers and increasingly more multipurpose process controllers were integrated into application implements for monitoring, application control, and data acquisition related to working time, working speed, application amounts, etc. Also, first proprietary data transmission tools from the controller to the on-farm PC were developed and used.

While IBM created the worldwide accepted standard DOS for PCs, the implement controllers still followed proprietary solutions with barriers in acceptance on the one hand, and with specified and additional demand in data communication procedures on the other.

Consequently, at the end of the 1980s, the demand for standardized communication systems in agriculture were discussed in regions where farmers preferred tractors and implements from different manufacturers in order to perform field work in the best way under the given conditions, and with the available infrastructure in service and maintenance of the used technology.

10.2.2 Standardized Electronic Communication

Early on, electronic communication in mobile agricultural equipment was considered either as a company-specific solution by dominating market leaders or as independent solutions by a commonly accepted standard.

One of the widely utilized process controllers in Europe (dlz Spezial, 1990), with more than 50,000 sold units since 1985, could be used for implement monitoring, implement control, and for data transfer to the on-farm computer using a chip card in a proprietary way (Figure 10.3).

Farmers widely accepted electronics coming from "one hand." A single controller used year round with its simple man-to-machine (M2M) interface facilitated the interaction. Depending on the mounted implement, the connector with its specific pin allocation ensured the required control software, and the chip card allowed for data transfer in both directions to and from the management computer. Similar products were developed worldwide and thousands are still in use.

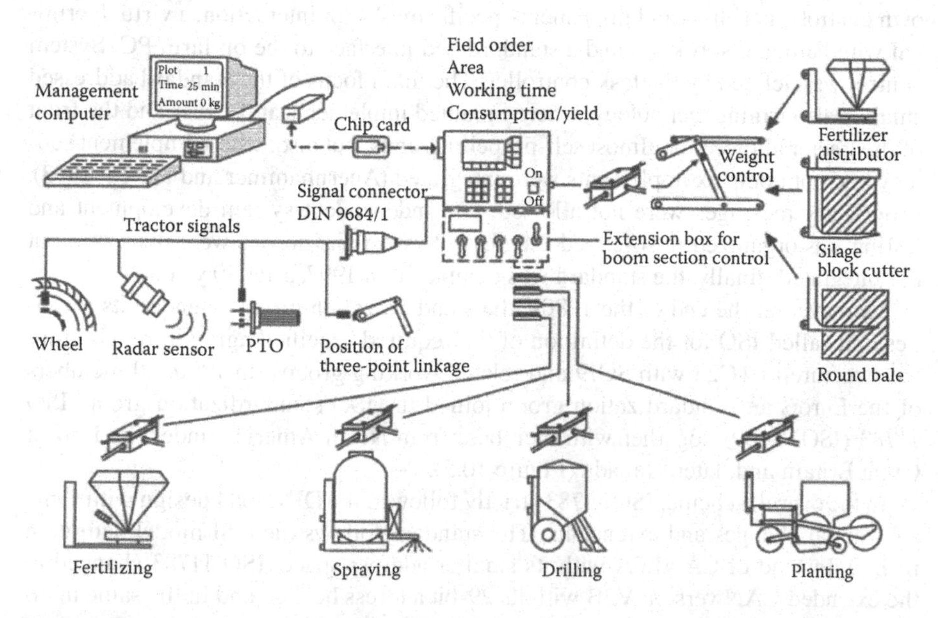

FIGURE 10.3 Mobile multipurpose agricultural process controller MÜLLER Unicontrol.

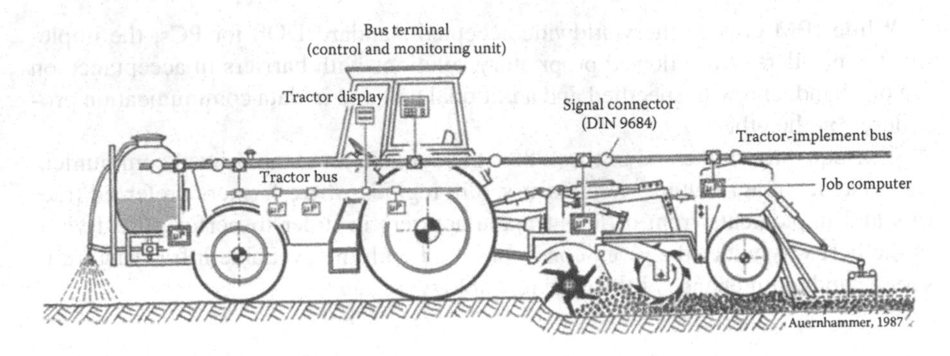

FIGURE 10.4 Agricultural BUS-System (Landwirtschaftliches BUS-System LBS) by DIN 9684.

With all these solutions, farmers as well as manufacturers began getting strongly dependent on electronic suppliers. The small- and medium-sized implement manufacturers in Europe were especially eager to get independence as soon as possible by creating a standard for electronic tractor implement communication (Auernhammer, 1989) with an interface to the farm management computer in DIN 9684 (DIN, 1997), as seen in Figure 10.4.

In a 1987 initiative chaired by the German DIN organization, representatives from Denmark, the Netherlands, France, and Great Britain worked together. The serial bus system controller area network (CAN) (BOSCH, 1987) from BOSCH was initially selected. The main characteristics of the standard are CAN1.0A protocol with 125 kB/s, electronic control units (ECU) in the tractor and implements, with their own control algorithms and implement-specific masks for interaction, a virtual terminal with hard and soft keys, and a standardized interface to the on-farm PC. System control was defined by the task controller. The main focus of the standard addressed small-scale farming technology, where mounted implements at the rear and the front of the tractor allowed for almost self-propelling units, but also, where implement subnetworks for complex implements were integrated (Auernhammer and Frisch, 1993). Proprietary messages were not allowed, and independent system development and testing was organized in so-called Plugfests. System diagnostics were discussed but not integrated; finally, the standard was completed in 1997, after 10 years.

Moreover, at the end of the 1980s, the standardization group extended its activities and called ISO for the definition of the required specified agricultural committee structure in TC23 with SC19 and related working groups. In 1990, all members of the European standardization group joined the ISO standardization group, ISO 11783 (ISO, 2009), together with members from North America under the lead of Great Britain and, later, Canada (Figure 10.5).

In the overall scheme, ISO 11783 strictly followed the DIN 9684 design with some substantial changes and extensions. The standard follows the OSI model definition in ISO. Instead of CAN1.0A with its smaller address space, ISO 11783 changed to the extended CAN version V2B with its 29-bit address header, and in the same manner, the bus transmission rate was doubled to 250 kB/s. More attention was given to

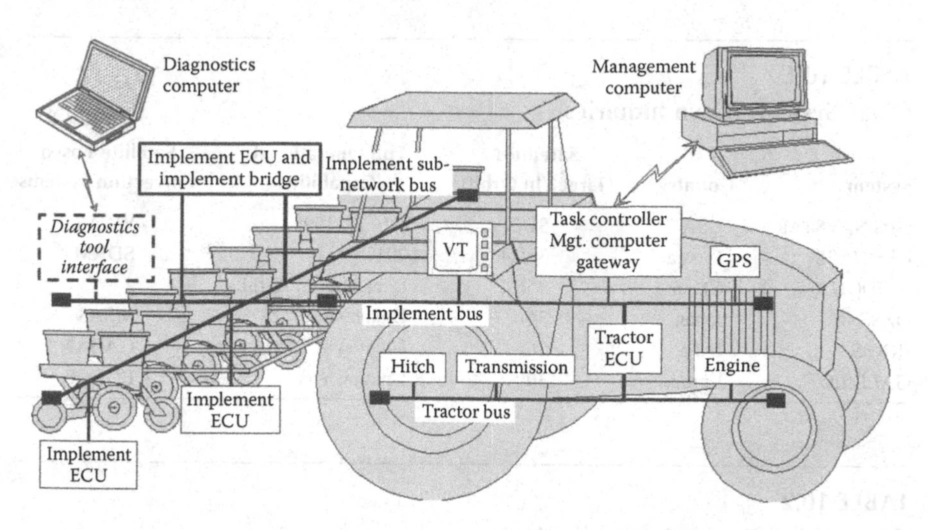

FIGURE 10.5 Agricultural BUS-System ISO 11783. (From Stone, M.L. et al. 1999. ISO 11783: An electronic communications protocol for agricultural equipment. ASAE St. Joseph, MI, USA. Modified with "diagnostics" by Stone, M. L. 2011. ISO 11783 Part 10 Task controller and management information system data interchange. ASABE AET. http://www.shieldedpair.net/downloads/ISO%2011783%20Part%2010.pdf. With permission.)

tractor and implement combinations used in the semi-mounted or trailed mode with large working width and hydraulic power supply. Therefore, implement subnetworks are of higher importance and more well defined. Diagnostics with interface and diagnostic tools are integrated. Most importantly, today, proprietary messages are allowed and may be used by anyone within the standard to get manufacturer-specific advancement within their own tractor implement production segment.

ISO 11783 may be called a "living standard" with no foreseen finalization at this time. Additional definitions are under development and will be added as required or changed by technical enhancements.

10.2.3 Location Sensing

From a technical point of view, the second pillar of precision farming evolved from a military initiative, with the development of the Global Navigation Satellite System (GNSS), NAVSTAR, and, in parallel, GLONASS, offering time and positioning signals (Auernhammer, 1994). Receivers are able to determine their own position by calculating the signal transit times from at least four visible satellites.

After the launch of the first GPS test satellite in 1971 and the test of an interim system, the predominantly used system, NAVSTAR (Navigational Satellite Timing and Ranging), reached its "full operational capability" (FOC) in June 17, 1995, with civilian usability for location sensing by all, at no cost. Most important for the agriculture sector, as one of the worldwide first users, are: availability all day, usability with no restrictions of daytime and visibility, no need for additional infrastructure when basic accuracy is sufficient, and higher accuracy and lower dependency as more satellite systems are globally available, and as more signal improvement tools are used.

TABLE 10.1
GNSS Systems (Own Inquiries)

System	Country	Satellites (Target/In Orbit)	Full Operational Capability	Satellite-Based Correction Systems
GPS NAVSTAR	USA	31/31	July 17, 1995	WAAS
GLONASS	Russia	30/24	2011	SDKM
BEIDOU	China	35/14	December 27, 2011	?
QZSS	Japan	3/1	Unknown	MSAS
IRNSS	India	7/3	Unknown	GAGAN
GALILEO	Europe	30/5	Unknown	EGNOS

TABLE 10.2
GNSS Signal Processing Technologies

System	Signal Correction	Examples	Accuracy (m)
Single-frequency GNSS	No	NAVSTAR, GLONASS	10–15
Differential GNSS	Postprocessing	–	0.5–1.0
Differential GNSS	Radio communication	IALA Radio Beacon	0.5–1.0
Differential GNSS	Satellite communication	WAAS, EGNOS	0.5–1.0
Double-frequency GNSS	Second frequency	–	2–5
Differential double-frequency GNSS	Radio or satellite communication + double frequency	Starfire II, Omnistar HP	0.1–0.2
RTK GNNS	Radio communication	–	0.02–0.05

Since the first tests, more than 200 satellites have been launched into orbit. Besides the two prime pioneering systems, GPS NAVSTAR and GLONASS, four more systems are under development (Table 10.1).

Advanced receivers are able to pick up signals from all visible satellites, even those from different systems, and select the best geometry for the highest possible accuracy. Additionally, different signal processing technologies (Table 10.2) can be used to achieve the required location sensing precision for navigation (± 10 m), field operations (± 1 m), vehicle guidance (± 0.10 m), and tool guidance (± 0.01 m), defined by Auernhammer and Muhr (1991).

10.3 FARM MANAGEMENT

Information-driven farm management needs data and algorithms to analyze, plan, and control farm processes, as well as to follow social and environmental conventions. Widespread databases guarantee any needed documentation, and allow for comprehensive analyses and predictions. At present, besides the traditional on-farm data storage and data processing, increasingly more off-farm services are offered and in use (Table 10.3).

TABLE 10.3
Data Handling Systems

System	Characteristics	Advantage	Disadvantage
On-farm	On-farm data storage On-farm-specific software	Private data ownership Preferred hardware Specialized software	Being alone in • Long-term data storage • Hardware exchange • Software update
Provider (fertilizer, plant protection, contractor, etc.)	Data storage Farm-related analyses and predictions	No on-farm infrastructure Analyses be demand Predictions by demand Specialized on-farm services	Only provider-specific predictions Data ownership? Data security?
Manufacturer	Data storage Remote machine analysis and service	No on-farm infrastructure Improved machinery service and repair Farm data contribute to machinery improvement	Only company related provider-specific predictions Data ownership? Data security?
Cloud	Offered by software companies and others Data storage Access of web-based apps and services	No on-farm infrastructure Great choice in required apps Ongoing development of new and more beneficial apps	Data ownership? Data security?

As a very rough assessment, it may be generalized that small- and medium-sized farms still work with on-farm systems as most of them have personal field-based knowledge and experience, whereas larger farms with more employees and less field-based information prefer either contractors for fertilizing or plant protection, or rely more on the on-farm dominant machinery supplier.

10.3.1 Data Acquisition and Data Analysis

Automatic process data acquisition is mainly used to establish comprehensive field records and provide cost element data for bookkeeping (Steinberger, 2012). Data acquisition systems differ widely and there is still no data definition standard (Table 10.4).

Data processing is going through a change. Smaller farms often retain special software packages with data history of cultivated fields tailored to simplified usage and with more or less no specialized analytical tools. By focusing on some methods of PA, improved software packages are used with a central database and with farm-specific analytical tools (Daberkow and McBride, 2003). Very often, those systems are offered with the cooperation of tractor manufacturers and contractual partners to simplify data transfer from mobile technology to the farm database and the software tools at the farm.

Data storage in the cloud at this time is an exception, mainly owing to questions of data ownership and data security. Also, concerns related to financial data and financial information may be seen as an obstacle when using this more beneficial and more powerful data handling possibility.

TABLE 10.4
Process Data Acquisition Systems

System	Sensor Data	Data Transmission	Standard
Specialized implement controllers	Implement, location, time, throughput (as applied)	None	No
Multipurpose process controllers	Implement, location, time, throughput (as applied)	Chip card, USB	No
Tractor terminals	Tractor, location, time, implement (if ECU) throughput (as applied)	USB, radio	No
ISOBUS task controllers	Tractor, location, time, implement (if ECU) throughput (as applied)	USB, radio	Partly following ISO XML
Controllers at self-propelled vehicles	Vehicle, location, time, throughput (gathered/as applied)	Paper print, USB, radio	No

10.3.1.1 Soil Mapping

Although they were not seen as part of precision farming in the past, very detailed soil maps were established and have been available on-farm for field-related measures for a long time. The resolution of these solely analog documents differs from region to region and from country to country, and they are increasingly being offered at no charge.

Today, soil mapping systems identify soil type and soil nutrients. The former are mainly detected on-the-go using electromagnetic or electroconductivity sensors such as EM38® or Veris®, whereas soil nutrient data are mainly gathered through soil sampling technologies and associated chemical analysis (Friedman, 2005; Ladoni et al., 2010; Sinfield et al., 2010). Often, environmental laws determine the time intervals of soil nutrient examinations.

Soil type and soil nutrient data are mainly used to establish field-specific homogeneous fertilizing strategies according to the base nutrients (once in a growing season) and nitrogen fertilization, either once or multiple times, in a growing season.

10.3.1.2 Yield Mapping

The yield monitor, used in combine harvesters, was one of the first widely adopted precision farming technologies (Schueller et al. 1985; Searcy et al., 1989; Reyns et al., 2002). In yield monitors, data acquisition is carried out with specific sensors and processors. Signal processing at the combine is performed in a company-specific way. Nearly all high-performance harvesters are equipped with yield monitors using different sensor types depending on the harvested crop. In grain harvesting technologies as well as in forage harvesters, moisture sensors are also state of the art (Table 10.5).

Data transfer to the farm management system (FMS) and mapping software is part of the yield monitoring system. Mapping with GIS software mostly differentiates yields into classes of one-metric ton and can be achieved by grid mapping or contour mapping (Figure 10.6).

TABLE 10.5
Yield Monitors in Harvesting Technologies

Harvester	Sensor Type	Moisture Sensor	Relative Accuracy (%)
Grain	Impact, light barrier	Capacitive	5–10
Cotton	Light barrier	–	5–10
Forage	Feed roller displacement	Capacitive and NIR	5–15
Root crops	Weigh cells	–	5–10
Sugarcane	Weigh cells, roller displacement	–	5–15

Source: Demmel, M. 2007. *Landtechnik*, 62 SH: 270–271; Vellidis, G. et al. 2003. *Applied Engineering in Agriculture*, 19:259–272; Ehlert, D. 2002. *Biosystems Engineering*, 83:47–53; Demmel, M. 2013. Site-specific recording of yields. In Heege, H.J. (Ed.), *Precision in Crop Farming*. Springer, Berlin, Germany, pp. 314–329; Molin, J.P. and L.A. Menegatti. 2004. Field-testing of a sugar cane yield monitor in Brazil. ASABE St. Joseph, MI, USA, Paper No. 041099. With permission.

Grid mapping simply puts all available yield measurement values, according to their position, into a grid. Grid sizes may be a single or multiples of the working width of the combine or may be related to the working width of the application technologies, again as a single or in multiples. Yield measurement values are averaged with their standard deviation to show the mean of a grid and the variation in it. Different colors represent yield zones with different colors in different yield monitor systems.

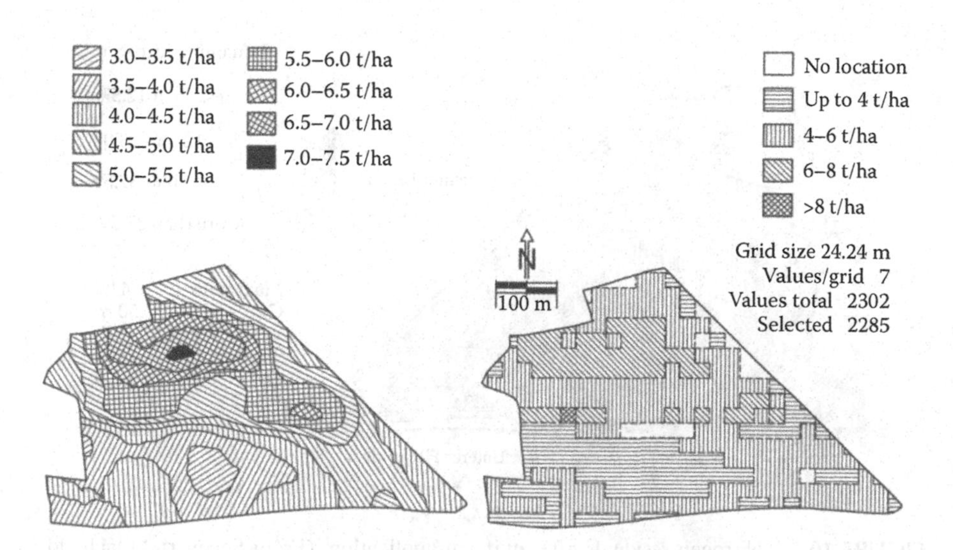

FIGURE 10.6 Yield maps in contour shape (left) and grid shape (right).

Contour mapping is based on geostatistical data processing operations and allows for very sophisticated yield differentiation. Research is ongoing to reduce or exclude measurement errors caused by neighborhood influences (up and back harvesting design) or by variations in the working velocities or interrupts in the work flow (Lyle et al., 2014).

Nevertheless, all mapping systems deal with unavoidable measurement errors caused at the field end by filling and emptying the material flow. Consequently, yield maps of small fields (short field length) show more faulty information compared to larger fields or plots.

In an overall assessment, it can be noticed that yield mapping is widely adopted at the farm level. First of all, it makes the main target of farming, the yield, and its variety, visible within the field and, in so doing, allows a more precise reaction for future growing seasons. As a first measurement procedure, it also provides existing knowledge and experience of the given yield performance of a field.

Taking those results into consideration, for example, a generated map of a field-based nitrogen balance, will then offer more "true information" as well as a better understanding of the findings in the map (Figure 10.7).

Regarding fertilizing, the often established component, classification may be improved by differentiation related to the "mean yield" of the field. A first map may be divided into two classes with yield above and yield below the average. Most fields may better fit into a system with three yield classes, for example, ±10% around the average and classes above and below. Finally, a system with five classes, first with ±10% around the average, another two classes with 20% above and below the average class, and a further two classes above and below, may represent the in-field yield variations in an operation-oriented way. In this way, farm-specific strategies based on control and accuracy of the available fertilization technology will be able to precisely apply the required amount of nutrients in a site-specific manner.

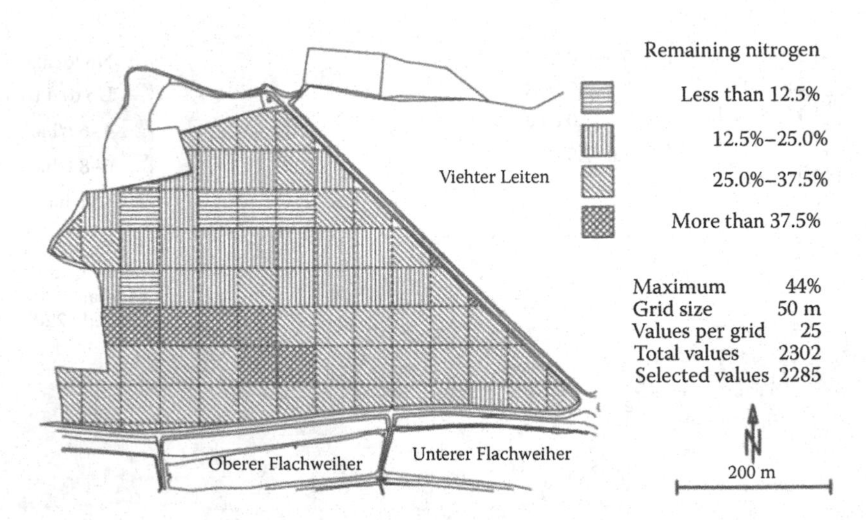

FIGURE 10.7 Nitrogen residuals after uniform application. (From Scheyern "Flachfeld," 1991. With permission.)

There are also many concerns against repeated yield monitoring:

- Yield maps generated from yield data of different manufacturers within a field or within a growing period may not be comparable as there is no standard in data harmonizing and data processing. Yield maps in this way are often colored pictures only.
- Repeated yield measurement is influenced by different weather conditions and by different crops in a crop rotation. Even with similar crop management procedures, it is very difficult to analyze and interpret the results. Also, crops for which there is no current practical yield measurement technology available do have an important influence.
- Finally, data handling in the long term is difficult, especially on smaller farms with no specialists in data storage and retrieval.

10.3.1.3 Weather Monitoring

Crop farming depends on climate conditions in regard to field measures or to produced yield of a field. In the same way, weather conditions at a farm may vary widely, especially in regard to wind speed, with implications for pest management and in rainfall, with an impact on fertilization and nutrient movements in the soil. Also, any operation scheduling depends on weather conditions and forecasts (O'Neal et al., 2004).

Larger farms as well as farms with a certain topographic differentiation of fields need more regional differentiated weather data obtained from their own weather stations. Standardized sensors with standardized signal processing algorithms allow the utilization of nearly all weather stations offered in the market. The data link can be either a wired interconnection or through radio communication to the FMS, where the first may cause problems due to lightning and thunderstorms, but offers independence through a parallel power supply.

Weather forecasting has improved during the last decades, by having the use of more powerful computers and more sophisticated models. The on-farm use of this information is often free of charge, while more detailed information needs specific contracts with suppliers or may be a free-of-cost amendment offered by other service providers or by leading agricultural machinery suppliers.

10.3.2 Administration

Farm management depends on or is increasingly influenced by laws and regulations formulated by legislation. Examples of restrictions in different parts of the world are

- *Nitrogen restrictions*, which allow a maximum amount per hectare documented through so-called farm-gate regimes. More diverse regulations already look to the field-gate and it might be expected that in the near future, together with improved site-specific technologies, the gate will come to the part-field.
- *Use of prohibited agents*, especially in pest protection scenarios.

- *Limited time intervals* in conjunction with applications of manure to frozen soil or during times of no plant growth, which keep lateral flow and ground water contamination in check.
- *Exclusion zones* of natural resources such as waterways, surfaces with an inclination higher than a given threshold and restricted areas of nonarable land.

In all these examples, precise data acquisition without gaps and verifiable documentation have to be guaranteed.

10.3.3 On-Farm Research

Plant treatments by fertilizing, pest management, and irrigation mainly follow common models and/or adviser-created recommendations. While the former suggestions are based on universal references, the latter may be more related to real farm conditions. Therefore, whenever local conditions should be integrated more intensively, on-farm research is crucial. Mainly focusing on fertilizing or pest management, different types of implementation can be chosen:

- *Untreated windows* are able to show the effect of any surrounding applications during fertilization or spraying.
- *Strips* with different application rates allow the evaluation of varying amounts or concentrations of agents.
- *Small test plots* with identical treatments and different varieties give genuine information about the performance of a particular variety.

All-in-all, on-farm research needs precise treatment as well as precise and specific data acquisition. The outcome contributes mainly to the farm management itself with the focus on increasing profit. It also contributes to more precise fieldwork in homogeneous treatments during fertilizing, pest management, and irrigation, avoiding over-applications as well as shortcomings resulting in reduced yields or in dangerous infestation.

10.3.4 Quality Management

Farms tend to grow and farm work often is transferred from family workers or well-trained farm workers to untrained laborers in full-time or part-time employment. Also, the transition of field work to contractors and/or machinery communities (joint ownership) is increasing, and in all these cases, monitoring will suffer.

More precise farm management, therefore, needs a well-defined work order and detailed data from any field activity, which might be included in existing field records or in quality management data pools (Kruize et al., 2013; Nawi et al., 2014).

10.3.4.1 Traceability and Good Agricultural Practices

When considering the farm management activities mentioned in this subchapter, it might become clear that high-quality data acquisition, together with application of different methods, which is how PA is defined, are fundamental requirements for more precise farming.

Data storage, in the same way, is the second challenge. Agricultural data will become very valuable over time. This information requires sophisticated data management systems and long-term data storage tools with adoption of new storage devices and enhanced formats. Special attention therefore should be given to the ownership and safety of data.

Large gaps, however, remain in the documentation storage scenario. These are mainly related to nonsensor-based or nonautomated data acquisition processes:

- *Worker identification* in common and especially in regard to hazardous agents or implements
- *Implements without electronics* in tractor–implement combinations following the ISOBUS
- *Agents* for seeding, planting, fertilizing, and plant protection
- *Yield sensors* in all of the used harvesting technologies
- *Soil stress and soil compaction* caused by field work under suboptimal conditions or created by improper tires or tire inflation pressure or by too high axle loads (Demmel et al., 2008; Hemmat and Adamchuk, 2008)

Consequently, the data where failures will occur are generally manually acquired, as humans are never perfect:

- *Forgotten* in times of heavy workload or at the end of a long working day
- *Wrong figures* either as wrong or unclear reception or willfully done to hide the right ones
- *No perception*, regarding certain items such as soil damage or soil crouching
- *Others*

Finally, thorough data records document attempts to perform farm work more precisely, to fulfill community laws and regulations, to achieve the required quality items in field operations, and to trace products back to the field and plot, if required (see Chapter 9 of this book).

10.4 CROP MANAGEMENT

Advanced agricultural technology, new sensors, data processing, and powerful software systems may be seen as the key elements of site-specific crop production and crop management (Auernhammer and Schueller, 1999; Schueller, 2002). These elements were first driven by profit maximization, mainly focusing on yield and fertilization; today, environmental issues gain higher importance (Bongiovanni and Lowenberg-Deboer, 2004). In this way, the ideas and possibilities of this concept are provided to conventional as well as organic farming systems, even though in the latter precision farming is still not the mainstream and many constraints can be observed.

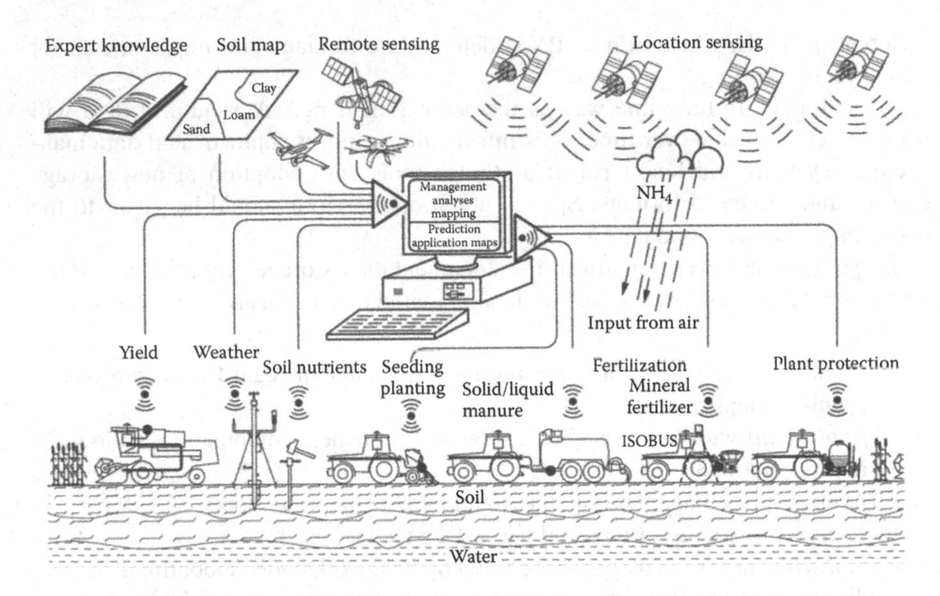

FIGURE 10.8 Site-specific crop management in precision agriculture. (Adapted from Sommer, C. and H.-H. Voßhenrich. 2004. *Managementsystem für den ortsspezifischen Pflanzenbau. Verbundprojekt pre agro*, Darmstadt, Germany, Chap. 4: 121–150, CD-ROM 43013, http://www.preagro.de/Veroeff/Liste.php3.)

Crop management covers the whole plant growing season with tillage at the beginning, and seeding or planting, followed by application measures of fertilization, plant protection and irrigation, and, finally, with the harvesting of grown plants (Figure 10.8).

However, the use of precision farming technologies today differs widely in crop management activities.

10.4.1 Tillage

Although tillage has not been long in the focus of precision farming applications, a number of utilities have been applied to tillage measures, and spatial variable tillage (by intensity and depth) has been investigated and discussed.

GNSS-based automated guidance of tractors has the strongest influence on the optimization of tillage. Its main goal is to avoid overlapping or gaps. Different investigations have shown that overlapping can be reduced by 5%–10%, and the relative figures increase with smaller working widths. Further, changed turning regimes—wide U-turns with skipping passes instead of swallow tail turns—will reduce turning times by one-third. This increases the field efficiency on short fields and with small working width (small-scale farming). The possibility of combining automatic steering with headland automation of tractors increases these effects and reduces the workload.

Some investigations have tried to evaluate the effects of site-specific primary tillage, varying the tillage depth according to soil type and soil moisture (Figure 10.9). While soil type does not change over time, soil moisture is a variable and, today,

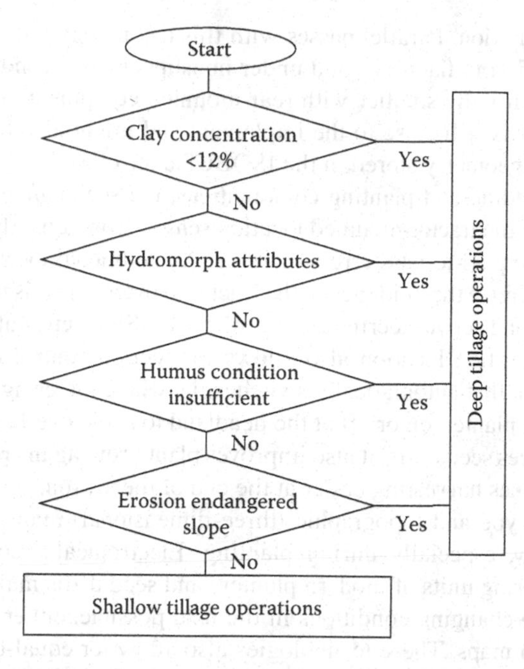

FIGURE 10.9 Algorithms of site-specific primary tillage. (Modified from Sommer, C. and H.-H. Voßhenrich. 2004. Soil cultivation and sowing. In KTBL (Ed.), *Managementsystem für den ortsspezifischen Pflanzenbau. Verbundprojekt pre agro*, Darmstadt, Germany, Chapter 4: 121–150, CD-ROM 43013, http://www.preagro.de/Veroeff/Liste.php3.)

hard to measure on-the-go. Therefore, the adoption of site-specific tillage first needs reliable soil moisture sensors.

The rise of controlled traffic farming, a farming strategy concentrating all field traffic on permanent tracks and for the first time discussed in the 1980s, was facilitated by minimum till and no-till technology and the availability of automatic guidance systems for agricultural machinery. Although it is an entire system, it will be mentioned here (Demmel et al., 2012a,b).

10.4.2 Seeding and Planting

From the very beginning, seeding and planting have been the processes with the highest requirement for precision. They constitute the fundamentals of a crop stand. Traditionally, markers have been used to accurately align passes while travelling the field. In advanced plant production systems, during seeding, track lines laid out by multiplying seeding implementation widths are established, matching the working width of the application technology. In manually guided seeding, those track lines show, in the average, overly narrow distances up to 8% of working width, and also show the higher overlapping caused by increasing slope.

Besides tillage, seeding and planting are increasingly being carried out with the assistance of GNSS guidance systems. This makes sense, with the increased working width of seeders and planters in both the rear-mounted and the trailed tractor implement configuration. Parallel passes with this technology will come to a precision of ±0.02–0.05 m in flat areas and under nonslippery soil conditions. Pass connection errors are usually smaller with rear-mounted equipment, as the position of the location sensor is still close to the implement and might also be reduced in the tractor implement geometry stored in the ISOBUS controller.

With trailed seeding and planting combinations, the seeding/planting tools are a long distance from the tractor-mounted location sensor. Consequently, under unfavorable conditions, the pass-to-pass errors increase. An additional location sensor at the implement can optimize the guidance of the tractor to overcome this problem. Another possibility is a second active steering system with a GNSS receiver at the implement.

GNSS-based position location also allows for section control of seed drills or planters, a function that automatically switches the whole metering unit, or sections or single rows of a planter, on or off at the headland to avoid overlapping. This function not only reduces seed costs, it also improves plant growing in critical areas at the headlands and makes harvesting easier at the end of the season.

Based on soil type and topographic (three-dimensional) maps, farmers tend to vary plant density, especially during planting. Electronically controlled electric drives of the metering units of modern planters and seed drills make the adaptation of the seed rate to changing conditions in the field possible, either manually by the driver or based on maps. These technologies also allow for equal-distance planting in areas with a triangular or rectangular geography. In the future, weeding and even pest control may simply be taken over by small autonomous field robots working across the field or following weed spots. Even single plant husbandry would become possible in this planting design.

10.4.3 Application in Fertilization

Following local yield measurements in combine harvesters and georeferenced soil sampling, site-specific fertilization was a very quickly adopted implementation (Auernhammer et al., 1999). From a systematic point of view, there are three different approaches to master this new challenge and possibility (Auernhammer et al., 1999), as shown in Figure 10.10.

10.4.3.1 Fertilizing by Balance

Based on local yield measurement from the previous harvest combined with soil nutrient sampling and analyses at the beginning of the vegetation, a highly reliable estimation of needed nutrients related to a nutrient balance can be achieved. It may be called "farming by balance" or, nowadays, "prescription farming." Additionally, long-term information from the historic data of a field can be taken into estimation. But whatever decisions are taken, this concept mainly focuses on "one-treatment only" applications, as any changes within the growing season cannot, or can only be included with low reliability, into the final determination. This approach is dealing with basic nutrients (P, K, Ca, etc.), one application of nitrogen only, the choice of the mostly beneficial crop variety, and the adjacent needed/preferred chemicals. Fertilizing operations then are based on application maps and appropriate application implements.

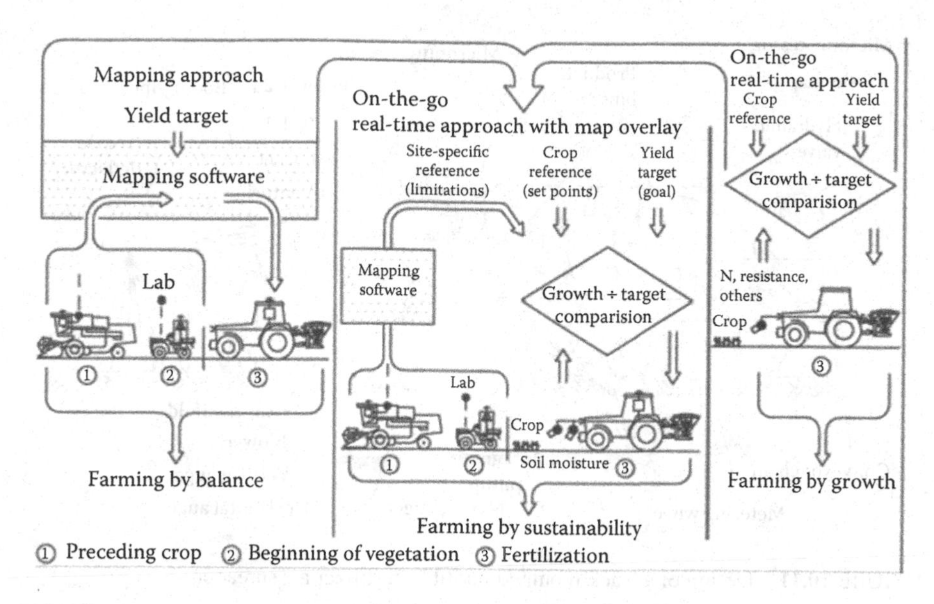

FIGURE 10.10 Theoretical approaches of site-specific fertilization.

Application maps are typically grid-based. The grid size is limited by either the working width of the spreader (it makes no sense if narrower) or by the section width when section control is available. Application maps may follow definitions in an ISOBUS system or might be in a proprietary format related to the used spreader controller. Application implements may be either spin spreaders or air spreaders, while spin spreaders usually have no section control units or a maximum of two section control units, and can deliver only one fertilizer type, either a single or a mixed nutrient agent, according to the required nutrient application. Control of the required distribution amount strictly follows the map and causes more rapid rather than smooth adjustments. Highly precise dosing is possible when the controller takes care of the time of flight in relation to the distributed material, adjusting the required amount of output prior to the map-based boundary. Nonuniform driving speeds cause less precision.

Air spreaders support section control but still follow the above-stated principles in distribution. Additionally, there is an extra influence from curved application tracks. This curve may be integrated into the control algorithms if path planning information is available to the application map. Otherwise, under- or over-application at the boom ends is unavoidable. Moreover, on-the-go nutrient mixing is an option available through a multibin design of the spreader (Peisl, 1993), shown in Figure 10.11.

10.4.3.2 Fertilizing by Growth

Whenever multiple applications deliver benefits, for example, avoiding overfertilization and allowing a reaction to unforeseen weather conditions, the growth factor has to be included in application management. This is the case in nitrogen fertilizing and especially under humid conditions with unpredictable rainfall. Under these conditions, the needed amount of fertilizer at a certain time results from the difference between the aspired growth target and the on-site growth situation.

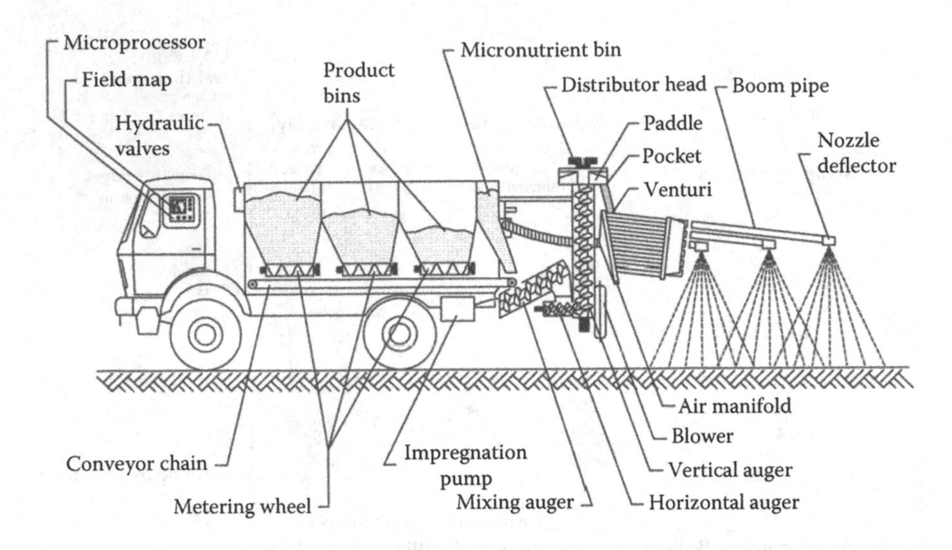

FIGURE 10.11 Design of a truck-mounted multibin fertilizer air spreader.

It is well known that the greenness of crops is an indication of the amount of chlorophyll and it is also well known that there is a very close correlation between chlorophyll content and nitrogen uptake, as chlorophyll mainly consists of nitrogen. Therefore, a greenness or chlorophyll sensor could give the needed information to determine the required amount of nitrogen derived from a standardized growth development curve of a specific variety with respect to the expected yield.

This type of precision fertilization is well adopted (Table 10.6), especially in Europe, with split nitrogen fertilizing strategies in three to four applications.

There are more than 1000 systems in operation in Europe (as of 2014) with an average field capacity per system of around 4,000 ha/year. The standard control procedure adds more nitrogen to those parts of the fields with lower biomass; this is often changed to the opposite for the last dressing. The majority of the systems take

TABLE 10.6
Crop Growth Sensors in Nitrogen Fertilizing Systems

Sensor Type	Principle	Vegetation Index	Wavelength	Algorithm	Fertilizing System
Yara N-Sensor	Passive	S1, S2	Confidential	Yes	Online
Yara ALS	Active	S1	730, 760	Yes	Online
CropCircle 430	Active	NDVI, WDVI	630, 730, 780	No	Online
GreenSeeker	Active	NDVI	656, 774	No	Online
Isaria	Active	IRMI	Confidential	Yes	Online with map overlay

Source: Adapted from Maidl, F.-X., A. Spicker, and K.-J. Hülsbergen. 2014. *LfL-Schriftenreihe Heft 7, Neue Techniken im Ackerbau,* Hrsg.: G. Wendl. Bayerische Landesanstalt für Landwirtschaft, Freising, Germany, pp. 63–74 (ISSN 1611–4159).

control of rear-mounted spin spreaders through proprietary data link interfaces. The sensors are typically mounted at the cabin roof of the tractor in a fixed position through the growing season with simplified cabling, and normally there is no location sensor.

A more detailed assessment of the aims using those online systems delivers four different strategies (Leithold, 2014):

- Yield maximization in more than 50% of use with the risk and the acceptance of overfertilization and also with extended fertilizer costs.
- Quality optimization in about 30% improving the protein content.
- Crop harmonization in about 10% to get higher harvesting performance.
- Manual overcontrol in about 5% when soil properties are known or visible on-the-go.

Today, manufacturer-specific control algorithms cover the main cereal crops, including maize and rapeseed, and long-term investigations into their use with potatoes and other root crops have also started.

Wireless data transmission, integration into the ISOBUS standard, and the use of the ISOBUS user terminals (UT), follow the efforts using standardized electronic communication systems and allow a simple sensor movement to other tractor–spreader combinations. Location sensing can be used with the ISOBUS to allow the generation of "as-applied maps" via the task controller (TC).

10.4.3.3 Fertilizing by Sustainability

Compared to manually controlled nitrogen application, any growth sensor replaces the "eye of a farmer" only where all his field experience and knowledge is sidelined. In a sustainable system, therefore, the "farmer's brain" must be included into the set-point definitions together with long-term spatial field data such as soil type, topography, local yields, soil resistance, rain fall, and others. This requires a sensor with a map-overlay system, which is able to make an adjustment of the growth sensor signals either given as a function of statistically well-confirmed dependencies or by well-established agronomical rules of interactions (Figure 10.12).

The final set point of the required local application can be derived through sensor fusion. When integrated into the ISOBUS standard, this technology could widely be used in mineral fertilizing as well as in organic fertilizing, pest management, and in seeding and planting (Ostermeier, 2013).

10.4.3.4 Organic Fertilization

Besides mineral fertilizing where farmers always try to make the most precise application, organic fertilization remains in the background. This might be acceptable in manure spreading as this type of organic fertilizer may be seen as a soil agent improving the organic matter in the soil and contributing to the stabilizing of humus. But in conjunction with slurry, known as a rapid effective nitrogen fertilizer, it is inadmissible. Besides, the indispensable requirement of highly precise pass-to-pass operation using two different solutions can be seen at this time:

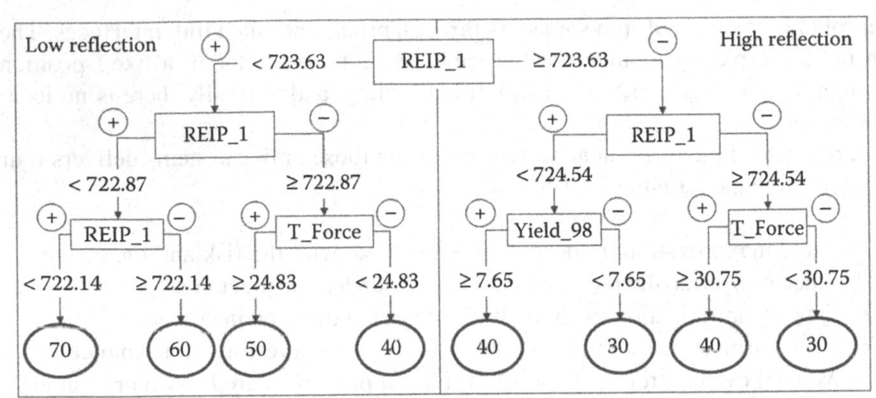

FIGURE 10.12 Decision tree formulating site-specific nitrogen requirement through data mining. (From Weigert, G. 2006. Data Mining und Wissensentdeckung im Precision Farming-Entwicklung von ökonomisch optimierten Entscheidungsregeln zur kleinräumigen Stickstoff-Ausbringung. Dissertation: Technische Universität München, Professur für Unternehmensforschung und Informationsmanagement, Freising-Weihenstephan, Germany, http://mediatum.ub.tum.de/?id=603736. With permission.)

- Uniform or even site-specific spreading of manure may be based on wagon platforms with weight sensors together with a variable hydraulic drive to feed the manure to the spreading unit according to the planned application rate. Either homogeneous or site-specific application may be addressed in this way.
- In slurry application, on-the-go measurement of the flow rate and the nitrogen content is essential. NIR sensors show good results (Reeves III and Van Kessel, 2000) and allow precise nitrogen application in the field, as is followed in the management of mineral nitrogen fertilizer. Also, P and K contents may be introduced into a more precise slurry application.

Again, also in organic fertilizing, location sensing is a must to be able to create "as-applied" maps for long-term improved field management.

10.4.3.5 All-in-All

Fertilization, especially site-specific nitrogen fertilization, can be seen as the most advanced precision farming technology at the field level today. Depending on given situations and related to nutrient requirements, "well predicted one application only" or "growth adjusted multiple" operations can respond precisely in a site-specific manner to avoid overfertilization with leaching, as well as to avoid underapplication, which causes yield losses. Nevertheless, some problems at the farm level can still be seen:

- Prediction of fertilizer amounts often follows simple balance attempts where historical field data are neglected, or due to data handling, are not available.

- In multiple application strategies, growth sensors offer a high potential through real-time crop state measurement but are still restricted to cereals. Application algorithms cover selected plant varieties only and in most cases a calibration at the field is required.
- Application implements are able to guarantee highly precise dosing with section control. Precise pass-to-pass operation prevents overlapping, but in nonlinear field structures, under and over supply takes place when no pass planning information or true look-ahead direction sensor is part of the system.
- Application control based on application maps normally generates stepwise set-point adjustments with no smooth and therefore no natural transitions.
- Sensor fusion with historic field data in systems with growth sensors are still an exception.
- Data communication in tractor–spreader combinations often is proprietary and forces the farmer to stay with the established system, having no choice to replace either the tractor or the spreader without new problems and open questions.

10.4.4 Application in Plant Protection

Contrary to fertilization with a wide range of specific strategies to rate the necessary amount of nutrients in relation to an established yield target, non-GMO plant protection measures can hardly be determined in advance. Treatments regularly take charge when monitoring shows an infestation greater than a defined threshold, or when spot-wise critical expansions of weeds, insects, or fungi are observed. In other words, plant protection either means monitoring in the first stage, or neglecting the given situation and following well-established recommendations with homogeneous whole-field treatments.

10.4.4.1 Monitoring

When farm and field sizes grow larger, thorough and accurate monitoring requires an exponentially increasing time when no dedicated aids or tools are available. Thus, comprehensive research activities have been carried out worldwide to close this gap, by looking for sensors and on-the-go weed detection methods on tractor–sprayer combinations or on autonomous vehicles or unmanned aerial vehicles (UAVs) (Thorp and Tian, 2004; Sankaran et al., 2010; Zhang and Kovacs, 2012; Pérez-Ruiz et al., 2015). At the field level, however, few of these solutions can be found.

Besides governmental organized or extension-based plant monitoring activities covering all major crops and all important plant infestations, on-farm monitoring with sensors in combination with the sprayer is starting to be adopted:

- Tractor mounted or spray boom-located NIR-based growth sensors are used to detect local biomass and apply more agents in dense crop standings and in the opposite way during whole-field treatments.
- Also, tractor-front-mounted plant-density sensors are used to react in a similar way.

10.4.4.2 Application

Owing to the development of powerful herbicides, fungicides, and insecticides, chemical plant protection has played and continues to play a key role in the development of modern agricultural plant production. Application technology has followed this trend. Its performance was increased by wider spraying booms, larger solution tanks, and faster working speeds. Application rates have been dramatically reduced, increasing the requirements on a precise allocation. Electronics became the most appropriate technology and today they are integrated in nearly all sprayers in "developed" plant production regions.

Electric three-way or by-pass valves regulate the liquid flow in a closed-loop control system according to the required spray pressure with regard to nozzle type and application rate. Radar velocity sensors or speed data from GNSS receivers provide slip-free true ground speed information. Position data from GNSS receivers also allow for electronic section control based on application maps. Boom control requires more attention with wider spraying booms in rough and hilly fields as well as when working at higher speeds. In this case, high-performance sprayers include distance sensors in the boom sections to facilitate active hydraulic or electrical boom distance control, and active boom suspension.

With trailed sprayers and nonlinear field structures, sprayer wheels might destroy additional plants when no action is taken to guide the sprayer within the tractor path. Precision sprayer track-guidance measures tractor steering and actuates with adjusted guidance activities in the sprayer drawbar. But nevertheless, even with sprayer track control, curved passes result in under- and over-application at both boom ends if the pass direction is not integrated into the control algorithms and/or driving speed is too high in relation to control time delay and nozzle adjustment time. An example is shown in Luck et al. (2011), where during spraying with a single nozzle control in irregular field shapes of a total of 185 ha, nearly 20 ha received less than 90% and nearly 13 ha more than 110% of the target rate.

And finally, as described above, NIR-based single nozzle control can become a part of the site-specific treatment that controls the output volume in accordance with the locally sensed crop biomass. Or the sensor signal can be used to create local overcontrol in accordance with predefined site-specific application set points.

10.4.4.3 Mechanical Weeding

In cereals, nearly half of the chemicals are used for weed control. From an environment-friendly standpoint, most of the pesticides could be replaced by mechanical weeding. On the one hand, vision-based implement guidance is available and mainly used in row crops or in cereals with wide row distances; typically, organic farmers use this technology more often than the conventional ones. On the other hand, GNSS-based RTK implement guidance is available. Both technologies allow for interrow weeding whereas in-row weeding still suffers.

10.4.4.4 All-in-All

In general, all of these electronic and sensor-based control prospects offer highest precision during application of chemicals at the field level. But in the future, these treatments have to be improved by more localized information. Ongoing development of in-field monitoring therefore is the key to environmentally sound plant protection measures with the following challenges:

- Any manual monitoring is time consuming and requires very special knowledge.
- Adapted sensor technologies related to a fast and clear identification of weeds, fungi, and insects are still not available.
- Autonomous air-based or land-based platforms for such sensors are not available or have problems in use at the farm level.
- Also, monitoring systems with cognitive capabilities to identify and maintain spots with different propagations in the right way may not be available today.

Further improvements in spray technology are required, such as

- Spray booms with a steady adjustment to the crop surface with no lateral and horizontal deviations even at higher driving speeds, on rough or hilly surfaces and also under nonlinear field conditions.
- Direct injection systems (Peisl et al., 1992) with infestation-specific treatment.
- Integration into the monitoring equipment with site-specific or plant-specific handling of microagents and newly developed physical or electrical treatment possibilities.

And finally, more attention should be given to mechanical weeding in newly designed field cropping systems (Demmel et al., 1999) together with autonomous vehicles in interrow, in-row, and also at single plant surroundings (Slaughter et al., 2008).

When taking all these different aspects into account, it might be concluded that, contrary to precision fertilization, future environmentally-friendly, sound, precise, site-specific or even plant-specific, plant protection is a long way off. And also, GMO crop farming in monocultures with specialized monoculture agents will not be able to overcome foreseeable problems.

10.4.5 Application in Irrigation

While just 15% of all arable land in the world is irrigated, this land generates nearly half of the value of all crops sold (UNESCO, 2007a). Worldwide agriculture (and horticulture) accounts for over 85% of water consumption (UNESCO, 2007b). It is expected that irrigated areas and water consumption will increase by 20% by 2025.

Irrigation systems can be divided into gravity-based (flooding furrows or entire fields) and pressure systems (sprinkler systems and drip irrigation). Worldwide, about 94% of irrigated land is under gravity irrigation. In the United States, about 50% of the irrigated land is under gravity irrigation; the other 50% is irrigated by pressure systems. Owing to the fact that agricultural irrigation accounts for the largest part of water consumption worldwide, different attempts have been made to increase water efficiency (Demmel et al., 2014). Based on the development and adoption of precision farming technologies in crop farming, ideas of precision irrigation and site-specific irrigation have also been discussed and investigated. Unfortunately, their application is limited to pressure irrigation systems. Two steps or levels can be distinguished, improved and automated irrigation control and site-specific irrigation.

10.4.5.1 Irrigation Monitoring and Control

In many cases, irrigation is controlled manually based on the experience of the farmer. Often, the applied amount of water is controlled only by the time the equipment (pump and sprinklers) is running. In a first attempt, the installation and use of a water meter significantly improves the accuracy and efficiency of irrigation. Furthermore, systems have been developed, investigated, and evaluated that increase water efficiency by using simulation models based on the Penman–Monteith equation (extended by local weather data from small electronic on-farm weather stations) or soil moisture sensors (Noborio, 2001), or a combination of both (Figure 10.13).

The development of fast-reacting, low-maintenance, and low-cost soil moisture sensors is still in process due to the fact that it is difficult to combine all three features. Besides, wireless networks are needed to cover the whole irrigation area with soil moisture sensors. In larger center-pivot or linear-move irrigation systems, these models are combined with ECUs for the pump, the valves, and the automation drive.

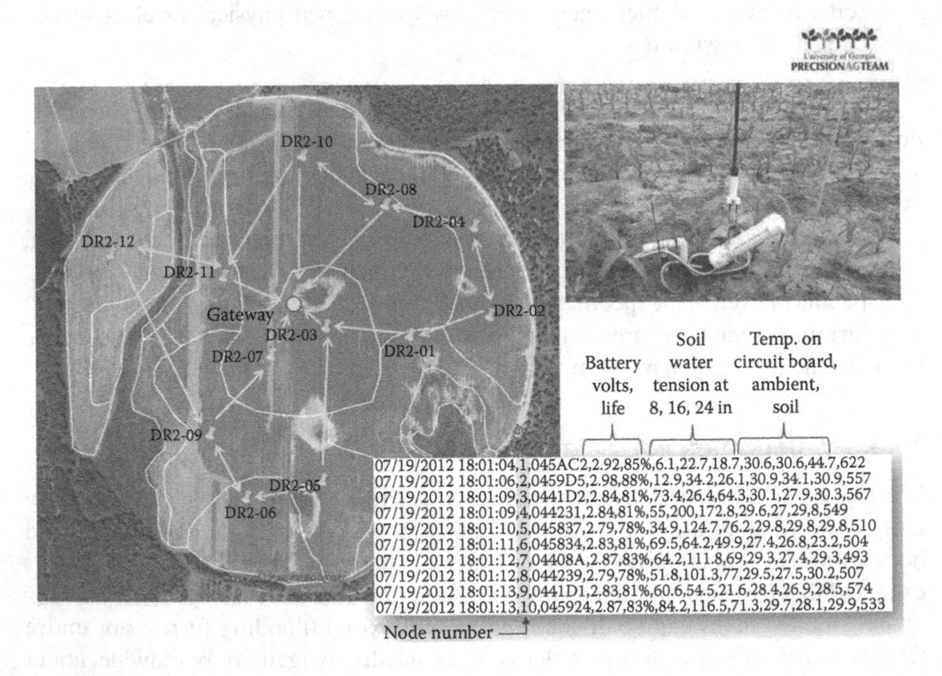

FIGURE 10.13 (**See color insert.**) Irrigation soil sensor network. (From Vellidis, G. 2015. Irrigation sensor network. Personal communication on teaching material, January 14, 2015. With permission.)

These control units include comprehensive monitoring and visualization functions that give the user instant information about the system, and the applied amount of water on the farm PC, or a smart phone, and also send an alarm if a malfunction is detected. The automation of center-pivot and large linear-move irrigation systems is characterized by continuous and intensive development.

10.4.5.2 Site-Specific Irrigation

Owing to the fact that soil heterogeneity also influences soil water balance and creates the need for irrigation, the observation of high soil heterogeneity within a center-pivot circle with a diameter of 400–500 m has been carried out and systems for spatial variable irrigation have been developed and investigated (Evans et al., 2013). This technology, called "site-specific variable rate irrigation" (SS-VRI) has been commercially available for center-pivots for several years, but its adoption by producers has been on a very low level. It is expected that higher costs for irrigation water, water scarcity, and the implementation of economic incentives for compliance with environmental or other regulations, will potentially provide the necessary incentives for much greater adoption of various advanced irrigation technologies.

10.4.6 Harvesting

All actions the farmer takes during the growing season are aimed at contributing to high yields, attaining the required quality, and delivering the outcome with marginal losses at the right time and at the lowest cost. To increase efficiency and to reduce costs, harvesting technology is increasing in size, tends to be very specific to the harvested crop, and becomes highly complex. To realize this progress in the field, there must be either higher qualified operators or more electronics and well-designed control capabilities with a high degree of automation. The increase in size and performance, and extended control and automation can best be implemented with self-propelled machinery concepts.

10.4.6.1 Guidance

Large harvesters enable highest performance when driven within the genuine working width or very precisely along given rows to avoid cutting losses or overlapping. Operators are challenged by this the whole day and sometimes even during the night, with no decrease in concentration.

Especially in row crops such as corn, silage maize, or sugar beet, mechanical row guidance sensors became standard equipment and are used more than 80% of the operation time. More recently, with the increasing header width of self-propelled combine harvesters, edge detectors were adopted to guide these machines along the edge of the standing crop with accuracy better than 10 cm. Similarly to row sensors, these sensors are also part of today's harvesting technology and need no calibration and no additional infrastructure in the field (Figure 10.14).

Today, high-accuracy GNSS-based guidance systems (RTK) are used on harvesting equipment as well. These guidance controls are not influenced by laying crops and also allow for skipping passes with reduced turning times and less soil compaction at headlands.

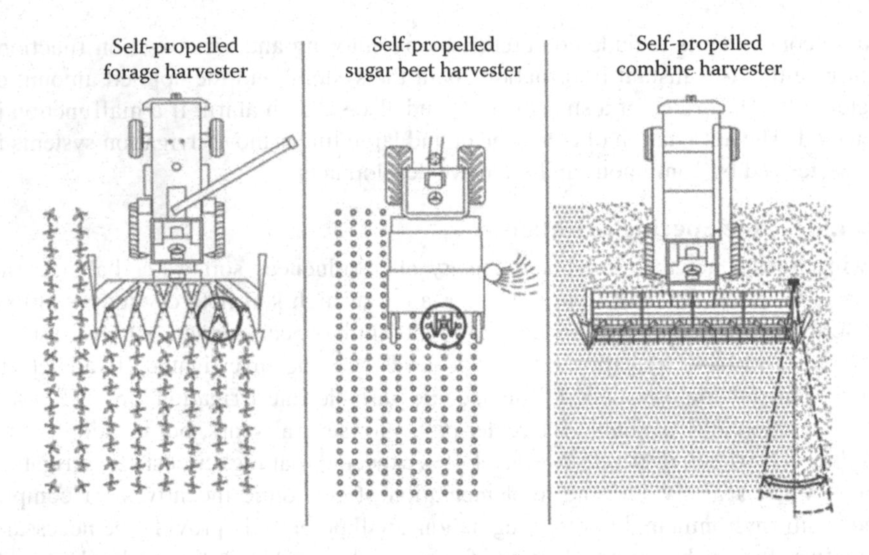

FIGURE 10.14 Row and edge guidance sensors in self-propelled harvesting machinery.

More precise work recently became possible by utilizing guidance systems in transport vehicles in silage maize harvest operations (Europe) and with grain carts unloaded on-the-go. Precise parallel speed adjustments of both vehicles in this case avoid material loss. Together with additional optical sensors measuring the filling situation an adjustment of the loading device (spout) always allows perfect loading of the transport unit.

10.4.6.2 Operation Control

Large-capacity harvesting machinery allows maximum performance and precise work with low losses only when internal process units are well adjusted (Schueller et al., 1986) and in harmony with their neighbors. Preprogrammed adjustment tools, mainly in combine harvesters, enable manually activated employment of the harvester related to the coming crop. Site-specific control improvements can then be stored over a period of time and used for future operations. More recently, these field-specific control adjustments have also been transferred in real time to combines from the same manufacturer and of the same type (members of a group or swarm in a leader–follower concept) through wireless communication.

Besides guidance systems that give more freedom to the driver and allow more time for machinery observation and control, constant material throughput with less variation is realized by so-called cruise control systems optimizing all separation and cleaning processes. This advancement can be seen as an important and helpful supplementary feature, either following a control strategy to get maximum throughput under time restrictions, to optimize the output quality, or to minimize losses.

10.4.6.3 All-in-All

Never before has such huge and mainly self-propelled harvesting machinery done as comparable a job as it has done today. A wide range of different sensors, high-performance control algorithms, and fast and finely adjustable actuators allow for this, and will also be able to improve further:

- In-machine unit optimization relating to previously collected site-specific preinformation from parallel tracks. Site-specific information may even come from spatial soil data or from inclination maps (Bishop and McBratney, 2002). This information may be used to optimize cleaning at the sieves in combine harvesters or to empower soil separation from root crops more intensively or more smoothly.
- Besides locally diverse variations in crop quality and even with an increasing working width, the quality of the harvested crops may differ more widely. Therefore, because of product as well as of cost reasons, selective harvesting and on-the-go division of low- and high-quality composition such as protein or starch would be an option.
- When harvesting under wet conditions, soil compaction is a major issue in environment protection. Optimized tires or rubber belt undercarriages are a first option.
- However, one limitation may be that overcollection of material in the hopper or tank in relation to an on-the-go operation measured by soil moisture value may be seen.
- Also, the continuing growth of harvesting technology might be questioned. Unmanned followers—all in smaller size—are able to increase the overall performance of an operator and may even be easily adapted to different operating conditions such as field size, field shape, topography, and others, but will still be under manual control.

10.5 MACHINERY MANAGEMENT

Cost-efficient utilization of farm machinery requires time-critical logistics to guarantee that any needed equipment with an appropriate performance will be at the right location at the right time with no breakdown. While this requirement has lower importance on family farms with widely seen overcapacities, it is a "knock-out criterion" in large-scale farming as well as in a contractor work organization.

10.5.1 Route Planning

Specialized very high-priced machinery tends to be increasingly used in cooperation as well as in so-called machinery rings or in most cases worldwide, by contractors. But independent from the organization, the main target is "to fulfill the required task in the right time," which is very difficult in areas with identical or very similar conditions.

A good example of this may be "planting of sugar beet," as planting has to be done as early as possible to achieve a long growing period. Also, beet will gain highest yields only in the best soils, which are typically found in a narrow area. Furthermore, there is no point of ripeness, which means beets should be in the soil as long as possible. From a farmer's business point of view, beet should grow as long as there is no frost, but from the sugar mill point of view, the request is to start as early as possible with processing to arrive at a long processing time with lowest costs per sugar unit. In other words, in a certain region, planting should be completed on one day and consequently all operations during the growing season will fit the same requirement, whereas any delay in any operation will lead to a decrease in profit and/or quality.

In similar operations, only very detailed route planning in agreement with all the involved farmers, including different pricing, will be successful (Figure 10.15).

Any mission depends on

- Available georeferenced field data with additional field metadata according to the required process or, if "unknown," specific GNNS-based field inventory and recording of field metadata. In both cases, the field traffic situation with road condition and restrictions is additionally required.
- Job schedule or flow diagram based on farmer's time preference, field sizes, in-field conditions, field-to-field distances, road conditions, working time per day, and many others.
- Single or multiple job execution, including either groups of similar machinery or the machine executing the leading task together with required transport units and associated facilities.

Several systems are available and used for optimization with highest possible precision in two different ways:

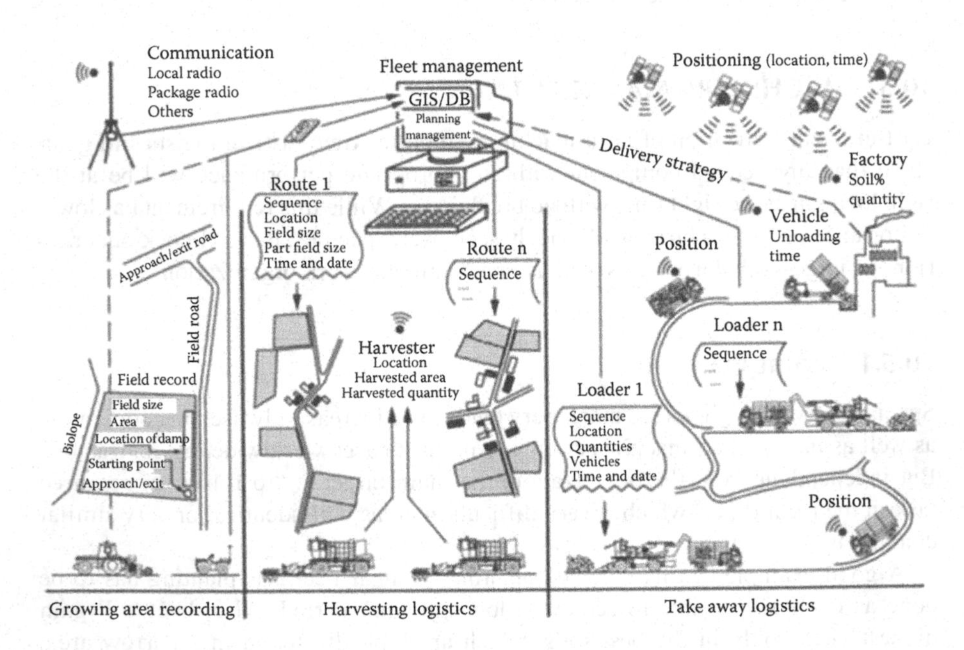

FIGURE 10.15 Route planning in sugar beet production.

- Operator-driven systems (contractor, cooperation, and large-scale farm) focusing on the optimized utilization of the available equipment to fulfill all required field tasks in a limited time with as little as possible idle time and minimized timeliness costs.
- Processing industry driven to keep the plant running, for example, in sugar beet, sugarcane, vegetables, starch potatoes, and others.

Also, route planning systems for combinable crops are of great importance following the maturing of small grains on strictly planned routes from south to north first and repeating this way for corn harvesting later in the northern hemisphere or vice versa in the southern one. Great attention has to be given to the end of the period as, often, weather conditions become insecure and fast changing.

10.5.2 Process Monitoring

Existing sensors in the machinery with GNSS receivers and telecommunication equipment together with GIS and on-screen software allow for centralized or decentralized real-time monitoring of any mechanized field work. Thus, at any time, machinery settings as well as the work progress can be observed. With well-adapted actions it will be possible to

- Confirm the instantaneous work situation being in a good fit with its machinery performance to the required field task and with the workflow in comparison with the previous established work schedule.
- Discover possible/necessary improvements in the machinery settings to gain more work progress and possibly compensate for a certain work flow delay.
- Be aware of an earlier fulfillment of the ongoing field operation with consequences to the work schedule of the day, the worker, or the engaged machinery.
- Detect and document any misuse or wrong use by time and location especially when untrained or less trustworthy workers are engaged.

Related to transportation tasks on rough field roads and/or heavy loaded public roads, the monitoring of multiple transportation units mainly during harvest of a huge amount of biomass (e.g., silage, sugarcane, and sugar beet) provide benefits as

- Transport capacity in number of units can be easily adapted to the required task.
- Unexpected traffic situations can be compensated for by altered navigation or to extended transportation units.
- Any idle time of the harvesting unit(s) may be eliminated to fulfill the field task within the scheduled time.

In summary, process monitoring of machinery used in a single task, or associated to clusters of similar machinery, or in combination with transport units, gains benefit through reduced idle time in the whole process as well as in any of the connected units. Also, unexpected situations can be easily seen, action can be immediately taken, and the total work can be undertaken in conformity with the work target and work schedule.

10.5.3 System Control

While route planning and process monitoring focus on "what should be done" and "what happens," system control focuses on "how it should be done." System control mainly concentrates on harvesting with a (large) fleet of harvesters, grain carts in the field, and a fleet of transport units. The most critical challenges occur while changing from one field or one harvest area to another. While the harvest has to be finished in a way so as not to run the whole fleet to an abrupt stop, the movement of the equipment has to be initiated and the harvesting of new fields has to be prepared and started in a way that the nonproductive time of all the equipment is minimized.

Still, this type of fleet management rests upon well-experienced human skills, but increasingly fundamental research and simulation is on-the-go and will soon come to the field level.

10.5.4 Remote Service

This section looks mainly at the machinery itself. Machinery manufacturers in particular show a very high interest in any solutions to problems or feedback with regard to their products. This can be achieved, first, by gathering field-level data according to a certain technology for further improvement and development, and second, by providing maintenance proactive services and repair, to avoid breakdown and idle time. The system approach is based on permanent or intermittent machinery data acquisition and transmission to a centralized database with access for manufacturers, dealers, service providers, and also contractors and/or farmers (Figure 10.16).

A system such as that described above is required by advanced engineering at the manufacturer's level as well as by service providers, as comprehensive and realistic field use data have not been available to

- Discover weak structure or design points in machines
- Provide the settings under different conditions
- Monitor different loads at specific in-machinery locations
- Analyze running time and load spectra
- Create more advanced maintenance information material and service instructions
- Develop improved and field-level designed replacement technology

On the other hand, farmers have some constraints when it comes to offering their process data to manufacturers, gaining no individual benefit and also considering data security concerns. Even when this system seems to be rewarding to all engaged parties, it is very difficult to bring it to the farm level with the required broad acceptance. There may be two steps to going forward.

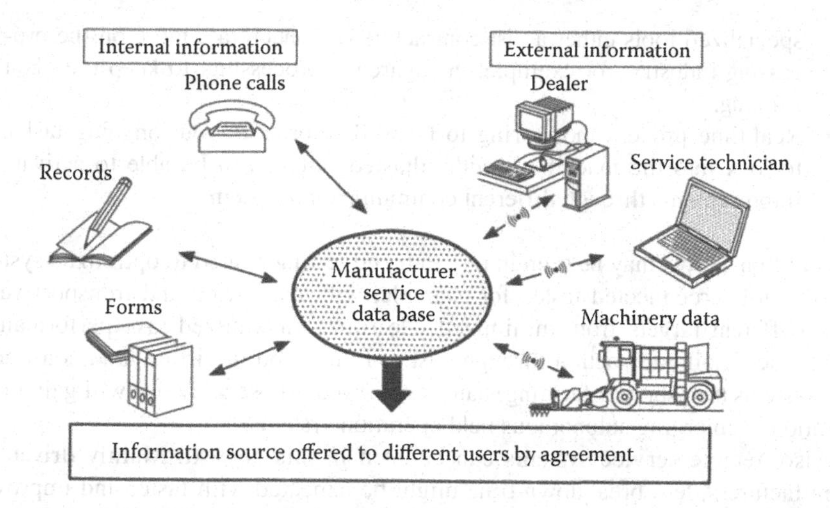

FIGURE 10.16 Remote service system with manufacturer service database. (Largely modified from Ahrends, O. 2003. Das Serviceformular—Ganzheitliche Umsetzung im Unternehmen. In: Eine gesamtheitliche Teleservice-Lösung für die Landmaschinenbranche, Kapitel 4: 39–53, CD-ROM, ISBN 3-00-011832-2. With permission.)

10.5.4.1 Service Hiring

If users are utilizing the machinery permanently at its maximum performance and are sure to have immediate service and/or repair after a breakdown, they may go into a special service lease contract. In this case, data are part of the contract and have to be transferred to the database without any restrictions. The disposition of the data will also be defined in the contract.

10.5.4.2 Machinery Hiring

Contractors may hire machinery mainly to get the most advanced technology and to avoid unexpected breakdowns through aging or wearing out. In the fulfillment of a contracted guaranteed operational readiness, data become part of the contract and can be obtained by the manufacturer or service provider to the extent and density required.

10.5.4.3 All-in-All

Only optimized machinery management guarantees the fulfillment of any task within time limits and with the required precision. As farmers predominantly tend to avoid risk, they often have overcapacities together with long years of experience in organization and utilization of the owned machinery. In contrast, contractors aim to get maximum performance out of their technology and normally operate with fleets of machines or integrated machinery systems. Besides long-term contracted clients, new clients with no historic information have to be served. Well-engineered management tools are able to support, at the field level:

- Route planning with included optimization algorithms to fulfill any required task at the right place and right time with the required performance. More

specialized tools either assist contractors or concentrate more on the processing industry, for example, in sugarcane processing, to keep the plant running.
- Real-time process monitoring to be well informed about ongoing tasks, to look into the machinery with adjusted settings, to be able to activate improvements through different communication systems.

Additional tools may be seen in research and testing related to optimized system control in interconnected tasks, for example, with harvesting and transport vehicles. Different targets from minimized idle times, maximized task performance, and reduced soil compaction through cost minimization in virtual land consolidation systems (transborder farming) can be observed. Those activities will gain more attention in emerging autonomous field operation systems.

Also, remote service systems can be seen in this context. Mainly driven by manufacturers, less breakdown time might be expected with faster and improved service and repair. Furthermore, field-level experience will lead to improved machinery design to better adjust next-generation developments to the needs of the farmer.

Machinery management was often not identified as an important section within precision farming. But as farm sizes, machinery sizes, and machinery performance grow, requirements in this area are increasing.

10.6 LABOR MANAGEMENT

Whatever is accomplished or will be accomplished in precision farming is related to labor, which means human beings. But human beings have different abilities, education, and knowledge; they differ in age; they have different motivations; and whatever work they are doing, in time they get tired. On the other hand, precision farming means doing everything more precisely, even with the differences in human beings and even when human beings get tired. Precise work therefore calls for improved tools and more automation.

10.6.1 TOOLS AND IMPLEMENTS

Worldwide, farms of less than 1 ha account for 72% and larger farms between 1 and 2 ha account for 12% of all agricultural operations. Together, these 84% of all farms control only 12% of all agricultural land (Big Picture Agriculture, 2014) and cultivate the land mainly by hand and/or using animals. As in highly mechanized regions with highly sophisticated implements for precise work, better adapted tools could lead to more precise work under these conditions as well. The following examples may focus on tasks for plant production:

- A wheel-guided plow allows for precise working depth in comparison to plow draw bars carried on animal shoulders.
- Seeding with wheel-driven metering devices guarantees uniformly distributed seeds compared to hand spread seeding.

- Fertilizing with any equipment driven by a wheel with material-adjusted dosing facilities minimizes fertilizer input with uniform distribution.
- Plant protection with hand-operated backpack sprayers will apply chemicals more precisely than hand-powdering.
- Finger knife mowers cut laterally more precisely than reaping hooks.
- Rotating mechanical threshing devices achieve better grain separation than flails.

Besides more precise work, all of these examples offer higher performance and often reduce failures, losses, and workload.

10.6.2 Guidance Systems

Modern agriculture is dominated by self-propelled machines such as tractors with implements, combine harvesters, and sprayers. All of them allow higher speeds, increased working width, and higher performance with more engine power. But they all have to be guided by drivers, often through long days as well as at night, in dust, fog, and often in very slippery soil surface conditions. So, any type of automation in guidance is very welcome to lighten the load for human beings and to allow a constant output of steering accuracy through a full working day. Available systems used on the farm level depend on the working situation (Table 10.7).

10.6.2.1 Mechanical Systems

Furrow guidance evolved first. Besides unsuccessful work with plows to increase working speed, the main focus was given to tractor rear-mounted interrow cultivators

TABLE 10.7
Guidance Systems

Sensing Type	System	Machinery	Acceptance at Field Level
Mechanical sensors	Furrow guidance	Plow Interrow cultivator	No
	Row guidance	Maize choppers Sugar beet harvesters	More than 80%
Optical sensors	Camera row guidance	Tractor Interrow cultivator	Less, mainly in special crops and in horticulture
	Laser edge guidance	Combine harvester	More than 80% of large harvesters
GNSS systems	Parallel swathing	Tractor–implement combination	Add-on
	Steering assistance	Tractor–implement combination Self-propelled harvesters	Add-on
	Auto-steering	Tractor–implement combination Self-propelled harvesters	High
	Auto-steering and headland management	Tractor with mounted (front, rear) or trailed implements	Low

to dispense with the cultivator guide. During planting, a v-shaped plate or a mole-blade established, in a simple way, the guidance furrow; a guidance plate at the inter-row cultivator followed and did a good job if there was no disturbance of the guiding furrow by rain, soil movement, etc. As no high-quality guidance was possible at all and upcoming tool carriers with interaxle-mounted cultivators could be used by the driver alone, furrow guidance systems never gained acceptance at the field level.

While in-field established furrows never guarantee robust conditions, strong-stem plants such as maize can do this. So, with a well-adjusted sensor, application guidance of harvesters along the rows became possible and offered freedom to the driver for either more supervision of the technique, for improving the work quality or to allow more accurate control of the material delivery into a parallel transport unit. A very high acceptance of this guidance system describes the benefit of both, the precise work with the higher performance of the harvester and the optimized loading of transport units.

10.6.2.2 Optical Systems

Camera-based systems still suffer as highly precise guidance is difficult in creating a robust centerline from one or more parallel crop rows independent from wind disturbances in higher crops, by differences in the growth habitat, or by deficiencies in one or all the rows. Furthermore, changing illumination caused by the altitude of the sun as well as shadows, dust, and fog may create problems.

Laser-based edge guidance in combine harvesters is well accepted on the field level. The main reasons from a labor point of view may be seen in fast and simple oversteering in special situations or in laying crops, and in more freedom for accurate control of the increasing complexity and increase in size of modern combine harvesters.

10.6.2.3 Satellite Systems

Today, GNNS-based guidance systems show a fast rising adoption rate due to their high accuracy independent from field conditions, time of day, and other influences. Besides direct integration into new machinery, many retrofitting systems are available and well accepted mainly in tractors. In-field usage is focused on parallel tracking following an A–B line in a linear or nonlinear shape. The overall usage in the field can be seen in three different types:

- Driver-assisted use refers to parallel tracking along linear A–B lines whereas unshaped areas as well as headlands are manually guided.
- Auto-steering means parallel tracking all over the field but not at the headlands.
- Extended auto-steering integrates headland management systems. Until now it is restricted to U-turns but can include down (up) shift, hitch, and hydraulic functions and power-take-off (PTO) engagement, and disengagement.

10.6.2.4 Overall Assessment

Today, the utilization of guidance systems is the most applied precision farming technology worldwide. Also, it might be expected that this trend will continue, very soon covering all larger tractors and all self-propelled agricultural machinery. Also, the trend to GNNS-based systems will continue.

From a labor point of view, all available systems offer secure handling with simple actuation and manual oversteering. Huge benefits may be seen in a formidable reduction of workload and a corresponding increase in comfort.

But headland management systems in small- and medium-scale farming systems with tractor–implement combinations are still challenging. More simplified systems should be used whenever possible to reduce the risk of accidents and damage when turning. However, manual management might be welcome in the field to break the monotony of auto-steering in long fields with less variations and low monitor and control requests.

10.6.3 Leader–Follower Systems

Auto-steering opens the door to driverless vehicles, at least inside fields, if a certain kind of observation and fast time intervention is possible. In a first and simple solution, this can be realized in a system where a manned leading vehicle is either followed by or following an unmanned one (Figure 10.17).

There may be two different solutions:

- The follower is of the same type. Through wireless communication, it takes over all settings from the leader. It may also be monitored in the reverse way by the leading driver who would also be able to immediately stop the follower in the event something went wrong.
- The follower is of a different type with a different duty such as a chaser bin working together with the combine and the transport unit. In this case, it is simple to call the follower to the best position, overloading in parallel pass. Later, the follower may go autonomously to the storage unit to unload and

FIGURE 10.17 Visionary leader–follower system for consecutive tillage and seeding where the leader carries out the most challenging task.

be waiting for the next call. Again, the leading driver takes care of the follower, but as distances will be larger, unexpected disturbances may occur with unforeseeable situations.

In comparison, a leader–follower system of identical type might be simpler, but has not been realized so far as

- Moving vehicles must have a driver on-board by the Vienna Agreement Law from 1948.
- Secure communication links do not exist (spies, hackers, etc.).
- Typical agricultural machinery is of high mass and of high power, so any uncontrollable situations would be hazardous especially when close to public roads or urban locations.
- Instead of less trained drivers on auto-steered vehicles, well-trained high-cost specialists would be required.

10.6.4 Field Robots

Leader–follower systems would be able to reduce the labor requirement at the field level by about 50% for one follower and more for multiple followers. The use of field robots could reduce the labor requirement to 10%–20% of the nonrobotic situation depending on the required service and field monitoring tasks and the number of robots. But nevertheless, field robots are moving vehicles and may not be allowed by law. Even if laws and regulations change, two more questions will arise:

- Who wants to take over the overall responsibility of running large and massive autonomous vehicles at the field level with more or less no human attendants?
- What might be the most appropriate target when only small autonomous vehicles are the solution for the mechanization of tomorrow?

From this point of view, a very clear answer can only focus on the circumstances related to the required operations in different fields (Tables 10.8 and 10.9).

Assuming that all of the most important pros and cons are considered, a threefold answer or prospect may be derived:

- There is no expectation to use large and massive autonomous field robots in the near future mainly owing to safety and responsibility reasons.
- Small autonomous field robots do not fit the high power requirements and also cannot carry large amounts.
- Small autonomous field robots equipped with specialized sensor applications or with very specific application tools requiring only easy-to-transport enhancements may be the first accepted solution at the field level.

10.6.4.1 All-in-All

Labor management is one of the most important activities in any business, including agriculture. Besides reduction in time consumption and in workload, more precise

TABLE 10.8
Large Autonomous Field Robots with Pros and Cons to the Required Tasks

Task	Pros	Cons
Tillage	Increased working time 24 h/day Immediately preparing field conditions to directly following seeding or planting No timeliness costs and minimized weather influences	Potential hazardous incidents Reduced site-specific interactions Relatively high soil compaction Large headland for trouble-free turnings Possible high timeliness costs through rainfall-stopped prepreparation
Seeding and planting	Increased working time 24 h/day	Potential hazardous incidents Reduced site-specific interactions Auto-refilling is challenging
Application of fertilizers and chemicals	Extended working time related to best wind and humidity conditions No exposure of operator to chemical agents	Potential hazardous incidents by vehicle or by agents Reduced site-specific interactions Auto-refilling is challenging
Harvesting	Extended working time related to crop and soil conditions	Potential hazardous incidents Reduced site-specific interactions in laying crops Auto-unload is challenging

TABLE 10.9
Small Autonomous Field Robots with Pros and Cons to the Required Tasks

Task	Pros	Cons
Tillage	Reduced soil compaction Most site-specific adaption in variable tillage effects Minimized headland need Minimized timeliness costs by prepreparation	Required power increases mass of the vehicle Less overall capacity Management and maintenance of "herds" asks for highly qualified people
Seeding and planting	Enlarged working time 24 h/day Minimized headland need Less payload required No auto-refilling in small fields required	Less overall capacity Management and maintenance of "herds" asks for highly qualified people "Herd supply" in refilling is challenging
Application of fertilizers and chemicals	Enlarged working time 24 h/day Minimized headland need No auto-refilling in small fields required	Less overall capacity Management and maintenance of "herds" asks for highly qualified people "Herd supply" in refilling is challenging
Harvesting	Extended working time related to crop and soil conditions Minimized headland need Possible selective harvesting in low-mass row crops	Less overall capacity Management and maintenance of "herds" asks for highly qualified people Auto-unload is challenging

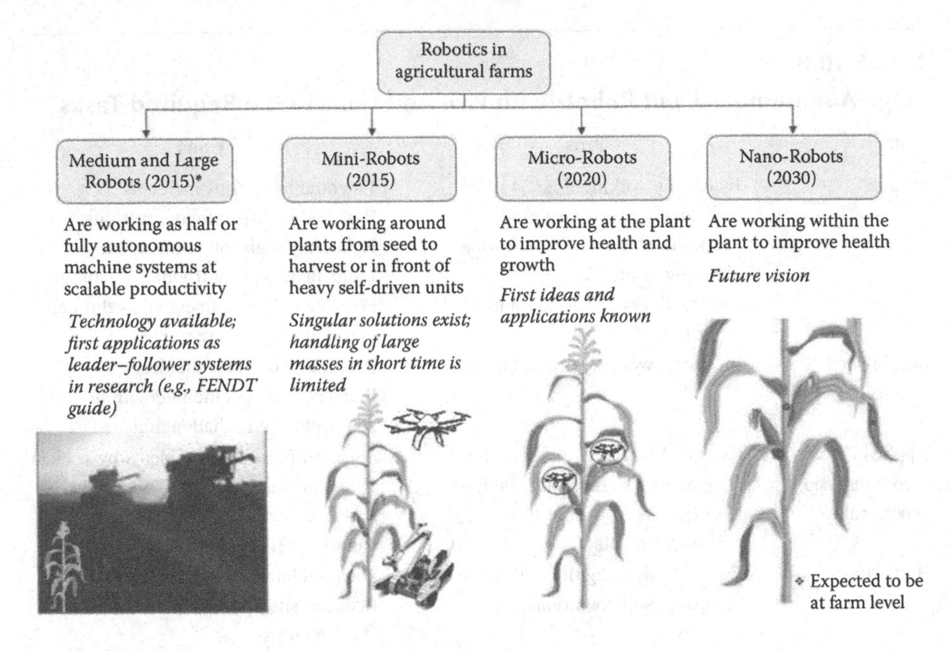

FIGURE 10.18 Expected field-robot development and usage in agriculture. (Adapted from Auernhammer, H. 2011. Twenty years of precision agriculture—More questions than answers? In *ACPA2011, The 4th Asian Conference on Precision Agriculture*, July 4–7, 2011, Tokachi Plaza, Obihiro, Hokkaido, Japan, CD-ROM, Keynote No. 4: 1–6.)

task fulfillment at the field level is challenging for sustainable land use in future. In this way:

- Auto-guidance in whatever shape is the most promising technology in labor management and will increase in numbers with nearly no restrictions.
- An increase in productivity may be seen in the use of leader–follower systems enabling agricultural equipment of smaller size.
- Small autonomous field robots may first come to the field level in field scouting, monitoring, and in conducting very specific treatments at the plant level.

In the future, miniaturization will continue and bring field robots to flowering plants and finally to larger-stem-sized plants (Figure 10.18).

10.7 FUTURE REQUIREMENTS

After more than 20 years of precision farming, the utilization at the farm level is still small compared to the huge potential in today's information-based land use (Auernhammer, 2011). While larger farms and farms with partial or major usage of contractor work progress, smaller and mid-sized farms are having problems in the switchover from familiar mechanical machinery to electronics-controlled equipment owing to the need for additional investments and specific knowledge, plus a certain mistrust of data acquisition, data storage, information management, and communication (Tey and Brindal, 2012).

Another reason may be in the transition of precision farming to the farm level. From both the research and the advisory angle, fertilization in site-specific farming only gained dominance with the expectation of making everything simpler, easier, and more profitable. As this happened when basic nutrients were addressed, it opened more questions than answers in nitrogen application measures. In this case, no answer or sometimes too many answers were available with regard to the best nutrition management:

- How should the most beneficial yield maps be established?
- How should management zones be defined?
- Should more nitrogen be used at the more fertile zones or should the opposite occur?
- Should on-the-go growth sensors and straightforward online control of the spreader follow integrated algorithms only?
- Do yield map-based measures influence plant protection?
- What is the overall value of year-by-year yield mapping?

There are many other open questions with no reliable, unreliable, or more philosophical answers. In other words, farmers, and in particular pioneering farmers, still stand alone and lose interest, motivation, and enthusiasm.

10.7.1 Big Data Challenge

Any type of site-specific farming means "information-driven farming." Farming in this context means data generation, data exploration, data modeling, and, finally, data-based operation control (Nash et al., 2009). In other words, precision farming approaches the big data challenge, and earlier precision farming models as seen in Figure 10.8 together with the above-defined situation "farmer left alone" approach the model of tomorrow (Figure 10.19).

Data of field operations will soon go to the cloud. Web services will then need to ensure that required data explorations or data modeling always use the newest, most advanced, and well-tested scientific algorithms. Farm management itself will evolve to the use of apps with twofold aims:

- Any financial data will still reside at the farm management system remaining as safe and limited-accessibility information.
- Field operation measures derived from big data will go in a timely manner from the cloud to in-field precision farming technology, and in parallel for safe documentation to the farm management system.

But all this will only become a reality in the future if big data in the cloud may be stored and handled in a highly secure way with farm-given access permission and with contractual and financial agreements to any data user.

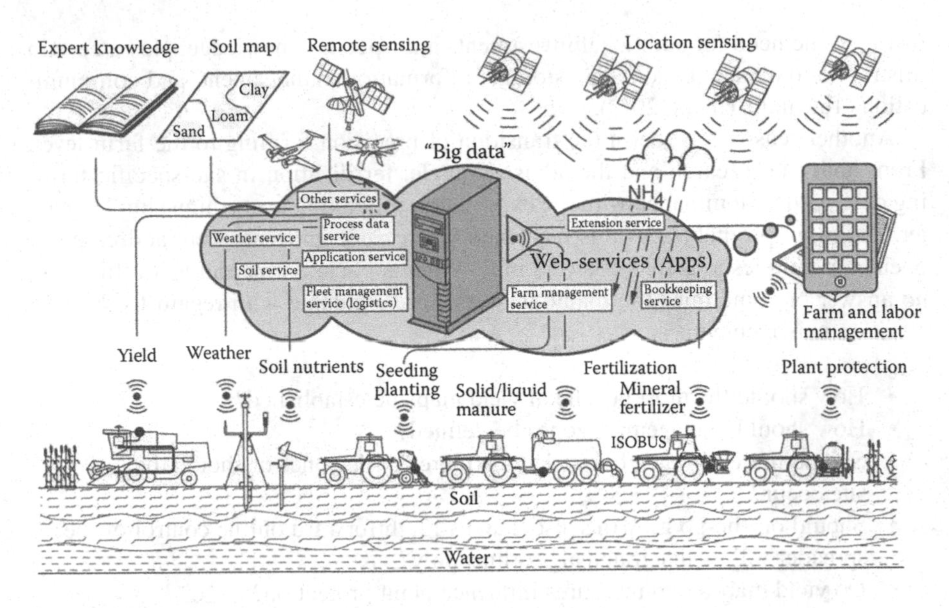

FIGURE 10.19 Cloud-based big data in precision cloud farming.

10.7.2 Models

In today's precision farming practice, existing models are mainly used, evaluated, and approved under homogeneous conditions in conventional farming systems. In the future, there is a big need for well-adapted new models related to the inhomogeneities of different farming types, given farm situations, farm-related rules and laws, and other factors (Table 10.10).

All these models may also be more detailed and subdivided into very specific farming systems. Feedback of any information on human overcontrol during operations or other influences or unexpected occasions into the cloud is essential. This allows for learning processes with more and better adjustments to the real world.

10.7.3 Automation

While site-specific farming suffers at small- and mid-sized farms in contrast to the overwhelming acceptance of guidance systems also in these farm types, farmers demonstrate a preference for more and higher automation. In future, auto-steering will be an internal application in tractors and self-propelled machines, and automation will integrate the whole tractor–implement combination. Control will move from the tractor to the implement in tractor implement management (TIM), as implements do the job, know the best conditions, and tell the tractor to react in a well-adapted manner. This will become possible if

- Communication between tractor and implement follows well-accepted standards open to any manufacturer, and specifically to small- and medium-sized producers. Safety control is a must; in this case to clearly avoid inappropriate breakdowns and also to allow dedicated maintenance and repair.

TABLE 10.10
Models in a Big Data Environment

Model	Related to	Main Influences
Nutrient balance	Application in only one operation	Yield map, soil map, soil nutrient map, cultivated variety, expected rainfall, etc.
Nutrient optimization	Multiapplication in multiple operations	Yield map, soil map, soil nutrient map, cultivated variety, expected rainfall, economic evaluation, etc.
Irrigation	Dry land farming	Soil type, inclination, cultivated variety, crop status, weather data, etc.
Soil fertility	Humus	Crop rotation, soil type, climate parameters, topography, tillage regime, working depth and working intensity, humus enrichment target, etc.
Soil compaction	Large-scale machinery usage	Type of machinery, soil moisture, tires and tire inflation, path planning, yields, yield moisture, etc.
Task quality	Contractor work and nonfamily workers	Task order, task documentation, settings, on-the-go measured task parameters, worker identification, etc.
Field size optimization	Transborder farming	Crop rotation, soil map, soil nutrient map, topography, yield targets, roads and logistics, etc.
Organic farming	Organic farming	Crop rotation, humus content, infiltration, distance to conventional field plots, etc.
Hillside farming	Upland farming	Topography, plot length, tillage system, crop rotation, residual coverage
Consumer-focused farming	Good agricultural practice	Data availability and compatibility along the food chain

- Communication between tractor and implement relays in the response of a so-called full-liner, where both machines come from the same manufacturer. Safety and responsibility in this solution will be clearly defined.

Tractor and implement(s) may then be seen as an integrated self-propelled unit. Adjusted settings given from implements to the tractor will be stop and go, left–right steering, forward speed adoption, PTO-revolution adjustment, hitch position changing, and flow control at auxiliary hydraulic valves.

Automation may also spread across field machinery, including cloud settings into the loop either related to a task operation or in real time during job execution. Automation in this case will evolve to the Internet of agricultural utilities.

10.7.4 Sustainability

As precision farming in whatever shape, type, and application is used at the farm level, it is always interacting with land and nature. But more often, it is primarily used to increase profit, accepting negative or even harmful and long-lasting impacts on nature, such as overfertilization with increased leaching, wasting of scarcely available water, and reducing number of crops in the rotation; or, far worse, going into monoculture crops.

For tomorrow, whatever term will be used instead of the well-known term of today, precision farming should be more than profit driven; it should move to sustainable agriculture in the right manner. Therefore, besides the dominating economic prospects and expectations, the ecological and the social implications should also be integrated targets:

- First of all, site-specific data in detail as well as from history should be part of any model and field operation.
- Major attention should also be given to crop rotation, including cover crops and their positive effects on soil fertility, plant growth and reduction, and suppression of weeds, plant diseases, and insects.
- Under more arid climate conditions, irrigation should focus on minimum consumption of water together with the most adapted plant varieties, whereas under more humid conditions, the reduction of soil compaction and soil erosion should be the dominant target.
- In all plant protection systems, the use of chemicals should be reduced or even eliminated, creating systems with physical weed control and with spatial or single plant-related highly specified plant protection measures or organic agents.
- Finally, most crop production will go in some way or another to the consumer. Traceability creates trust in the produce and allows secure seek-backs when problems occur.

With all these facets, it is clear that PA means more than site-specific application. PA means sustainable agriculture with no equivalent alternative in either conventional or in organic farming systems.

REFERENCES

Ahrends, O. 2003. Das Serviceformular—Ganzheitliche Umsetzung im Unternehmen. In: Eine gesamtheitliche Teleservice-Lösung für die Landmaschinenbranche, Kapitel 4: 39–53, CD-ROM (ISBN 3-00-011832-2).

Auernhammer, H. 1989. The German Standard for Electronical Tractor Implement Data Communication. In Sagaspe, J.P. and A. Villeger (Eds.), *AGROTIQUE 89, Proceedings of the Second International Conference*, September 26–28, 1989, Bordeaux, France, pp. 395–402 (ISBN 2-87717-012-8).

Auernhammer, H., Ed. 1994. GPS in agriculture. *Computers and Electronics in Agriculture*, 11, No. 1, Special Issue (ISSN 0168-1699).

Auernhammer, H. 1999. Precision farming for the site-specific fertilization. *Zeitschrift für Agrarinformatik*, 3:58–67.

Auernhammer, H. 2011. Twenty years of precision agriculture—More questions than answers? In *ACPA2011, The 4th Asian Conference on Precision Agriculture*, July 4–7, 2011, Tokachi Plaza, Obihiro, Hokkaido, Japan, CD-ROM, Keynote No. 4: 1–6.

Auernhammer, H., M. Demmel, F.X. Maidl, U. Schmidhalter, T. Schneider, and P. Wagner. 1999. An on-farm communication system for precision farming with nitrogen real-time application. ASAE St. Joseph, MI, USA, Paper No. 991150 (http://mediatum.ub.tum.de/?ID=1238205).

Auernhammer, H. and J. Frisch, Eds. 1993. Landwirtschaftliches BUS-System LBS (Agricultural BUS-System LBS). *KTBL-Schriftenvertrieb im Landwirtschaftsverlag*

GmbH, Münster-Hiltrup, Germany, Arbeitspapier 196 (ISBN 3-7843-1841-X, http://mediatum.ub.tum.de/?id=683691).

Auernhammer, H. and T. Muhr. 1991. GPS in a basic rule for environment protection in agriculture. In *Proceedings of the 1991 Symposium "Automated Agriculture for the 21st Century"*, December 16–17, 1991, Chicago, Illinois, USA, ASAE St. Joseph, MI, USA, pp. 394–402 (ISBN 0-92935-521-0).

Auernhammer, H. and J.K. Schueller. 1999. Precision agriculture. In *CIGR-Handbook of Agricultural Engineering. Vol. III: Plant Production Engineering*. ASAE St. Joseph, MI, USA, pp. 598–616.

Big Picture Agriculture. 2014. Statistics on Global Family Farm Size—Shares of Farms, By Land Size Class. Available at http://www.bigpictureagriculture.com/category/farm-statistics, accessed on January 27, 2015.

Bishop, T.F.A. and A.B. McBratney. 2002. Creating field extent digital elevation models for precision agriculture. *Precision Agriculture*, 3:37–46.

Bongiovanni, R. and J. Lowenberg-Deboer. 2004. Precision agriculture and sustainability. *Precision Agriculture*, 5:359–387.

BOSCH. 1987. CAN—Controller Area Network. Stuttgart, Germany.

Daberkow, S.G. and W.D. McBride. 2003. Farm and operator characteristics affecting the awareness and adoption of precision agriculture technologies in the US. *Precision Agriculture*, 4:163–177.

Demmel, M. 2007. Mass flow and yield measurements in harvesting machines. In *Landtechnik*, 62 SH: 270–271.

Demmel, M. 2013. Site-specific recording of yields. In Heege, H.J. (Ed.), *Precision in Crop Farming*. Springer, Berlin, Germany, pp. 314–329.

Demmel, M., H. Auernhammer, G. Kormann, and M. Peterreins. 1999. First results of investigations with narrow row equal space planting of corn for silage. ASAE St. Joseph, MI, USA, Paper No. 997051 (http://mediatum.ub.tum.de/?ID=1238206).

Demmel, M., R. Brandhuber, and R. Geischeder. 2008. Effects of heavy agricultural machines for sugar beet harvesting on soil physical properties. ASABE St. Joseph, MI, USA, Paper No. 084719.

Demmel, M., R. Brandhuber, and H. Kirchmeier. 2012a. Strip Tillage for corn and sugar beet—results of a three year investigation on three locations. *CIGR/AgEng Conference*, July 6–10, 2012, Barcelona, Spain, Paper No. 1990.

Demmel M., R. Brandhuber, H. Kirchmeier, M. Mueller, and M. Marx. 2012b. Controlled traffic farming in Germany—Technical and organizational realization and first results. *CIGR/AgEng Conference*, July 6–10, 2012, Barcelona, Spain, Paper No. 1987.

Demmel, M., S. Kupke, R. Brandhuber, B. Blumental, M. Marx, A. Kellermann, and M. Mueller. 2014. Drip irrigation for potatoes in rain fed agriculture—Evaluation of drip tape/drip line position and irrigation control strategies. *AgEng Conference*, July 6–10, 2014, Zurich, Switzerland, Paper No. C0174.

DIN 9684/2-5 1997. *Landmaschinen und Traktoren—Schnittstellen zur Signalübertragung*. Beuth Verlag, Berlin.

dlz Spezial 1990. Elektronik für Profis (A publication of the research and development program "Introduction of Electronics into Farm Field Work"). Landwirtschafts Verlag, München, Germany (ISSN 0340-787X, http://mediatum.ub.tum.de/?id=683796).

Ehlert, D. 2002. Advanced throughput measurement in forage harvesters. *Biosystems Engineering*, 83:47–53.

Evans, R.G., J. LaRue, K.C. Stone, and B.C. King. 2013. Adoption of site-specific variable rate sprinkler irrigation. Publication from USDA-ARS/UNL Faculty. Paper No. 1245 (http://digitalcommons.unl.edu/usdaarsfacpub/1245).

Friedman, S.P. 2005. Soil properties influencing apparent electrical conductivity: A review. *Computers and Electronics in Agriculture*, 46:45–70.

Gorsek, E.J. 1983. Grain sensor using a piezoelectric element. United States Patent Application, US 4401909 A.

Göhlich, H. 1978. Pflanzenschutztechnik 1978. *Landtechnik*, 33:310–312.

Heiser, J. and W. Kobald. 1979. Einrichtung zur Hubwerksregelung. German Patent Application, DE 2731164 C2.

Hemmat, A. and V.I. Adamchuk. 2008. Sensor systems for measuring soil compaction: Review and analysis. *Computers and Electronics in Agriculture*, 63:89–103.

International Organization for Standardization. 2009. ISO 11783 Tractors and machinery for agriculture and forestry—Serial control and communications data network, Parts 1–14, Geneva, Switzerland.

Kruize, J.W., R.M. Robbemond, H. Scholten, J. Wolfert, and A.J.M. Beulens. 2013. Improving arable farm enterprise integration—Review of existing technologies and practices from a farmer's perspective. *Computers and Electronics in Agriculture*, 96:75–89.

Ladoni, M., H.A. Bahrami, S.K. Alavipanah, and A.A. Norouzi. 2010. Estimating soil organic carbon from soil reflectance: A review. *Precision Agriculture*, 11:82–99.

Leithold, P. 2014. Use of YARA-N-Sensors. Personal information, September 19, 2014.

Luck, J.D., S.K. Pitla, R.S. Zandonadi, M.P. Sama, and S.A. Shearer. 2011. Estimating off-rate pesticide application errors resulting from agricultural sprayer turning movements. *Precision Agriculture*, 12:534–545.

Lyle, G., B.A. Bryan, and B. Ostendorf. 2014. Post-processing methods to eliminate erroneous grain yield measurements: Review and directions for future development. *Precision Agriculture*, 15:377–402.

Maidl, F.-X., A. Spicker, and K.-J. Hülsbergen. 2014. Mit Sensoren Bestände besser führen? In *LfL-Schriftenreihe Heft 7, Neue Techniken im Ackerbau,* Hrsg.: G. Wendl. Bayerische Landesanstalt für Landwirtschaft, Freising, Germany, pp. 63–74 (ISSN 1611–4159).

Molin, J.P. and L.A. Menegatti. 2004. Field-testing of a sugar cane yield monitor in Brazil. ASABE St. Joseph, MI, USA, Paper No. 041099.

Nash, E., P. Korduan, and R. Bill. 2009. Applications of open geospatial web services precision agriculture: A review. *Precision Agriculture*, 10:546–560.

Nawi, N.M., G. Chen, and T. Jensen, 2014. In-field measurement and sampling technologies for monitoring quality in the sugarcane industry: A review. *Precision Agriculture*, 15:684–703.

Noborio, K. 2001. Measurement of soil water content and electrical conductivity by time domain reflectometry: A review. *Computers and Electronics in Agriculture*, 31:213–237.

O'Neal, M., J.R. Frankenberger, D.R. Ess, and J.M. Lowenberg-Deboer. 2004. Profitability of on-farm precipitation data for nitrogen management based on crop simulation. *Precision Agriculture*, 5:153–178.

Ostermeier, R. 2013. Data Fusion in einem mobilen landtechnischen BUS-System für die Realtime Prozessführung in sensorgestützten Düngesystemen). Dissertation: Technische Universität München, Lehrstuhl für Agrarsystemtechnik, Freising-Weihenstephan, Germany (http://mediatum.ub.tum.de/?ID=1113617).

Peisl, S. 1993. Technische Entwicklung und verfahrenstechnische Einordnung eines Gerätes zur mobilen Herstellung von Mineraldüngermischungen mit variablen Nährstoffanteilen (Mehrkammerdüngerstreuer). Dissertation: Technische Universität München, Institut für Landtechnik, Freising-Weihenstephan, Germany (ISBN 3-7843-1908-4, http://mediatum.ub.tum.de/?ID=984249).

Peisl, S., M. Estler, and H. Auernhammer. 1992. Direkteinspeisung von Pflanzenschutzmitteln—Ein systemvergleich. *PSP*, 4:24–27.

Pérez-Ruiz, M., P. Gonzalez-de-Santos, A. Ribeiro, C. Fernandez-Quintanilla, A. Peruzzi, M. Vieri, S. Tomic, and J. Agüera. 2015. Highlights and preliminary results for autonomous crop protection. *Computers and Electronics in Agriculture*, 110:150–161.

Reeves III, J.B. and J.S. Van Kessel. 2000. Near-infrared spectroscopic determination of carbon, total nitrogen, and ammonium-N in dairy manure. *Journal of Dairy Science*, 83:1829–1836.

Reyns, P., B. Missotten, H. Ramon, and J. De Baerdemaeker. 2002. A review of combine sensors for precision farming. *Precision Agriculture*, 3:169–182.

Ryder, J.P. and S. Victor. 1966. Electronic seed monitor. United States Patent Application, US 3527928 A

Sankaran, S., A. Mishra, R. Ehsani, and C. Davis. 2010. A review of advanced techniques for detecting plant diseases. *Computers and Electronics in Agriculture*, 72:1–13.

Schueller, J.K. 2002. Advanced *Mechanical and Mechatronic Engineering Technologies and their Potential Implementation on Mobile Agricultural Equipment*. ASAE St. Joseph, MI, USA, Paper No. 021064.

Schueller, J.K., M.P. Mailander, and G.W. Krutz. 1985. Combine feedrate sensors. *Transactions of the ASABE*, 28:0002–0005.

Schueller, J.K., R.M. Slusher, and S.M. Morgan. 1986. An expert system with speech synthesis for troubleshooting grain combine performance. *Transactions of the ASABE*, 29:342–344.

Searcy, S.W., J.K. Schueller, Y.A. Bae, S.C. Borgelt, and B.A. Stout. 1989. Mapping of spatially variable yield during grain combining. *Transactions of the ASABE*, 32:826–829.

Sinfield, J.V., D. Fagerman, and O. Colic. 2010. Evaluation of sensing technologies for on-the-go detection of macro-nutrients in cultivated soils. *Computers and Electronics in Agriculture*, 70:1–18.

Slaughter, D.C., D.K. Giles, and D. Downey. 2008. Autonomous robotic weed control systems: A review. *Computers and Electronics in Agriculture*, 61:63–78.

Sommer, C. and H.-H. Voßhenrich. 2004. Soil cultivation and sowing. In KTBL (Ed.), *Managementsystem für den ortsspezifischen Pflanzenbau. Verbundprojekt pre agro*, Darmstadt, Germany, Chapter 4: 121–150, CD-ROM 43013 (http://www.preagro.de/Veroeff/Liste.php3).

Steinberger, G. 2012. Methodische Untersuchungen zur Integration automatisch erfasster Prozessdaten von mobilen Arbeitsmaschinen in ein Informationsmanagementsystem "Precision Farming". Dissertation: Technische Universität München, Lehrstuhl für Agrarsystemtechnik, Freising-Weihenstephan, Germany (http://mediatum.ub.tum.de/?ID=1096419).

Stone, M.L. 2011. ISO 11783 Part 10 Task controller and management information system data interchange. ASABE AET. http://www.shieldedpair.net/downloads/ISO%2011783%20Part%2010.pdf

Stone, M.L., D.M. Kee, C.W. Formwalt, and R.K. Benneweis 1999. ISO 11783: An electronic communications protocol for agricultural equipment. ASAE St. Joseph, MI, USA.

Tey, Y.S. and M. Brindal. 2012. Factors influencing the adoption of precision agricultural technologies: A review for policy implications. *Precision Agriculture*, 13:713–730.

Thorp, K.R. and L.F. Tian. 2004. A review on remote sensing of weeds in agriculture. *Precision Agriculture*, 5:477–508.

UNESCO. 2007a. Summary of the Monograph. *World Water Resources at the Beginning of the 21st Century*, Prepared in the framework of IHP UNESCO, 1999. www.espejo.unesco.org.uy/summary/html, accessed on October 09, 2007.

UNESCO. 2007b. Water Use in the World: Present Situation/Future Needs, 2000. www.unesco.org/science/waterday2000/water_use_in_the_world.htm, accessed on October 09, 2007.

Vellidis, G. 2015. Irrigation sensor network. Personal communication on teaching material, January 14, 2015.

Vellidis, G., C.D. Perry, G.C. Rains, D.L. Thomas, N. Wells, and C.K. Kvien. 2003. Simulataneous assessment of cotton yield monitor. *Applied Engineering in Agriculture*, 19:259–272.

Weigert, G. 2006. Data Mining und Wissensentdeckung im Precision Farming-Entwicklung von ökonomisch optimierten Entscheidungsregeln zur kleinräumigen Stickstoff-Ausbringung. Dissertation: Technische Universität München, Professur für Unternehmensforschung und Informationsmanagement, Freising-Weihenstephan, Germany (http://mediatum.ub.tum.de/?id=603736).

Zhang, C. and J.M. Kovacs. 2012. The application of small unmanned aerial systems for precision agriculture: A review. *Precision Agriculture*, 13:693–712.

Index

A

B

D

M

N

O

P

Q

R

S

T

U

V

Printed in the United States
by Baker & Taylor Publisher Services